T0331878

ASTROPHYSICS FROM ANTARCTICA

IAU SYMPOSIUM No. 288

COVER ILLUSTRATION: THE SOUTH POLE TELESCOPE

The 10 m diameter South Pole Telecope (or SPT), seen under the light of the full Moon on a clear night, just before sunrise at the Pole. The SPT is located in the Dark Sector Laboratory (DSL) at the Amundsen–Scott South Station, 1 km away from the geographic South Pole.

The SPT is being used to observe the Cosmic Microwave Background Radiation, surveying a 2,500 square degree area of sky. The science objectives include searches for galaxy clusters using the Sunyaev Zel'dovic effect and making the highest angular resolution measurements of the CMB power spectrum. The results are being used to constrain models for Dark Energy and other cosmological parameters.

Photograph taken by Daniel Luong-Van in September 2010, wintering scientist for the SPT during 2010 and 2011. The picture was taken with a Canon 7D camera about 1.5 km away from the telescope in order to make the Moon and the SPT comparable in scale. It is interesting to note that the band of white is not cloud but is a mirage, being the reflection off the snow surface.

IAU SYMPOSIUM PROCEEDINGS SERIES

Chief Editor

THIERRY MONTMERLE, IAU General Secretary
Institut d'Astrophysique de Paris,
98bis, Bd Arago, 75014 Paris, France
montmerle@iap.fr

Editor

PIERO BENVENUTI, IAU Assistant General Secretary
University of Padua, Dept of Physics and Astronomy,
Vicolo dell'Osservatorio, 3, 35122 Padova, Italy
piero.benvenuti@unipd.it

INTERNATIONAL ASTRONOMICAL UNION

UNION ASTRONOMIQUE INTERNATIONALE

ASTROPHYSICS FROM ANTARCTICA

PROCEEDINGS OF THE 288th SYMPOSIUM OF THE INTERNATIONAL ASTRONOMICAL UNION HELD IN BEIJING, CHINA AUGUST 20–24, 2012

Edited by

MICHAEL G. BURTON
University of New South Wales, AUSTRALIA

XIANGQUN CUI
Nanjing Institute for Astronomical Optics and Technology, CHINA

and

NICHOLAS F. H. TOTHILL
University of Western Sydney, AUSTRALIA

CAMBRIDGE
UNIVERSITY PRESS

Shaftesbury Road, Cambridge CB2 8EA, United Kingdom

One Liberty Plaza, 20th Floor, New York, NY 10006, USA

477 Williamstown Road, Port Melbourne, VIC 3207, Australia

314–321, 3rd Floor, Plot 3, Splendor Forum, Jasola District Centre, New Delhi – 110025, India

103 Penang Road, #05–06/07, Visioncrest Commercial, Singapore 238467

Cambridge University Press is part of Cambridge University Press & Assessment, a department of the University of Cambridge.

We share the University's mission to contribute to society through the pursuit of education, learning and research at the highest international levels of excellence.

www.cambridge.org
Information on this title: www.cambridge.org/9781107033771

First published 2013

A catalogue record for this publication is available from the British Library

ISBN   978-1-107-03377-1   Hardback
ISSN   1743-9213

This journal issue has been printed on FSC-certified paper and cover board. FSC is an independent, non-governmental, not-for-profit organization established to promote the responsible management of the worlds forests. Please see www.fsc.org for information.

Table of Contents

Preface

Overview

The remarkable environment of Antarctica offers many advantages for astronomical observations. Over the past two decades this field of scientific endeavour has developed dramatically and Antarctic-based observatories now regularly contribute to front line astrophysical research, in particular in the measurement of the cosmic microwave background radiation. Scientific stations where astronomy is a major focus now exist at five locations on the Antarctic plateau, as well as on the Antarctic coast, and are operated as international facilities. This Symposium examined the contributions to astrophysics that Antarctic telescopes have made. It reviewed our understanding of the Antarctic environment, in particular that on the high Antarctic plateau. It considered the developments taking place across the continent. It also looked at the parallel opportunities offered by the Arctic environment. Finally, the Symposium examined the science that may be best addressed by future facilities on the Antarctic continent.

The Science

Over the past decade astronomy has matured as one of the fields of science that is being pursued in Antarctica, notably through high angular resolution studies of the cosmic microwave background, including the demonstration that the Universe is flat (de Bernardis *et al.* 2000), the first measurements of polarization of the CMB E–mode (Kovac *et al.* 2002), measurement of the kinematic and thermal SZ–effect in galaxy clusters (Staniszewski *et al.* 2009) and the demonstration that at high-l the background radiation is dominated by emission from dusty galaxies (Leuker *et al.* 2010). Several experiments are now being conducted, worldwide, to search for the B–modes of the CMB, with a large part of this effort taking place in Antarctica, using both ground-based and balloon-borne experiments. Over the same time period there have been science programs conducted in Antarctica from the optical to the sub-millimetre wavebands, and also a range of facilities constructed for high energy astrophysics, particularly for neutrino detection. These science achievements of astronomy in Antarctica have been summarised in the recent review of the field by Burton (2010).

The Stations

Concurrently, major infrastructure has been developed over the Antarctic plateau, undertaken by several different nations. The US have completed a major upgrade to the Amundsen–Scott South Pole station, as well as the construction of the first neutrino telescope (IceCube), in addition to an on-going CMB program. France and Italy have completed the building of the Concordia Station at Dome C, and have wintered personnel there since 2005. Dome A was first visited by humans that year and China has begun the construction of Kunlun station at the site. Dome Fuji station at Dome F has had personnel winter over for ice-core drilling, and astronomical site testing has recently been initiated there. Ridge A was first visited this very year, and a fully robotic telescope installed by the USA and Australia. Long durations balloons equipped with astronomical payloads are now being regularly launched from the US McMurdo station on the Antarctic coast with further extended missions under development.

The Environment

Our understanding of the Antarctic environment as it pertains to the conduct of astronomical observations has improved dramatically over the past decade. Through the use of autonomous observatories at the South Pole, Dome C and Dome A, considerable quantitative information is now available on the properties and behaviour of the sky background and atmospheric transparency, as well as its stability over the high plateau. In particular, the special properties of the narrow surface boundary layer are being revealed. The median thickness of the boundary layer is now known to be less than 15 m thick in winter at the highest places on the plateau, and is often only half that height (e.g. Bonner *et al.* 2010). Further development of automated observatories at Dome A, Dome F and Ridge A is underway to expand this knowledge of the Antarctic environment. Test-bed astronomical facilities are also being operated at Domes A and C, prior to the completion of full infrastructure at these sites.

The Astronomy

Astrophysics has now been conducted in Antarctica in the optical, infrared, terahertz (or far-infrared) and sub-millimetre portions of the spectrum with a variety of small-scale facilities, undertaking studies mostly of stars, gas and dust in the Galaxy. Telescopes such as the 60 cm infrared SPIREX and the 1.7 m sub-millimetre AST/RO have demonstrated the ability to undertake new science in these bands in Antarctica. Of particular note is the recent demonstration of the opportunity for time series investigations in the optical domain with an extremely high cadence and duty cycle (the CSTAR experiment; Yang *et al.* 2010) at Dome A, and at Dome C for precision photometry to be undertaken (sIR-AIT; Strassmeier *et al.* 2008) and the monitoring of exoplanet transits (ASTEP-South; Crouzet *et al.* 2010). Thermal infrared large scale repeated surveys, such as proposed for the PLT (Dome C) and KDUST (Dome A) telescopes, would also greatly benefit from Antarctic atmospheric conditions.

The Role of the IAU

The IAU has a special role in furthering the development of astronomy in Antarctica because of the unique political situation on the continent. No country owns Antarctica. International collaboration and co-ordination is the modus operandi for Antarctic science. This is facilitated through SCAR, a fellow ICSU body with the IAU with specific interests for Antarctic science. IAU recently affiliated with SCAR, the ninth scientific union to do so. Astronomy is not, however, a "traditional" Antarctic science. It faces particular difficulty in attracting funding for infrastructure as a result of many national funding arrangements. The IAU can play an important role in furthering astronomy in Antarctica through supporting the holding of a Symposium in the field, so giving recognition to both the maturing of the field as well as to the science potential it offers.

Chinese Plans for Antarctica

China, the host country for the IAU's XXVIII GA in 2012, also has a special interest in the development of astronomy in Antarctica at its newly established Kunlun station, following the first human visit to the site by China in just 2005. The first astronomical experiments are now taking place there during the construction phase of the station. Under the Chinese government's 12th 5-year plan two major facilities would operate at

Dome A when the station is completed: for the optical / infrared the 2.5 m KDUST telescope and for the THz waveband the 5 m DATE5 telescope.

Opportunities in the Arctic

The Arctic regions also offer several sites with comparable characteristics to Antarctica, notably in Greenland and in Ellesmere Island in Canada. These are high ($> 2,500$ m), dry locations on the summits of large islands located in the Arctic Ocean. Several sites are now under investigation there regarding their suitability for astronomical observations. Sub-orbital programs are also being planned for the Arctic regions, so as to achieve the full-sky coverage needed for large-scale polarization surveys of the CMBR. The science opportunities presented by Arctic observatories was also a theme at the meeting.

Looking to the Future

"Astrophysics from Antarctica" examined all the above themes. The Symposium reviewed our current understanding of the Antarctic environment as it pertains to astronomical observations. It looked at the major achievements of astrophysics in Antarctica to date, in particular with the CMB, but also in high energy astrophysics (neutrinos), radio astronomy (sub-millimetre and THz) and optical/IR astronomy. The Symposium also examined international developments in Antarctica, in particular at the high plateau stations, and the science themes that are emerging for them. In turn, these have many common needs for infrastructure, as well as in challenges that need to be met. To facilitate their furtherance, the Symposium also overviewed this international endeavour, including both the science programs being conducted through national Antarctic programs and the facilities being built or considered for their pursuit. Large international projects are increasingly driving developments in astronomy, and Antarctica and its method of government provides a locale where a "World Observatory" might be built. A particular focus was given to where Antarctic facilities can best contribute to addressing the big science questions posed for astronomy in many national reviews. The Symposium ended by seeking a vision for the future development of international facilities on the continent.

This Volume presents the papers delivered on all these subject matters at IAU Symposium 288, "Astrophysics from Antarctica", held in the China National Convention Center in Beijing from 20–24 August, 2012.

Michael Burton and Xiangqun Cui,
Co-chairs SOC, IAU Symposium 288
Sydney & Nanjing, November 1, 2012

References

Bonner, C. S., *et al.* 2010, *PASP*, 122, 1122
Burton, M. G. 2010, *A&A Rev.*, 18, 417
Crouzet, N., *et al.* 2010, *A&A*, 511, A36
de Bernardis, P., *et al.* 2000, *Nature*, 404, 955
Kovac, J. M. *et al.* 2002, *Nature*, 420, 772
Lueker, M., *et al.* 2010, *ApJ*, 719, 1045
Staniszewski, Z., *et al.* 2009, *ApJ*, 701, 32
Strassmeier, K. G., *et al.* 2008, *A&A*, 490, 287
Yang, H., *et al.* 2010, *PASP*, 122, 490

ORGANISATION AND ACKNOWLEDGEMENTS

Scientific Organising Committee

Michael Burton (Australia) (co-chair)
Xiangqun Cui (China) (co-chair)
Leo Bronfman (Chile)
Nicolas Epchtein (France)
Peter Gorham (USA)
Takashi Ichikawa (Japan)
Doug Johnstone (Canada)
John Kovac (USA)
Silvia Masi (Italy)
Young Minh (Korea)
Klaus Strassmeier (Germany)
Ji Yang (China)
Zhaohui Shang (China)

Editors of the Proceedings

Michael Burton (Australia)
Xiangqun Cui (China)
Nicholas Tothill (Australia)

Acknowledgements

The Symposium was sponsored and supported by the IAU's Divisions IX (Optical and Infrared Techniques), X (Radio Astronomy) and XI (Space and High-Energy Astrophysics), as well as by the IAU Working Group for the Development of Antarctic Astronomy. In addition, the IAU's cognate ICSU body SCAR (Scientific Committee for Antarctic Research), the Chinese Astronomical Society and the National Research Council of Canada's Herzberg Institute of Astrophysics, gave their support for the Symposium.

The Symposium Dinner was sponsored by the Nanjing Institute of Astronomical Optics and Technology together with Purple Mountain Observatory in China.

The Symposium website, at www.phys.unsw.edu.au/IAUS288, was supported by the School of Physics at the University of New South Wales in Australia.

CONFERENCE PHOTOGRAPH

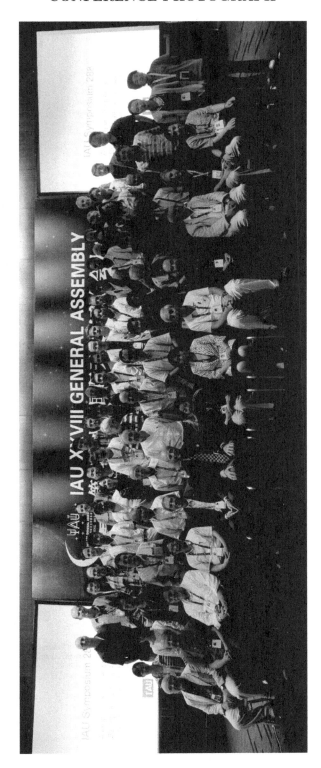

Participants at IAU Symposium 288, "Astrophysics from Antarctica", held from 20–24 August 2012, in the China National Convention Center in Beijing's Olympic Park, as part the the XXVIII General Assembly of the International Astronomical Union.

Participants

Yl Ai	ayl@ynao.ac.cn
Karle Albrecht	karle@icecube.wisc.edu
Tilquin Andre	tilquin@cppm.in2p3.fr
Moore Anna	amoore@astro.caltech.edu
Challinor Anthony	a.d.challinor@ast.cam.ac.uk
Stark Antony	aas@cfa.harvard.edu
Eric Aristidi	aristidi@unice.fr
Steve Barwick	barwick@HEP.ps.uci.edu
Sabogal Beatriz	bsabogal@uniandes.edu.co
Kumthekar Bhagvat	bkumthekar@gmail.com
Thide Bo	bt@irfu.se
Benson Bradford	bbenson@kicp.uchicago.edu
Danielle Briot	danielle.briot@obspm.fr
ten Brummelaar Theo	theo@chara-array.org
Kuo Chao-Lin	clkuo@stanford.edu
Tao Charling	tao@cppm.in2p3.fr
Huang Chen	hc@shao.ac.cn
Liu Chengzhi	lcz@cho.ac.cn
Vasconez Christian	cvasconez@gmail.com
Martin Christopher	Chris.Martin@oberlin.edu
Pun Chun Shing Jason	jcspun@hku.hk
Sarah Church	schurch@stanford.edu
Pryke Clem	pryke@physics.umn.edu
Bischoff Colin	cbischoff@cfa.harvard.edu
Kulesa Craig	ckulesa@email.arizona.edu
Licai Deng	licai@bao.ac.cn
Mekarnia Djamel	mekarnia@oca.eu
Bock Douglas	Douglas.Bock@csiro.au
Steve Durst	info@iloa.org
Pascale Enzo	enzo.pascale@astro.cf.ac.uk
Steinbring Eric	Eric.Steinbring@nrc-cnrc.gc.ca
Fossat Eric	Eric.Fossat@oca.eu
Wang Gang	wg@lamost.org
Zhao Gang	gzhao@nao.cas.cn
Sims Geoff	g.sims@unsw.edu.au
Lombardi Gianluca	glombard@eso.org
Novak Giles	g-novak@northwestern.edu
Durand Gilles	durandgs@cea.fr
Zhao Gong-Bo	gong-bo.zhao@port.ac.uk
Murray Graham	g.j.murray@durham.ac.uk
Liu Guoqing	liuguoqing@tsinghuaeducn
Francis Halzen	francis.halzen@icecube.wisc.edu
Zinnecker Hans	hzinnecker@sofia.usra.edu
Zuo Heng	hengz@niaot.ac.cn
Okita Hirofumi	h-okita@astr.tohoku.ac.jp
Matsuo Hiroshi	h.matsuo@nao.ac.jp
Wang Hongchi	hcwang@pmo.ac.cn
Sylwester Janusz	js@cbk.pan.wroc.pl
Hamilton Jean-Christophe	hamilton@apc.univ-paris7.fr
Rivet Jean-Pierre	jean-pierre.rivet@oca.eu
Yang Ji	jiyang@pmo.ac.cn
Wu Jianghua	jhwu@bao.ac.cn
Fu Jianning	jnfu@bnu.edu.cn
Rui Jin	sgjin@shao.ac.cn
Nam Jiwoo	namjiwoo@gmail.com
Greiner Jochen	jcg@mpe.mpg.de
Dickey John	john.dickey@utas.edu.au
Storey John	j.storey@unsw.edu.au
Lawrence Jon	jl@aao.gov.au
Vernon Jones	w.vernon.jones@nasa.gov
Navarrete Julio	jnavarre@eso.org
Hoffman Kara	kara@umd.edu
Strassmeier Klaus	kgoetz@aip.de
John Kovac	jmkovac@cfa.harvard.edu
Guoliang Li	guoliang@pmo.ac.cn
Xiaoyan Li	xyli@niaot.ac.cn
Yuansheng Li	lysh@pric.gov.cn
Wang Lifan	wang@physics.tamu.edu
Liu Liyong	liuly@nao.cas.cn
Damé Luc	luc.dame@latmos.ipsl.fr
Abe Lyu	Lyu.Abe@unice.fr
Inoue Makoto	inoue@asiaa.sinica.edu.tw
Ruiqing Mao	rqmao@pmo.ac.cn
Sarazin Marc	msarazin@eso.org
Sergio Marenssi	smarenssi@dna.gov.ar
Wolfire Mark	mwolfire@astro.umd.edu
McCaughrean Mark	mjm@esa.int
McAuley Mark	mark.mcauley@astronomyaustralia.org.au
Reina Maruyama	rmaruyama@wisc.edu
Seta Masumichi	seta@physics.px.tsukuba.ac.jp
Colless Matthew	colless@aao.gov.au
Ashley Michael	m.ashley@unsw.edu.au
Burton Michael	m.burton@unsw.edu.au
Young Minh	minh@kasi.re.kr
Jeremy Mould	jmould@swin.edu.au
Imae Naoya	imae@nipr.ac.jp
Tothill Nicholas	n.tothill@uws.edu.au
Crouzet Nicolas	crouzet@stsci.edu
Sharp Nigel	nsharp@nsf.gov

Halverson Nils	nils.halverson@colorado.edu
Jay Pasachoff	jay.m.pasachoff@williams.edu
Rojo Patricio	pato@oan.cl
Koch Patrick	pmkoch@asiaa.sinica.edu.tw
Jones Paul	pjones@phys.unsw.edu.au
Chong Pei	cpei@niaot.ac.cn
Qingyu Peng	tpengqy@jnu.edu.cn
Carlberg Raymond	carlberg@astro.utoronto.ca
Matsushita Satoki	satoki@asiaa.sinica.edu.tw
Lee Seong-Jae	seong@chungbuk.ac.kr
Sheng-Cai Shi	scshi@pmo.ac.cn
Kazayuki Shiraishi	kshiraishi@nipr.ac.jp
Muchovej Stephen	sjcm@astro.caltech.edu
Ivan Syniavskyi	syn@mao.kiev.ua
Cheung Sze-leung	cheung.szeleung@hku.hk
Gaisser Thomas	gaisser@bartol.udel.edu
Zhang Tianmeng	zhangtm@nao.cas.cn
Stevenson Toner	toners@phm.gov.au
Mroczkowski Tony	tonym@astro.caltech.edu
Pascal Tremblin	pascal.tremblin@gmail.com
Peter Tuthill	p.tuthill@physics.usyd.edu.au
Nam Uk-Won	uwnam@kasi.re.kr
Papitashvili Vladimir	vpapita@nsf.gov
Lingzhi Wang	lzwang520213@mail.bnu.edu.cn
Jones William	wcjones@princeton.edu
Cui Xiangqun	xcui@nioat.ac.cn
Jiang Xiaojun	xjjiang@bao.ac.cn
Guan Xin	guan@ph1.uni-koeln.de
Zhou Xu	zhouxu@bao.ac.cn
Wu Xuefeng	xfwu@pmo.ac.cn
Chen Xuepeng	xpchen_yale@hotmail.com
Akira Yamaguchi	yamaguch@nipr.ac.jp
Gong Yan	ygong1@uci.edu
Zhang Yanxia	zyx@bao.ac.cn
Kang Yong Woo	byulmaru@kasi.re.kr
Omura Yoshiharu	omura@rish.kyoto-u.ac.jp
Chu You-Hua	yhchu@illinois.edu
Xiangyan Yuan	xyyuan@niaot.ac.cn
Motizuki Yuko	motizuki@riken.jp
Shang Zhaohui	zshang@gmail.com
Zhaowen	wzhao7@zjut.edu.cn

Editorial

One hundred years ago, on a bright and clear day on December 5, 1912 three young explorers – Francis Bickerton, Alfred Hodgeman and Leslie Whetter – were traversing across the sastrugi in the coastal highlands of Adelie Land in Antarctica, dragging a sledge along behind them. They had just abandoned an 'air tractor' which they were using to pull their load on their journey of discovery, and were struggling across the rough surface. They saw ahead of them a shiny black object, partially buried in the snow. About the size of one's hand, they immediately recognised it to be a meteorite, for how else could it have got there, all alone on top of an immense visage of ice?! They picked it up for their collection and later study. Now known as the Adelie Land Meteorite, and displayed in the Australian Museum in Sydney, it was the first meteorite to be found in Antarctica, though it was to be another 11 years before this find was written up as a scientific paper (Bayly & Stillwell, 1923). It was the start of astronomy in Antarctica. However, it was to be almost half a century later before another meteorite was found. Indeed, it was not until 1969, when several different types of meteorites were found close together (Nagata, 1975), that it was realised that Antarctica provides a superb platform for finding meteorites on account of the great flows of the ice, gathering objects that fall upon them and taking them to blue-ice fields where the snow is ablated by the wind, leaving them easy to spot and so collect.

Astronomy in Antarctica has had a slow gestation. Cosmic ray studies were the first astronomical science programs, being instigated around the time of the International Geophysical Year (1956–57). It was not until 1979 that the first astronomical science was carrried out at the South Pole, using an optical telescope to measure oscillations in the interior of the Sun (Grec, Fossat & Pomerantz, 1980). This was followed by a variety of experiments at millimetre-wavelengths looking for interstellar dust and the cosmic microwave background radiation. It was not until the 1990's, with the instigation of CARA, the Center for Astrophysical Research in Antarctica, at the South Pole, that astronomy in Antarctica began to be undertaken in earnest†. Since then, the growth in the field has been rapid.

Today astronomical investigations are being carried out at five locations on the Antarctic plateau (South Pole, the ice Domes A, C and F, and at Ridge A), as well as long duration ballooning from McMurdo station on the coast. The images on the following pages provide a flavour of the activity at each of these sites. A rich variety of different types of astrophysics is now being being accomplished using data gathered in Antarctica.

The IAU recognised the potential for astronomy in Antarctica in 1991 at the XXI General Assembly in Buenos Aires, creating the 'Working Group for the Development of Astronomy in Antarctica', under the chair of Peter Gillingham. It was set up as an inter-divisional working group under Division IX (Optical/IR Techniques) and Division X (Radio Astronomy). Since then meetings on Astronomy in Antarctica have been held at the GA's in The Hague, Kyoto, Sydney, Prague and Rio de Janeiro‡. The size and context of these meetings has grown over time, with the last three being designed "Special Sessions", and their proceedings appearing in the IAU's *Highlights of Astronomy*.

---

† For a history of astronomy in Antarctica during these early years see Indermuehle, Burton & Maddison, 2005.

‡ See www.phys.unsw.edu.au/jacara/iau/.

With the XXVIII GA in Beijing in 2012, widespread interest internationally in the prospects for astronomy in Antarctica, and with China beginning the construction of Kunlun Station at the highest point on the Antarctic plateau (Dome A), the time was ripe for the proponents of astronomy in Antarctica to seek to hold an IAU Symposium on this field, and so receive recognition from their peers of the maturity of this field of endeavour. The IAU agreed, and hence was born IAU Symposium 288, "Astrophysics from Antarctica"¶. In the pages ahead you will find papers written by the presenters at this Symposium describing a wide variety of activities in a vibrant field.

IAUS288 is the first, but it also may be the last, IAU Symposium devoted to the full subject matter of astronomy in Antarctica. For, having entered the mainstream, the results from Antarctic experiments will be reported more and more at discipline-specific meetings in the future, alongside results from the multitude of facilities that other domains of astronomy bring to the science.

We hope you enjoy reading these Proceedings!

Michael Burton, University of New South Wales
Xiangqun Cui, Nanjing Institute for Astronomical Optics and Technology
Nicholas Tothill, University of Western Sydney
Editors, IAU Symposium 288, "Astrophysics from Antarctica".

References

Bayly, P. G. W. & Stillwell, F. L. 1923, The Adelie Land Meteorite, *Australasian Antarctic Expedition 1911–14*, Scientific Reports Series A, Vol. IV: Geology, Sydney. Publisher A.J. Kent (Sydney)

Grec, G., Fossat, E. & Pomerantz, M. A. 1980, *Nature*, 288, 541

Indermuehle, B., Burton, M. G. & Maddison, S. T. 2005, *PASA*, 22, 73

Nagata, T., ed. *Mem. Nat. Inst. Polar Res.*, 5

Top: The US Amundsen–Scott Station at the South Pole. Four CMB telescopes can be seen: SPT, BICEP, ACBAR and QUAD/DASI, with the IceCube neutrino telescope under the ice to left. An LC130 aircraft is taking off. Credit: Steffan Richter. Bottom: The French–Italian Concordia Sation at Dome C. Labelled are the cluster of experiments that make up the Concordiastro site testing program. The twin towers of the station are to the rear. Credit: Karim Agabi.

The 0.5 m AST3 optical telescope at the Chinese Kunlun station at Dome A. The yellow and green buildings behind are the Australian-built PLATO laboratory. Credit: NIAOT.

The Tohoku-DIMM placed on the mount for the AIRT 40 cm infrared telescope at the Japanese Dome Fuji Station at Dome F. Credit: Hirofumi Okita

Launch of the 2 m BLAST THz telescope from the Long Duration Balloon Facility (LDBF) at McMurdo Station. The Mount Erebus volcano is in the background. Credit: Mark Halpern.

The 60 cm HEAT THz telescope deployed at the SCAR international station at Ridge A by the USA and Australia. Credit: Craig Kulesa.

Astrophysics from Antarctica
Proceedings IAU Symposium No. 288, 2012
M. G. Burton, X. Cui & N. F. H. Tothill, eds.

© International Astronomical Union 2013
doi:10.1017/S1743921312016596

Review of Antarctic astronomy

John W. V. Storey

School of Physics, University of New South Wales,
Sydney, NSW 2052, Australia
email: j.storey@unsw.edu.au

Abstract. Astronomers have always sought the best sites for their telescopes. Antarctica, with its high plateau reaching to above 4,000 metres, intense cold, exceptionally low humidity and stable atmosphere, offers what for many forms of astronomy is the ultimate observing location on this planet. While optical, infrared and millimetre astronomers are building their observatories on the ice, particle physicists are using the ice itself as a detector and exploration of the terahertz region is being conducted from circumpolar long-duration balloons. Remarkable astronomical discoveries are already coming out of Antarctica, and much, much more is just around the corner.

Keywords. Antarctica, Arctic

1. Introduction: why Antarctica?

Looking at Antarctica from a human perspective it may seem that this would be the last place on earth to build a astronomical observatory. Mention "Antarctica" and the horrible prospect of slowly freezing or starving to death may well come to mind. However, from an astronomical perspective, it also happens to be one of the best places on earth to build an observatory. And, thanks to a century of development in logistics capability, anyone can now work there. Like the air and the ocean, Antarctica is unforgiving of the foolhardy or unprepared. However, with several countries now operating well-established stations and providing excellent transport capabilities, anyone who loves the continent can now live and work there in safety.

Relative to the best temperate sites:
- The infrared sky is 20–50 times darker
- The water vapour content is ∼10 times lower
- The free-atmosphere seeing is ∼2 times better
- The coherence volume is 10–100 times greater
- The scintillation noise is ∼2.5 times less
- The atmospheric boundary layer is only 10–20 metres high
- The aerosol content is ∼50 times lower
- The sky is continuously dark for months

For particle physicists, the vast volumes of the purest ice on earth provide an unparalleled medium for detecting neutrinos, muons, and other subatomic beasties. For folk wishing to observe above (most of) the atmosphere, the circumpolar vortex invites long-duration ballooning and indeed ultra-long duration ballooning. Antarctica is not only beautiful, it is the only place on earth set aside for science by international treaty.

There are also of course some disadvantages and discouragements for the Antarctic astronomer:
- It is exceedingly cold. However, this from a human, not an engineering perspective. It is not as cold as the insides of a dewar, nor indeed space. A competent engineer can design for any temperature. While "cold" might sound daunting, this is probably the least difficult issue to cope with.

- There is longer twilight. The sun simply spends more time close to the horizon than at temperate sites. This can be good or bad, depending on the science, but is usually bad.
- The sky coverage is less. This is the flip-side of the fact that those objects that are visible are generally continuously visible.
- The relative humidity is high. This is perhaps surprising—the absolute humidity is lower than that in a cylinder of commercial dry nitrogen. However, given the intense cold, objects close to the ground are susceptible to slowly icing up if they are allowed to fall below (or even sit at) ambient temperature.
- The temperature in the boundary layer is unstable. This presents an engineering challenge to the realisation of the exception image quality that the site can potentially deliver. "Mirror seeing" must be avoided, and this has traditional required that the mirror temperature closely track the ambient temperature.
- At optical wavelengths, aurorae are at best a nuisance and, for some science, potentially a strong disadvantage.
- The accessibility is less than ideal, as even the strongest advocate for Antarctica would have to admit. Still, it beats L2.

This brings us to the inevitable comparison with space. Why bother with Antarctica, when space invariably offers even better observing conditions? There are two principal reasons: deployment to Antarctica is one thousand times cheaper than space (~$10/kg versus ~$10,000/kg) and—whatever your views are on spending several hours in a Hercules C130—it is much more accessible. Experiments can be developed and deployed quickly, and can be maintained and upgraded on annual basis during the summer months.

An extensive review of Antarctic astronomy was published by Michael Burton just two years ago (Burton 2010), so in this introductory paper to the Symposium only the briefest of brief overviews of the field will be given. In mentioning various experiments, no attempt is made to be comprehensive: I have simply chosen a few illustrative examples to whet the reader's appetite for these proceedings.

2. The existing stations

Antarctica is large—roughly 50% larger than China, Europe, or the USA. It should therefore not be surprising that the climates and, with them, the observing conditions, are very different at the different sites. Most astronomical work has been carried out from the inland stations, as these are higher, drier, and calmer than the coast.

2.1. South Pole

First established by the USA in 1957, the Amundsen-Scott Station at the geographic south pole has been host to a wide range of highly successful telescopes. It is a unique site, offering continuous observations of celestial objects at constant zenith distance. Although true geostationary satellites are not visible from South Pole (nor indeed any site south of about $-80°$ latitude, the National Science Foundation is nevertheless able to provide high bandwidth communications through "wobbly" geostationary satellites and through TDRSS.

Several major cosmic microwave background experiments at South Pole take advantage of the unique location. These include the imaginatively-named South Pole Telescope, a 10 metre diameter sub-millimetre telescope (Carlstrom et al. 2011) that has carried out extensive surveys using the Sunyaev-Zel'dovich effect.

Particle experiments at South Pole take advantage of the massive quantities of pure ice to detect neutrinos and muons. The largest of these is Ice Cube (Ahrens et al. 2004), and several new experiments (eg., the Askaryan Radio Array, Hanson *et al.* 2012) are under development.

2.2. *McMurdo*

Sited on the coast, the US McMurdo station is generally uncompetitive as an astronomical site. Nevertheless , it has proved to be an ideal location from which to launch high altitude balloons, which float around the continent and, after a period of typical 14 days, often return more or less directly overhead again. A recent well-publicised example is BLAST (Devlin *et al.* 2009).

2.3. *Dome A*

At an elevation of 4,050 metres, Dome A is the highest point on the plateau. Kunlun station was established there by China in 2008. Although only inhabited over a few weeks each summer, Kunlun hosts experiments that operate year-round using an autonomous power, heat and communications facility called PLATO (Yang *et al.* 2009).

Data from Kunlun indicate that the site offers exceptionally good terahertz transmission (Yang *et al.* 2010) and, of particular importance for optical/infrared astronomy, a boundary layer thickness of typically less than 14 metres (Bonner *et al.* 2010). The CSTAR wide-field optical telescopes have operated robotically at Kunlun for several years now, and have returned a wealth of data (Zhou *et al.*, 2010, Wang *et al.* 2011).

2.4. *Dome C*

Completed in 2005, the French/Italian Concordia station at Dome C offers extremely good free-atmosphere seeing (eg., Lawrence *et al.* 2004, Aristidi *et al.* 2005, Giordano *et al.* 2012), a low boundary layer and very good cloud-cover statistics (Crouzet *et al.* 2010). The site hosts several important astronomical facilities, including the wide-field optical survey telescope ASTEP-400 (Crouzet *et al.* 2011).

At millimetre wavelengths, Battistelli *et al.* (2012) have used the BRAIN Pathfinder Experiment to study the atmospheric transmission and polarisation properties, prior to deployment of a planned interferometer to study the cosmic microwave background.

Dome C also offers very favourable conditions for solar astronomy, with very low atmospheric scattering (eg., Faurobert *et al.* 2012).

2.5. *Dome F*

The Japanese station at Dome F (Dome Fuji) is an another outstanding observing site. Although it operated as a year-round station for a number of years, it currently is open only during the summer. As with Dome A, autonomous power systems will be used—at least for the near future—to allow instruments to operate through the winter months. A 40 cm infrared telescope is planned for installation there at the end of 2012 (Okita *et al.* 2010).

2.6. *Ridge A*

Modelling of site conditions in Antarctica has been carried out by a number of researchers (eg., Lascaux *et al.* 2011), and has the advantage that predictions can be made that cover longer periods and a greater range of locations than are available from in situ observational data. By analysing available site-testing, satellite and meteorological data and applying appropriate models, Saunders *et al.*(2009) set out to determine what observing conditions would be like across the entire Antarctic plateau. Their conclusion was that there was no unique "best" site, and that each of the existing stations offered a particular set of advantages. However a new location, dubbed Ridge A, was predicted to not only have the lowest precipitable water vapour (and hence best THz transmission) but also to be an exceptionally good site on all criteria.

To test this hypothesis, and and the same time take advantage of the THz windows opened up, a robotic 800 GHz telescope was deployed there at the beginning of 2012. Initial indications (Kulesa 2012) are that Ridge A (and, by implication, other high plateau sites) offers unparalleled opportunities for observations in this hitherto inaccessible part of the spectrum.

3. New projects

Some sense of the breadth and depth of Antarctic astronomy can be gauged from the large number of major new telescopes that are either under construction or in the advanced planning stages. These include AFSIIC, AIRT-40, ARA, AST3, DATE5, EBEX, IRAIT, KDUST, PILOT, POLAR, QUBIC and Super-TIGER, and there are many more!

4. The Arctic

While Antarctica is a continent surrounded by ocean, the Arctic is an ocean surrounded by continents. The north pole is typically at an elevation of only a few metres above sea level, and is thus not an attractive site for astronomy. However, close to the north pole are several sites that offer many of the advantages of their Antarctic counterparts. For example Summit station on Greenland is at the same elevation as Dome C, and at the same (absolute) latitude. A 12 metre dish is to be installed there over the coming year for use in sub-millimetre interferometery.

Site testing in the Arctic has focussed on Summit, and at sites on Ellesmere Island in northern Canada. Both locations appear to have superb potential. On Ellesmere Island, site studies over the past few years (Steinbring et al.(2012)) have demonstrated excellent conditions for optical astronomy.

5. Conclusion

With this symposium, Antarctic astronomy has come of age. The scientific results now routinely coming from Antarctic observatories demonstrate not only the unique qualities of that continent, but also that it offers a practical and cost-effective way to conduct science that would be impossible, or much more expensive, to conduct elsewhere.

In future IAU meetings, Antarctic astronomy might no longer be seen as so obviously "different". Major scientific results from Antarctica are now having an impact in many different fields, and gain their prominence not by being "Antarctic", but by the pivotal contributions they make to our understanding of astronomy. May such science continue to flow from Antarctica, and continue to bear witness to the advantages of observing from this beautiful, unique continent.

References

Ahrens, J., Bahcall, J. N., Bai, X., et al. 2004, New Astr., 48, 519
Aristidi, E., Agabi, A., Fossat, E., Azouit, M., Martin, F., Sadibekova, T., Travouillon, T., Vernin, J., et al. 2005, A&A, 444, 651
Battistelli, E. S., Amico, G., Baù, A., et al. 2012, MNRAS, 423, 1293
Burton, M. G. 2010, A&A Rev., 18, 417
Carlstrom, J. E., Ade, P. A. R., Aird, K. A., et al. 2011, PASP, 123, 568

Crouzet, N., Guillot, T., Agabi, A., Rivet, J., Bondoux, E., *et al.* 2010, *A&A*, 511, 36

Crouzet, N., Guillot, T., Agabi, K., Daban, J.-B., Abe, L., Mekarnia, D., Rivet, J.-P., Fante-Caujolle, Y., Fressin, F., Gouvret, C., Schmider, F.-X., Valbousquet, F., Blazit, A., Rauer, H., Erikson, A., Fruth, T., Aigrain, S., Pont, F., & Barbieri, M. 2011, *EPJ Web of Conferences*, 11, 06001

Devlin, M. J., Ade, P. A. R., Aretxaga, I., *et al.* 2009, *Nature*, 458, 737

Faurobert, M., Arnaud, J., & Vernisse, Y. 2012, *EAS Publications Series*, 55, 365

Giordano, C., Vernin, J., Chadid, M., Aristidi, E., Agabi, A., & Trinquet, H. 2012, *PASP*, 124, 494

Hanson, K., ARA Collaboration 2012, *Journal of Physics Conference Series*, 375, 052037

Kulesa, C. 2012, private communication

Lascaux, F., Masciadri, E., & Hagelin, S. 2011, *MNRAS*, 411, 693

Lawrence, J. S., Ashley, M. C. B., Tokovinin, A., & Travouillon, T. 2004, *Nature*, 431, 278

Okita, H., Ichikawa, T., Yoshikawa, T., Lundock, R. G., & Kurita, K. 2010, *Proc. SPIE*, 7733,

Saunders, W., Lawrence, J. S., Storey, J. W. V., *et al.* 2009, *PASP*, 121, 976

Steinbring, E., Ward, W., & Drummond, J. R. 2012, *PASP*, 124, 185

Wang, L., Macri, L. M., Krisciunas, K., *et al.* 2011, *AJ* 142, 155

Yang, H., Allen, G., Ashley, M. C. B., *et al.* 2009, *PASP*, 121, 174

Yang, H., Kulesa, C. A., Walker, C. K., Tothill, N. F. H., Yang, J., Ashley, M. C. B., Cue, X., Feng, L., Lawrence, J. S., Luong-Van, D. M., Storey, J. W. V., Wang, L., Zhou, X., & Zhu, Z. 2010, *PASP*, 122, 490

Zhou, X., Fan, Z., Jiang, Z., *et al.* 2010, *PASP*, 122, 347

Astrophysics from Antarctica
Proceedings IAU Symposium No. 288, 2012   © International Astronomical Union 2013
M. G. Burton, X. Cui & N. F. H. Tothill, eds.   doi:10.1017/S1743921312016602

Autonomous observatories for the Antarctic plateau

J. S. Lawrence[1], M. C. B. Ashley[2], and J. W. V. Storey[2]

[1] Australian Astronomical Observatory, North Ryde,
NSW 2113, Australia,
email: jl@aao.gov.au

[2] School of Physics, University of New South Wales,
Sydney, NSW 2052, Australia
email: m.ashley@unsw.edu.au, j.storey@unsw.edu.au

Abstract. Antarctic astronomical site-testing has been conducted using autonomous self-powered observatories for more than a decade (the AASTO at South Pole, the AASTINO at Dome C, and PLATO at Dome A/Dome F). More recently autonomous (PLATO) observatories have been developed and deployed to support small-scale scientific instruments, such as HEAT, a 0.6 m aperture terahertz telescope at Ridge A, and AST3, a 0.5 m optical telescope array at Dome A. This paper reviews the evolution of autonomous Antarctic astronomical observatories, and discusses the requirements and implications for observatories that will be needed for future larger-scale facilities.

Keywords. Antarctic astronomy, site testing, autonomous observatory

1. Introduction

Conditions on the Antarctic plateau are very promising for astronomy. The high altitude, low surface winds, and lack of a tropospheric jet stream lead to a beneficial turbulence profile. The cold and dry atmosphere leads to low thermal emission and high throughput. The cloud free conditions and the high site latitudes are ideal for high-cadence imaging programs. The initial verification of these conditions, and of the feasibility of conducting observations that take advantage of these conditions, has required the development of astronomical observatories that could operate in extreme environmental conditions with a high level of automation.

Potential sites for astronomy are at permanently manned stations such as the US Amundsen Scott South Pole station operated since the late 1950s, and the French/Italian Dome C station operated year round since 2005; mostly summer-only stations such as the Japanese base at Dome F operated since 1995, the Chinese Kunlun station at Dome A since 2007, and the Russian Vostok station operated since 1957; or field stations such as Ridge A operated since 2012.

Manned observatories at permanently-manned and summer-only Antarctic plateau stations have been vital for the collection of site testing and science data. Facilities at the South Pole have supported a range of telescopes: SPT, BICEP, Viper-ACBAR, DASI, AST/RO and SPIREX. At Dome C, facilities have supported the AstroConcordia program, A-STEP, COCHISE and IRAIT.

Collection of winter-time data from summer-only stations or from field locations has required the development of fully autonomous self-powered observatories, which have now been deployed to a number of sites on the high plateau. These observatories have significantly progressed in capability and reliability over the last fifteen years.

2. Autonomous Observatory Requirements

The primary requirements for a remote self-powered autonomous observatory for astronomy on the Antarctic plateau include:

(*a*) The observatory must provide electrical power to instruments. The total amount of power depends on the type and range of instrumentation but is usually in the range 10-100 W for a single experiment and 1-3 kW for an instrument suite or a moderate size science experiment.

(*b*) The observatory must provide heat to keep instrument components above ambient temperature. The total heat required depends on the instrument elements and the instrument size but generally the heat required is of similar magnitude to the electrical power required.

(*c*) The observatory must provide an autonomous and reliable control system. This system includes both the power distribution system which must have multiple redundancies and the computer control system.

(*d*) The observatory must provide an autonomous and reliable communications system. Communication is required for observatory housekeeping and control and for data and control of instrumentation.

(*e*) The observatory must comply with appropriate logistical constraints. Constraints are imposed by the scale of the observatory and its intended location and are primarily dictated by the method of transportation (e.g., Twin Otter, LC130 aircraft, ice sled, helicopter, icebreaker, etc).

3. AASTO: South Pole

The Automated Antarctic Site Testing Observatory (AASTO; Fig. 1) was deployed to South Pole in Jan 1997 (Storey *et al.* 1996). This was a collaboration between Australian and US institutes under the umbrella of the Center for Astrophysical Research in Antarctica (CARA) and the Joint Australian Centre for Astrophysical Research in Antarctica (JACARA). The AASTO was fabricated from a single fibreglass module based on a modified Automated Geophysical Observatory (AGO; Dudeney *et al.* 1998). Similar to the AGO program, the AASTO was designed to be transported to remote locations on the plateau via ski-equipped LC130 aircraft. The AASTO was initially powered via a Thermo-Electic Generator (TEG) that produced ~50 W electrical power and ~2 kW of heat. As was also the case in the AGO program, the TEG proved unreliable for unmanned operation and the AASTO was switched to South Pole station power.

The AAO control system consisted of a series of PC104-based computers communicating with instruments via 1-wire-bus and/or a serial ports (RS232/485). Communication was via the South Pole station network linked to a series of inclined geosynchronous satellites.

The AASTO was initially used to support a suite of site-testing instruments, including a mid-infrared sky monitor (MISM; Chamberlain *et al.* 2000), a near-infrared sky monitor (NISM; Lawrence *et al.* 2002), a sonic radar (SODAR; Travouillon *et al.* 2003), a sub-millimetre tipping radiometer (SUMMIT; Calisse *et al.* 2004), and a telescope fibre-coupled to an optical spectrograph (AFOS; Dempsey *et al.* 2004). It was later also used as a platform for science experiments, such as the Vulcan South Pole transit search experiment. The AASTO was decommissioned in 2005.

Figure 1. The AASTO at the South Pole from 1997.

Figure 2. The AASTINO at Dome C in 2003.

4. AASTINO: Dome C

The Automated Antarctic Site Testing INternational Observatory (AASTINO; Fig. 2) was deployed to the Italian French Dome C station in Jan 2003 (Lawrence *et al.* 2005). It was designed to support a series of site-testing instruments during the winter months prior to the station opening for full winter manned operation. The AASTINO was a collaboration between the University of New South Wales and the Australian, French and Italian national Antarctic agencies (AAD, IPEV and PNRA). Institutes from Australia, France, Italy and the US contributed to the instrument suite.

The AASTINO was constructed from an outer fibreglass casing with thick internal sprayfoam insulation. The entire module was assembled from a series of 16 panels on-site. Transport for the AASTINO was via icebreaker (from Tasmania to the Antarctic coastal station Dumont Durville) then via overland tractor traverse to Dome C.

Power and heat for the AASTINO were provided via two Stirling cycle engines running on JetA1 aviation fuel. They provided ~400 W peak electrical power and several kW of heat. This was augmented with a solar panel bank that provided 400 W during the summer months.

The AASTINO used a similar (though upgraded) 1-wire communications bus, control architecture and PC104 computer system to the AASTO. For communication the AASTINO used the Iridium satellite system.

The instrument suite for AASTINO included a multi-aperture scintillation sensor (MASS; Lawrence *et al.* 2004), the SUMMIT sub-millimetre tipping radiometer (Calisse *et al.* 2004), and an acoustic radar (SODAR; Travouillon *et al.* 2003).

The AASTINO ran for ~6 months in both 2003 and 2004, providing valuable data on the atmospheric qualities of the Dome C site well into the winter months of these years. The observatory was decommissioned in 2005, when the Concordia station was opened for year-round operation.

5. PLATO: Dome A

The PLATeau Observatory (PLATO; Fig. 3) was deployed to Dome A in Jan 2008 via an overland traverse conducted by the Polar Research Institute of China (PRIC). PLATO (Lawrence *et al.* 2009) was designed and constructed by UNSW in consultation with a number of teams from China and the US who provided site testing instruments (Yang *et al.* 2009).

PLATO departs from earlier observatories in a number of design aspects. A bank of 6 single cylinder diesel engines are used as the primary power source for PLATO, with by default, a single engine running at any one time. The engines, fuel (4000 litres of Jet A1), and power control electronics, are installed inside a heavily insulated customised 10 foot steel shipping container, "the engine module". A second "instrument" module (of similar construction) is installed ~50 metres away, near a solar-panel array. The PLATO instrument module houses the computer control system, instrument control computers, and a series of site-testing instruments mounted through the module wall or ceiling. The PLATO control system features many improvements over earlier observatories, in particular, it uses a CAN-bus network for subsystem control.

PLATO has collected data during the winter months at Dome A from a number of instruments, including: the CSTAR telescope array (Zhou *et al.* 2010), the Snodar ground layer turbulence profiler (Bonner *et al.* 2010), a sub-millimetre Fourier Transform spectrometer, the Gattini sky camera, the Pre-HEAT terahertz telescope (Yang *et al.* 2010), and the Nigel fibre optical spectrograph (Sims *et al.* 2012).

Figure 3. PLATO installed at Dome A in 2008.

Figure 4. PLATO-A installed at Dome A in 2012.

The PLATO observatory has been extremely successful: running autonomously for more than 200 days in its first winter of 2008, and running continuously (with yearly servicing missions) for more than 1230 days since 2009.

6. PLATO-A: Dome A

In Jan 2012, the PLATO-A (Fig. 4) observatory was deployed to Dome A, just 250 m from the original PLATO. The sole purpose of PLATO-A is to provide a power, control, and communications platform for the AST3 telescope array (Yuan & Su 2012), developed by a consortium of Chinese institutes.

The PLATO-A observatory is similar in most aspects to the original PLATO observatory. It is powered by a bank of five single-cylinder diesel engines with a solar panel array, capable of providing up to 3kW of heat and peaks of 3kW of electrical power. It comprises two separate modules, one for instrument control and one for power generation (though these are fibreglass construction rather than steel as in the original PLATO). Modifications have also been made to the power system control architecture to enhance reliability and operability.

The first telescope in the AST3 array was installed with PLATO-A in Jan 2012. At time of writing, the PLATO-A observatory is still functional though a power supply problem caused the telescope operation to cease. A servicing mission for the observatory will coincide with the deployment of the second AST3 telescope in Jan 2013.

7. PLATO-F: Dome F

With a view to collecting data on the site qualities of the Dome F location, Japanese and Australian institutes and polar research agencies collaborated on the development of the PLATO-F observatory. PLATO-F (Fig. 5) was deployed to Dome F in Jan 2011.

The PLATO-F observatory is similar to the PLATO-A (second generation) observatory at Dome A. The power generation system for PLATO-F consists of a bank of 5 single cylinder diesel engines plus a solar panel array. The observatory comprises two modules (one for the engines and one for the instruments and control system). These modules were designed with a focus on minimising weight so that the observatory could be assembled from individual sections at the Japanese coastal station after transfer from the icebreaker via helicopter.

The instrument suite for PLATO-F comprises a Snodar ground-layer turbulence profiler, the Twincam optical telescope experiement, and a metrology tower. PLATO-F ran unattended for more than half a year in 2011. The next servicing mission is planned for Jan 2013.

8. PLATO-R: Ridge A

The site at Ridge A was identified as likely to experience lower water vapour content than Dome A (Saunders *et al.* 2009), motivating the development of the PLATO-R observatory (Fig. 6). PLATO-R was designed specifically to provide power and control for the HEAT terahertz telescope. Both were deployed to Ridge A in Jan 2012.

Significant constraints were placed on the PLATO-R design because the Ridge A location had no infrastructure as it had previously not been visited. Transport was viable only via ski-equipped Twin Otter aircraft. This meant that all components of PLATO-R had to fit within the limited cargo bay doors and load capacity for this aircraft. The observatory architecture for PLATO-R was similar to previous PLATO versions, i.e., with

Figure 5. PLATO-F installed at Dome F in 2011.

Figure 6. PLATO-R installed at Ridge A in 2012.

separate engine and instrument modules and external solar panels. For PLATO-R these modules are both relatively small. Both are constructed from fibreglass. The engine module houses two diesel engines, the instrument module houses the iridium satellite system and observatory supervisor computer and electronics.

PLATO-R ran for 127 days in 2012, limited by the available Jet-A1 fuel supply. The next servicing mission is planned for the Austral summer of 12/13.

9. Conclusion

Autonomous self-powered observatories have been crucial for collecting data on the site conditions at unmanned Antarctic plateau locations. These observatories have progressed in capability and reliability over the last fifteen years. The field has evolved such that unmanned observatories now being deployed are powering instruments targeting astronomical science rather than atmospheric site qualification. Additionally, experiments are now possible in new locations that have no existing infrastructure.

Current generation observatories are scalable, though not indefinitely so. Future telescopes of diameter greater than $\sim$2 metres will likely need to be deployed at permanently manned locations (so the design costs are not prohibitively expensive). However, many of the subsystems used in self-powered observatories (such as power control and communications systems) will be highly applicable to such systems. It is likely to be some time before permanently manned stations are operational at all high plateau sites and thus there are many years for which self-powered autonomous observatories will be essential for the collection of wintertime data.

References

Bonner, C. S., Ashley, M. C. B., Cui, X., Feng, L., Gong, X., Lawrence, J. S., Luong-Van, D. M., Shang, Z., Storey, J. W. V., Wang, L., Yang, H., Yang, J., & Zhou, X. 2009, *P.A.S.P*, 122, 1122

Calisse, P. G., Ashley, M. C. B., Burton, M. G., Phillips, M. A, Storey, J. W. V., Radford, S. J. E. & Peterson, J. B. 2004. *P.A.S.A*, 21, 256

Chamberlain, M. A., Ashley, M. C. B., Burton, M. G., Phillips, A., Storey, J. W. V., & Harper, D. A. 2000, *AP.J.*, 535, 501

Dempsey, J. T., Storey, J. W. V., Ashley, M. C. B., Burton, M. G., Calisse, P. G. & Jarnyk, M. 2004. *Proc. SPIE*, 5492, 811

Dudeney, J. R., Kressman, R. I., & Rodger, A. S. 1998, *Antarct. Sci.*, 10, 192

Lawrence, J. S., Ashley, M. C. B., Burton, M. G., Calisse, P. G., Everett, J. R., Pernic, R. J., Phillips, A., & Storey, J. W. V. 2002, *P.A.S.A*, 19, 328

Lawrence, J. S., Ashley, M. C. B., Kenyon, S., Storey, J. W. V., Tokovinin, A., Lloyd, J. P., & Swain, M. 2004, *Proc. SPIE* 5489, 174

Lawrence, J. S., Ashley, M. C. B., & Storey, J. W. V. 2005, *Aus. J. Electic. Electron. Eng.*, 2, 1

Lawrence, J. S., Ashley, M. C. B., Hengst, S., Luong-Van, D. M., Storey, J. W. V., Yang, H., Zhou, X., & Zhu, Z. 2009, *Rev. Sci. Instrum.*, 80, 064501

Saunders, W., Lawrence, J. S., Storey, J. W. V., Ashley, M. C. B., Kato, S., Minnis, P., Winker, D. M., & Liu, G., Kulesa C. 2009, *P.A.S.P*, 121, 976

Sims, G., Ashley, M. C. B., Cui, X,. Everett, J. R., Feng, L., Gong, X., Hengst, S., Hu, Z., Kulesa, C., Lawrence, J. S., Luong-Van, D. M., Ricaud, P., Shang, Z., Storey, J. W. V., Wang, L., Yang, H., Yang, J., Zhou, X., & Zhu, Z. 2012, *P.A.S.P*, 124, 74

Storey, J. W. V., Ashley, M. C. B., & Burton, M. G. 1996, *P.A.S.A*, 13, 35

Travouillon, T., Ashley, M. C. B., Burton, M. G., Storey, J. W. V., & Loewenstein, R. F. 2003, *A & A*, 400, 1163

Yang, J., Allen, G., Ashley, M. C. B., Bonner, C. S., Bradley, S., Cui, X., Everett, J. R., Feng, L., Gong, X., Hengst, S., Hu, J., Jiang, Z., Kulesa, C. A., Lawrence, J. S., Li, Y., Luong-Van, D., McCaughrean, M. J., Moore, A. M., Pennypacker, C., Qin, W., Riddle, R., Shang, Z., Storey, J. W. V., Sun, B., Suntzeff, N., Tothill, N. F. H., Travouillon, T., Walker, C. K., Wang, L., Yan, J., Yang, J., York, D., Yuan, X., Zhang, X., Zhang, Z., Zhou, X., & Zhu, Z. 2009, *P.A.S.P.*, 121, 174

Yang, H., Kulesa, C. A., Walker, C. K., Tothill, N. F. H., Yang, J., Ashley, M. C. B., Cui, X., Feng, L., Lawrence, J. S., Luong-Van, D. M., Storey, J. W. V., Wang, L., Zhou, X., & Zhu, Z. 2010, *P.A.S.P*, 122, 490

Yuan, X. & Su, D. 2012, *MNRAS*, 424, 23

Zou, H., Zhou, X., Jiang, Z., Ashley, M. C. B., Cui, X., Feng, L., Gong, X., Hu, J., Kulesa, C. A., Lawrence, J. S., Li, G., Luong-Van, D. M., Ma, J., Moore, A. M., Pennypacker, C. R., Qin, W., Shang, Z., Storey, J. W. V., Sun, B., Travouillon, T., Walker, C K., Wang, J., Wang, L., Wu, J., Wu, Z., Xia, L., Yan, J., Yang, J., Yang, H., Yao, Y., Yuan, X., York, D. G., Zhang, Z., & Zhu, Z. 2010, *A.J.*, 140, 602

Astrophysics from Antarctica
Proceedings IAU Symposium No. 288, 2012
M. G. Burton, X. Cui & N. F. H. Tothill, eds.

© International Astronomical Union 2013
doi:10.1017/S1743921312016614

Site characteristics of the high Antarctic plateau

Michael C. B. Ashley

School of Physics, University of New South Wales, Sydney NSW 2052, Australia
email: `m.ashley@unsw.edu.au`

Abstract. A brief review is given of the major results from the last twenty years of astronomical site-testing in Antarctica. Suggestions are made for how to resolve some outstanding questions, such as the infrared sky background at Antarctic sites other than South Pole station.

Keywords. site testing, atmospheric effects, techniques: photometric

1. Introduction

The last twenty years has seen observational confirmation that the unique atmospheric conditions over Antarctica hold many benefits for astronomy. This brief review does not attempt to be all-encompassing. I have concentrated on a few site characteristics, and with a bias towards those relevant for optical, infrared, and terahertz astronomy. For each site characteristic I also briefly comment on what measurements remain to be made.

In the early 1990s it was realised that there were clear theoretical reasons for Antarctica to be an excellent observatory site. For example, the cold atmosphere, low precipitable water vapor, low aerosol concentration, and relatively high altitude of the Antarctic plateau, should lead to very low infrared and sub-mm backgrounds, and high sky transparency.

There were also some not-so-obvious advantages, such as a predicted "cosmological window" at 2.27–2.45 μm resulting from a natural gap in airglow emission (Lubin 1988, Harper 1989), and the possibility of "superseeing" following from the unique Antarctic atmospheric conditions (Gillingham 1993).

And there were some major uncertainties, most notably the fraction of cloudy skies over Antarctica in winter.

Quantifying the site conditions has occupied many groups and produced many dozens of scientific papers over the past twenty years (see Burton 2010 for a review).

2. Astronomical seeing and atmospheric turbulence

The movement of the air above the Antarctic plateau is dominated by very low velocity katabatic winds at ice level, and the polar vortex higher up; the high altitude jet streams that are common in mid-latitudes are absent. These unique conditions lead to a highly turbulent, but thin, boundary layer near the ice, with the prospect of potentially excellent "free-atmosphere" conditions above it.

A direct measurement of the seeing is difficult since it requires a relatively large telescope above the boundary layer, hence we rely on measurements from microthermal sensors on balloons and towers, and instruments such as the DIMM, SCIDAR, SODAR, and MASS. The key parameters we need to determine are the boundary layer thickness, the free atmosphere seeing, the isoplanatic angle and the coherence time.

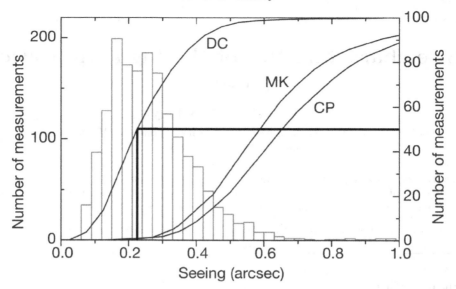

Figure 1. Histogram and cumulative distributions of Dome C seeing above 30 m from MASS combined with SODAR, and cumulative distributions of seeing at Dome C (DC), Mauna Kea (MK), and Cerro Paranal (CP). The median Dome C seeing is 0.23″. From Lawrence *et al.* (2004).

The turbulent boundary layer height at the South Pole was first measured by Neff (1981) with a SODAR (sonic radar). Marks *et al.* (1996, 1999) used microthermal sensors at the South Pole to show that the boundary layer height was typically ∼220 m and that above this the median free atmosphere seeing was 0.32″. On the ice at South Pole, the seeing was a disappointing 1.6″ (Loewenstein *et al.* 1998).

The poor seeing at ice level saw many astronomers lose interest in Antarctica for optical astronomy. However, interest was again ignited when Lawrence *et al.* (2004) used a MASS and a SODAR at Dome C during wintertime, to demonstrate 0.23″ median seeing (Figure 1) above a boundary layer that was less than 30 m in height—a height that is less than that of many mid-latitude telescopes. Note that there is an error in the abstract of Lawrence *et al.* (2004): they state the *median* seeing was 0.27″; this should be *mean* seeing; the median is 0.23″ as is clear from the text of their paper and Figure 1.

Given the stunning result from Lawrence *et al.* (2004), replication of their results was vital, and this has now been done through several independent means. There have been hundred of thousands of DIMM measurements from Dome C, and extensive balloon-borne microthermal campaigns (see, e.g., Agabi *et al.* 2005, Aristidi *et al.* 2005, 2009 and Trinquet *et al.* 2008); these papers show ∼0.36″ median seeing above a 25–40 m boundary layer. It is worth noting that DIMM measurements are always upper limits on the seeing, since DIMMs are affected by local turbulence. At Dome C, when the DIMM is on the top edge of the Concordia building, there are good reasons to expect turbulence created by the building itself, both from the tens of kilowatts of heat from the building, and the effect of the building on the airflow.

Giordano (2012) analysed single-star SCIDAR data from Dome C in 2006 to show that the median seeing was less than 0.3″, the isoplanatic angle was greater than 6.9″, and the coherence time was greater than 10 msec.

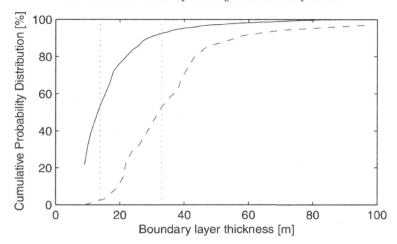

Figure 2. Cumulative probability distributions of the boundary-layer thickness over Dome A using a sonic radar during 2009 (solid line), and Dome C during 2005 (dashed line). Data for Dome C are from Trinquet *et al.* (2008). The median boundary-layer thicknesses for Dome A and Dome C are 13.9 m and 33 m, respectively. Above this height, a telescope would be in the free-atmosphere for half the time. From Bonner *et al.* (2010).

The most definitive measurements of the height of the boundary layer on the Antarctic plateau have been made by the Snodar instrument at Dome A (Bonner *et al.* 2010, see Figure 2). Snodar is a sonic radar that was purpose built to probe turbulence close to the ice with 1 m resolution (Bonner *et al.* 2009). The results showed a sharp ($<$ 1 m) transition between turbulence and non-turbulence. Winterover scientists at Dome C have also anecdotally reported that the transition zone is as thin as centimetres. These observations gives confidence that the height of the boundary layer can be unambiguously defined to better than a metre, and that a sonic radar is the best instrument to measure it. DIMMs and balloons are not able to measure the boundary layer height to this accuracy: DIMMs rely on a statistical analysis of thousands of observations above and below the boundary layer; balloons are affected by wake and limited spatial resolution. A sonic radar can make the measurement in one second, and produce detailed high time resolution plots of the behaviour of the layer.

Theoretical predictions of the boundary layer heights and free-atmosphere seeing over the entire continent were made by Swain and Gallée (2006). Lascaux *et al.* (2011) have used a mesoscale (1 km resolution) model of atmospheric circulation to predict seeing, and they showed a good agreement with observations. Interestingly, they predict that the seeing at Dome A will be better than that at Dome C by a factor of about 0.75.

In summary then, we have good agreement between theory and independent measurements taken with microthermals, DIMM, SCIDAR, Snodar, and MASS.

One important consequence of the atmospheric turbulence distribution is that the ultimate photometric limit from scintillation is significantly improved over mid-latitude sites. For example, for 60 s observations on a 4 m telescope Kenyon *et al.* (2006) predicted a scintillation limit of 52 μmag at Dome C, and $\sim$200 μmag for Chile and Mauna Kea.

For the future, we need to measure the seeing at the other high plateau sites: Domes A and F. These measurements are best made with a MASS. The outer-scale of turbulence could be usefully measured. It would also be worthwhile to send a high-resolution sonic radar (e.g., Argentini *et al.* 2012) to Dome C. Seeing measurements with a DIMM are not ideal since they are upper limits and are affected by local turbulence from the site.

3. Precipitable water vapour

The total column of precipitable water vapour (PWV), and its stability, is of crucial importance since it strongly affects the atmospheric transmission and background radiation. PWV is particularly important in the terahertz, since new windows can be opened up. Even in the optical, PWV has a significant effect in the red, as shown in Figure 3.

Simply speaking, the reason that the air above Antarctica has extremely low PWV is that the air is so cold that most of the water has frozen out. At 100% relative humidity, the Antarctic air in wintertime has about the same water content (a few parts per million) as a cylinder of commercial dry nitrogen.

There have been many measurements of PWV from Antarctica over many years, primarily from the South Pole. Recently, Yang *et al.* (2010) measured PWV above Dome A at 661 GHz (453 μm) and showed excellent agreement with satellite data (Figure 4). Tremblin *et al.* (2011) reported on three years of sub-mm sky opacity measurements from Dome C, and obtained reasonable agreement with the IASI interferometer on the METOP-A satellite. Sims *et al.* (2012) went on to show good agreement for Dome A between three different satellite instruments (Figure 5). A good case can now be made that further measurements of PWV from the ground can be replaced with satellite data, which have the added advantage of covering the whole continent. Ground-based measurements are still useful if you need higher temporal resolution that the satellites can give.

In many ways, the absolute PWV level is not as important as its stability. In both respects, Antarctica has far superior conditions to the best mid-latitude sites (e.g., Cerro Chajnantor).

One area in which Cerro Chajnantor wins is in "dry air opacity", i.e., the atmospheric opacity when no water vapour is present. Dry air opacity comes from collisionally induced multipole moments in nitrogen and oxygen molecules. The increased altitude of Cerro Chajnantor (5600 m) compared with Dome A (4100 m) results in less nitrogen and oxygen above the site, and so the dry air opacity is less. However, dry air opacity is of little practical consequence since it is highly stable: its only effect will be to slightly increase integration times. Variations in opacity over short timescales are far more important, and these are dominated by fluctuations in PWV.

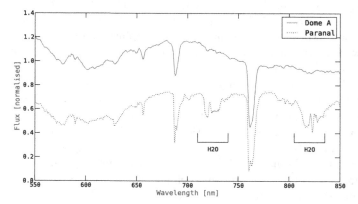

Figure 3. Comparison of Dome A and Paranal twilight sky spectra. The lack of H_2O absorption in the Dome A spectrum is evident. Both spectra were obtained while the Sun was $\sim$1.4° below the horizon. The Dome A spectrum has been offset by +0.3. From Sims *et al.* (2012).

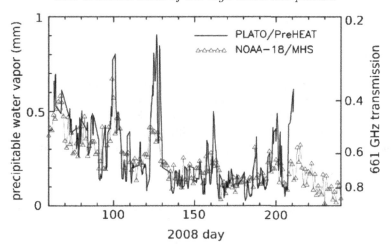

Figure 4. A comparison between satellite measurements of precipitable water vapour above Dome A (the MHS instrument on board NOAA-18) with ground-based of 661 GHz transmission from the Pre-HEAT/PLATO instrument, for five months in 2008. From Yang *et al.* (2010).

For the future there is much work to be done in analysing the large amount of available satellite data showing PWV over the continent. The data need careful analysis to consider the line of sight of the satellite over the continent.

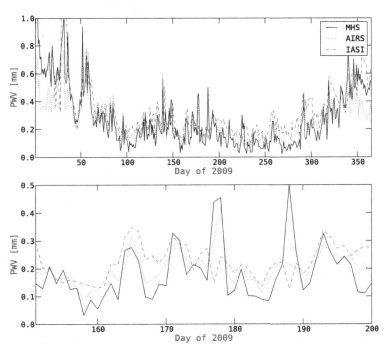

Figure 5. Daily averages of precipitable water vapour for Dome A in 2009 from three different satellite instruments (MHS, AIRS, and IASI), showing good agreement. Top: Spans the entire year. Bottom: Zooms in on a 50 day period during winter. From Sims *et al.* (2012).

4. Infrared sky brightness

As predicted, the infrared sky over Antarctica has been found to be very dark. The following measurements, apart from one, refer to data from the US South Pole station.

At J ($1.25\,\mu$m) and H ($1.65\,\mu$m), Phillips (1999) found a factor of 2–3 times improvement over Siding Spring Observatory (which is a very dark mid-latitude site); at these wavelengths, the sky background is dominated by airglow.

In the "cosmological window" at 2.27–$2.45\,\mu$m ("Kdark") Ashley *et al.* (1996) and Nguyen *et al.* (1996) reported a factor of 100 improvement over mid-latitude sites. This is not as dark as expected, by at least a factor of two, and has yet to be satisfactorily explained.

In the L band (2.9–$4.1\,\mu$m) Ashley *et al.* (1996) and Phillips (1999) found a factor of 20–40 times improvement over Siding Spring.

From 4 to $14\,\mu$m Chamberlin *et al.* (2000) found a factor of 10 improvement over Mauna Kea.

From 8.5 to $17\,\mu$m Smith and Harper (1998) measured a factor of 10 improvement over Mauna Kea, and greatly improved temporal stability at $10\,\mu$m.

Away from the South Pole, the only infrared observations I am aware of are from Walden *et al.* (2005) who used an FTS to make summertime measurements at 3–$20\,\mu$m from Dome C. The results were comparable to South Pole wintertime measurements from Smith and Harper (1998).

Interestingly, Lawrence (2004) predicted that Domes C and A will be substantially darker than the South Pole—factors of 2 to 100 from the near to far-IR—and with significant differences between the sites.

For the future we clearly need comprehensive infrared sky background measurements at 1–$20\,\mu$m from Domes A, C, F. We also need further measurements at Kdark, ideally at high enough resolution to resolve any residual airglow lines.

It is worth considering that some of the earlier work from the South Pole may have been affected by aerosol variations in the atmosphere caused by volcanic eruptions (see Figure 6 and the discussion below in section 9).

5. Atmospheric transmission variations arising from gases

For precision optical photometry, the stability of the transmission of the atmosphere is an important consideration. Various molecules contribute, e.g., from 0.3–$2.5\,\mu$m we need to examine H_2O, O_3, O_2, CO_2, CH_4, NO_2, N_2O, and CO. We have already discussed H_2O in the section on precipitable water vapour.

Ozone is not often thought about in this context, but it turns out to be the dominant absorber through much of the optical. Apart from the Hartley bands that absorb in the ultraviolet, the Chappuis bands produce up to $\sim$10% absorption from 400–540 nm. Ozone concentration is measured in Dobson Units (DU), where one DU is equivalent to a $10\,\mu$m layer of pure ozone. Allen and Reck (1997) give a table showing the day-to-day variations in ozone concentration as a function of latitude on the Earth's surface. Mauna Kea and Paranal have variations of $\sim$4–5 DU (this is the standard deviation of the absolute value of the difference from one day to the next). Sites further from the equator have larger variations, e.g., Siding Spring is $\sim$9 DU. Allen and Reck's data do not extend to Antarctica. While ozone does fluctuate on monthly timescales in Antarctica—most notably the ozone hole that is prominent in spring—the day-to-day fluctuations at South Pole in wintertime are relatively small, although still greater than that at Mauna Kea and Paranal, at $\sim$10 DU (Evans 2012).

The concentration of CO_2 is steadily rising from anthropogenic fossil fuel use. It is quite well mixed throughout the atmosphere, and at all latitudes. There is a clear annual fluctuation of $\sim2\%$ peak-to-peak at Mauna Loa, resulting from vegetation growth cycles in the northern hemisphere. The variation is still present at the South Pole, but is significantly less at $\sim0.4\%$ peak-to-peak.

I have not considered the effect of CH_4, NO_2, N_2O, and CO. It would be interesting to do this, and to quantify the stability of the concentrations of all these molecules. This is relevant to understanding the ultimate limits of ground-based photometry.

6. Optical sky brightness and aurora

The optical sky background from Antarctica appears to be comparable to the best mid-latitude observatories, although there is much data that has been taken and not yet published. Zou *et al.* (2010) reports i-band sky brightness, and estimates that aurora affect $\sim2\%$ of observations, based on observations with CSTAR at Dome A in 2008. Other measurements are available from ASTEP data at Dome C (Crouzet *et al.* 2010, 2011) and Nigel at Dome A (Sims *et al.* 2012). Data will soon be published from the Gattini experiments at Domes C, A and South Pole, and the HRCAM instruments on PLATO (Lawrence *et al.* 2009) at Domes A, F, and Ridge A.

For the future, our priority should be to publish the data that has been taken, and to continue to acquire uniform datasets. The strength of the airglow lines could be more thoroughly investigated, from the optical out to $2.4\,\mu$m. In common with other observatories, we expect a variation in sky brightness with the solar cycle, so long-term monitoring is important.

7. Clouds

Measuring the cloud cover over Antarctica during wintertime is complicated by the fact that few stations have a winterover crew; satellite instruments have difficulty in distinguishing cloud from the surface ice. Human "eye-ball" observations of eighths of cloud cover during winter are somewhat unreliable.

The first wintertime cloud observations from the high plateau were reported by the ICECAM experiment at Dome C (Ashley *et al.* 2005). This, and subsequent observations from Domes C and A (e.g., Zou *et al.* 2010, Crouzet *et al.* 2010, 2011), show significantly larger percentages of photometric conditions than mid-latitude sites. E.g., Zou *et al.* (2010) report 67% photometric (<0.3 mag extinction) conditions at Dome A during winter, compared with $\sim50\%$ for Mauna Kea.

For the future, both the Gattini and HRCAM experiments have acquired a large dataset of cloud statistics, and this is awaiting publication. Continuing to acquire a uniform record of the photometric conditions is very important, since this will help define the window functions which quantify the expected performance for, e.g., extrasolar planet transit and variable star campaigns.

8. Meteorological properties

We have a reasonable knowledge of the wind speed and direction, as a function of height, across the plateau. There are also balloon measurements of temperature as a function of height at South Pole and Dome C. Relative humidity measurements are difficult to make. A 15 m meteorological tower on PLATO acquired nine months of high quality data (temperature and wind speed/direction) from Dome A during 2011.

For the future, we need long-term meteorological towers of at least 15 m in height, instrumented with standard sensors, at Dome A, Dome F and Ridge A.

9. Aerosols

The "thumb test" (where you block out the sun using your thumb and look at the brightness of the surrounding sky) is an obvious indication to anyone that visits Antarctica that the aerosol concentrations are very low.

Recently, this has been quantified at Dome C by Dame *et al.* (2012) and Faurobert *et al.* (2012), showing that the sky is a factor of 2–4 times darker than Mauna Kea in the visible, and is dominated by Rayleigh scattering from molecules, not aerosols.

Tomasi *et al.* (2007) reviewed the observations of aerosols over Antarctica. The optical depth from aerosols at the South Pole is about 0.015 at 500 nm, but increases dramatically following volcanic eruptions, as shown in Figure 6.

For the future, we should take aerosol optical depth into consideration when measuring sky transmission, stability, and the infrared sky brightness. Some of the historical measurements may need redoing during periods well away from volcanic activity.

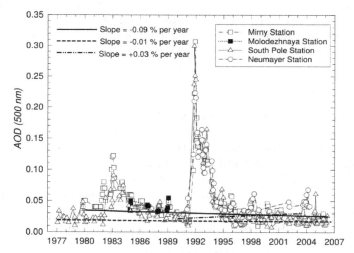

Figure 6. Time variation of the monthly mean aerosol optical depth (AOD) at 500 nm measured from four Antarctic stations. The peaks come immediately following volcanic eruptions: El Chichon in 1982 and Pinatubo and Cerro Hudson in 1991. From Tomasi *et al.* (2007).

10. Conclusion

In some respects, the Antarctic high plateau (Figure 7) is one of the best characterised astronomical sites. The measurements have in most cases confirmed Antarctica as having the best locations on Earth for ground-based observations. There are some important measurements still to be made, notably the infrared sky backgrounds at sites other than the South Pole, and the seeing at Dome A. In both of these examples, there are theoretical reasons for expecting significant improvements that are still to be revealed.

Figure 7. The Antarctic plateau, 3 km from the US Amundsen-Scott South Pole Station, January 2012. M. Ashley.

References

Agabi, A., Aristidi, E., Azouit, M., Fossat, E., Martin, F., Sadibekova, T., Vernin, J., & Ziad, A. 2006, *PASP*, 118, 344

Allen, D. R. & Reck, R. A. 1997, *JGR*, 102, 13603–13608

Argentini, S., Mastrantonio, G., Petenko, I., Pietroni, I., & Viola, A. 2012, *Boundary-layer Meteorology* 2012, 143, 177–188

Aristidi, E., Agabi, A., Fossat, E., Azouit, M., Martin, F., Sadibekova, T., Travouillon, T., Vernin, J., *et al.* 2005, *A&A*, 444, 651

Aristidi, E., Fossat, E., Agabi, A., Mkarnia, D., Jeanneaux, F., Bondoux, E., Challita, Z., Ziad, A., Vernin, J., *et al.* 2009, *A&A*, 499, 955

Ashley, M. C. B., Burton, M. G., Storey, J. W. V., Lloyd, J. P., Bally, J., Briggs, J. W., & Harper, D. A. 1996, *PASP*, 108, 721–723

Ashley, M. C. B., Burton, M. G., Calisse, P. G., Phillips, A., & Storey, J. W. V. 2005, *Highlights of Astronomy, ASP Conference Series*, 13, 936–938

Bonner, C. S., Ashley, M. C. B., Lawrence, J. S., Luong-Van, D. M., & Storey, J. W. V. 2009, *Acoustics Australia*, 37, 47–51

Bonner, C. S., Ashley, M. C. B., Cui, X., Feng, L., Gong, X., Lawrence, J. S., Luong-van, D. M., Shang, Z., Storey, J. W. V., Wang, L., Yang, H., Yang, J., Zhou, X., & Zhu, Z. 2010, *PASP*, 122, 1122–1131

Burton, M. G. 2010, *A&ARev*, 18, 417–469

Chamberlain, M. A., Ashley, M. C. B., Burton, M. G., Phillips, A., Storey, J. W. V., & Harper, D. A. 2000, *ApJ*, 535, 501–511

Crouzet, N., Guillot, T., Agabi, A., Rivet, J., Bondoux, E., *et al.* 2010, *A&A*, 511, 36

Crouzet, N., Guillot, T., Agabi, K., Daban, J.-B., Abe, L., Mekarnia, D., Rivet, J.-P., Fante-Caujolle, Y., Fressin, F., Gouvret, C., Schmider, F.-X., Valbousquet, F., Blazit, A., Rauer, H., Erikson, A., Fruth, T., Aigrain, S., Pont, F., & Barbieri, M. 2011, *EPJ Web of Conferences*, 11, 06001

Dame, L., Abe, L., Faurobert, M., Fineschi, S., Kuzin, S., Lamy, P., Meftah, M., & Vives, S. 2012, *EAS Publications Series*, 55, 359–364

Evans, R., 2012, *Private communication*, based on 2011 South Pole data, NOAA.

Faurobert, M., Arnaud, J., & Vernisse, Y. 2012, *EAS Publications Series*, 55, 365–367

Gillingham, P. 1993, *Optics in Astronomy*, Proceedings of the 32nd Hestmonceux Conference held in 11993. Edited by J.V. Wall. Cambridge, UK: Cambridge University Press, p. 244

Giordano, C., Vernin, J., Chadid, M., Aristidi, E., Agabi, A., & Trinquet, H. 2012 *PASP*, 124, 494–506

Harper, D. A. 1989, *Astrophysics in Antarctica, American Institute of Physics*, 123–129

Kenyon, S. L., Lawrence, J. S., Ashley, M. C. B., Storey, J. W. V., Tokovinin, A.,& Fossat, E. 2006, *PASP*, 118, 924–932

Lascaux, F., Masciadri, E., & Hagelin, S. 2011, *MNRAS*, 411, 693–704

Lawrence, J. S. 2004, *PASP*, 116, 482–492

Lawrence, J. S., Ashley, M. C. B., Tokovinin, A., & Travouillon, T. 2004, *Nature*, 431, 278–281

Lawrence, J. S., Ashley, M. C. B., Hengst, S., Luong-Van, D. M., Storey, J. W. V., Yang, H., Zhou, X., & Zhu, Z. 2009, *Rev. Sci. Inst.*, 80, 064501-1–064501-10

Loewenstein, R. F., Bero, C., Lloyd, J. P., Mrozek, F., Bally, J., & Theil, D. 1998, *ASP Conf Series*, 141, 296

Lubin, D. 1988, *Masters thesis, University of Chicago*

Marks, R. D., Vernin, J., Azouit, M., Briggs, J. W., Burton, M. G., Ashley, M. C. B., & Manigault, J.-F. 1996, *A&A Suppl*, 118, 385–390

Marks, R. D., Vernin, J., Azouit, M., Manigault, J. F., & Clevelin, C. 1999, *A&A Suppl*, 134, 161–172

Neff, W. D. 1981, *PhD thesis*, Wave Propagation Laboratory (Boulder Colorado, USA)

Nguyen, H. T., Rauscher, B. J., Harper, D. A., Loewenstein, R. F., Pernic, R. J., Severson, S. A., & Hereld, M. 1996, *PASP*, 109, 718

Phillips, A., Burton, M. G., Ashley, M. C. B., Storey, J. W. V., Lloyd, J. P., Harper, D. A., & Bally, J. 1999, *ApJ* 527, 1009–1022

Sims, G., Ashley, M. C. B., Cui, X., Everett, J. R., Feng, L., Gong, X., Hengst, S., Hu, Z., Kulesa, C., Lawrence, J. S., Luong-van, D. M., Ricaud, P., Shang, Z., Storey, J. W. V., Wang, L., Yang, H., Yang, J., Zhou, X., & Zhu, Z., 2012 2012, *PASP*, 124, 74–83

Sims, G., Ashley, M. C. B., Cui, X., Everett, J. R., Feng, L., Gong, X., Hengst, S., Hu, Z., Lawrence, J. S., Luong-Van, D. M., Moore, A. M., Riddle, R., Shang, Z., Storey, J. W. V., Tothill, N., Travouillon, T., Wang, L., Yang, H., Yang, J., Zhou, X., & Zhu, Z. 2012, *PASP*, 124, 637–649

Smith, C. H. & Harper, D. A. 1998, *PASP*, 110, 747

Swain M. & Gallée, H. 2006, *PASP*, 118, 1190

Tomasi, C., Vitale, V., Lupi, A., Di Carmine, C., Campanelli, M., Herber, A., Treffeisen, R., Stone, R. S., Andrews, E., Sharma, S., Radionov, V., von Hoyningen-Huene, W., Stebel, K., Hansen, G. H., Myhre, C. L., Wehrli, C., Aaltonen, V., Lihavainen, H., Virkkula, A., Hillamo, R., Strm, J., Toledano, C., Cachorro, V. E., Ortiz, P., de Frutos, A. M., Blindheim, S., Frioud, M., Gausa, M., Zielinski, T., Petelski, T., & Yamanouchi, T. 2012, *JGR*, 112, D12205

Tremblin, P., Minier, V., Schneider, N., Durand, G. A., Ashley, M. C. B., Lawrence, J. S., Luong-van, D. M., Storey, J. W. V., Durand, G. A., Reinert, Y., Veyssiere, C., Walter, C., Ade, P., Calisse, P. G., Challita, Z., Fossat, E., Sabbatini, L., Pellegrini, A., Ricaud, P., & Urban, J. 2011, *A&A*, 535, A112

Trinquet, H., Agabi, A., Vernin, J., Azouit, M., Aristidi, E., & Fossat, E. 2008, *PASP*, 120, 203

Walden, V. P., Town, M. S., Halter, B., & Storey, J. W. V. 2005, *PASP*, 117, 300–308

Yang, H., Kulesa, C. A., Walker, C. K., Tothill, N. F. H., Yang, J., Ashley, M. C. B., Cue, X., Feng, L., Lawrence, J. S., Luong-Van, D. M., Storey, J. W. V., Wang, L., Zhou, X., & Zhu, Z. 2010, *PASP*, 122, 490–494

Zou, H., Zhou, X., Jiang, Z., Ashley, M. C. B., Cui, X., Feng, L., Gong, X., Hu, J., Kulesa, C. A., Lawrence, J. S., Liu, G., Luong-Van, D. M., Ma, J., Moore, A. M., Pennypacker, C. R., Qin, W., Shang, Z., Storey, J. W. V., Sun, B., Travouillon, T., Walker, C. K., Wang, J., Wang, L., Wu, J., Wu, Z., Xia, L., Yan, J., Yang, J., Yang, H., Yao, Y., Yuan, X., York, D. G., Zhang, Z., & Zhu, Z. 2010, *AJ*, 140, 602–611

Astrophysics from Antarctica
Proceedings IAU Symposium No. 288, 2012
M. G. Burton, X. Cui & N. F. H. Tothill, eds.

© International Astronomical Union 2013
doi:10.1017/S1743921312016626

Dome Fuji Seeing -the Summer Results and the Future Winter-over Observations

Hirofumi Okita[1], Naruhisa Takato[2], Takashi Ichikawa[1], Colin S. Bonner[3], Michel C. B. Ashley[3], John W. V. Storey[3] and the 51st and 52nd JARE Dome Fuji teams

[1] Astronomical Institute, Tohoku University,
6-3 Aramaki, Aoba-ku, Sendai, Japan
email: h-okita@astr.tohoku.ac.jp

[2] Subaru Telescope, National Astronomical Observatory of Japan,
650 North A'ohoku Place, Hilo, Hawaii

[3] School of Physics, University of New South Wales,
NSW 2052, Australia

Abstract. We carried out the first seeing measurements at Dome Fuji in the 2010–2011 austral summer. From these observations, we found that the summer seeing at Dome Fuji was 1.2″ (mean), 1.1″ (median), 0.83″ (25th percentile) and 1.5″ (75th percentile), respectively. We also found that the seeing changed continuously and had a minimum around 0.7″ at ∼18:00 hours daily. We compared the seeing with some weather parameters obtained from the 16 m mast, and found that the seeing had good correlations with atmosphere temperature and wind shear. These results suggest that the seeing is degraded by turbulence near the surface boundary layer. Because the data were obtained only over a short duration in summer, the general characteristics of Dome Fuji's seeing could not be evaluated. We plan to observe the seeing in winter with a stand-alone DIMM telescope. This new DIMM, which we named the Dome Fuji Differential Image Motion Monitor (DF–DIMM), will be installed at Dome Fuji in January 2013.

Keywords. site testing, atmospheric effects, telescopes

1. Introduction

The Antarctic plateau is considered as the last frontier for ground-based astronomy. The extremely dry atmosphere, which is related to its low temperature and high altitude of the Antarctic plateau, provides high atmospheric transmittance. To confirm this, site-testing has been carried out at several Antarctic plateau sites (Peterson *et al.* 2003, Tomasi *et al.* 2008, Ishii *et al.* 2010, Sims *et al.* 2012). Furthermore, this site-testing showed extremely good seeing on the Antarctic plateau (Loewenstein *et al.* 1998, Travouillon *et al.* 2003a, 2003b, Aristidi *et al.* 2003, 2005, 2009, Lawrence *et al.* 2004, Bonner *et al.* 2010). Seeing is a parameter that describes how blurry a star image will be. Poor seeing decreases spatial resolution and detection limits. It is important to carry out astronomical observations at good seeing sites. Table 1 shows the measured seeing and the height of the surface boundary layers at various sites on the Antarctic plateau.

Dome Fuji is located at 77°19′S, 39°42′E. The altitude of Dome Fuji is about 3,810m, which is a local maxima, the second highest next to Dome A. Some simulations predict that the seeing and the height of the boundary layer at Dome Fuji is even better than at Dome A or at Dome C (Swain & Gallée 2006, Saunders *et al.* 2009). The first seeing measurements at Dome Fuji were carried out in the 2010–2011 austral summer by the 51st and 52nd Japan Antarctic Research Expedition (JARE).

Table 1. Seeing and the height of the surface boundary layer on the Antarctic plateau.

Site	Elevation	Free atmosphere seeing	Height of the surface boundary layer
South Pole[1]	2,835 m	0.37 arcsec	270 m
Dome A[2]	4,093 m	-	13.9 m
Dome C[3]	3,250 m	0.36 arcsec	23~27 m
Dome Fuji	3,810 m	-	-

Notes: [1] Travouillon *et al.* (2003a); [2] Bonner *et al.* (2010); [3] Aristidi *et al.* (2009).

2. Summer seeing in 2010-2011

We carried out the first seeing measurements at Dome Fuji from January 25 to January 28, 2011. We used a 40 cm primary mirror telescope for Differential Image Motion Monitor (DIMM) observations at Dome Fuji. DIMMs are now widely used for seeing measurements. This telescope was installed on the snow surface, and the DIMM entrance pupils were about 2 m above snow surface.

Figure 1 is the time series data of the DIMM seeing at Dome Fuji. The mean seeing was 1.2″, and the median was 1.1″, the 25th percentile and the 75th percentile were 0.83″ and 1.5″, respectively.

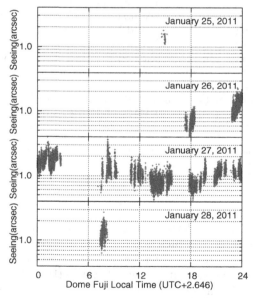

Figure 1. DIMM seeing at the wavelength λ=0.55μm from January 25 to January 28, 2011 as a function of the Dome Fuji local time, UTC+2.646. All measurements were carried out during daytime (when the Sun does not set). We chose exposure times of 1/1,000 second. The height of the entrance pupils of the DIMM apertures was ∼2 m above the snow surface.

We examined the time dependence of the seeing. Figure 2 (i) is the seeing averaged in one-hour intervals. The seeing changed continuously and had a minimum around 0.7″ at ∼18:00 hours. This trend is similar to the day-time seeing at Dome C (Aristidi *et al.* 2005). From comparison with the seeing and some meteorological data measured by the 16 m weather mast, we found that the seeing time dependence correlates with the atmospheric temperature and the wind shear near the snow surface. Figure 2 (ii) and (iii) show these results, averaged for each one-hour observation for the temperature (°C) and wind speed (m/s).

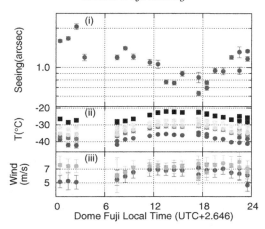

Figure 2. (i) Average seeing in one-hour intervals versus Dome Fuji local time for all seeing data in the 2010–2011 campaign. The error bars indicated the standard error. (ii) and (iii) are the one-hour averaged data measured at the 16 m weather mast; (ii) temperature with arbitrary offsets, (iii) wind speed. In (ii), the blue, magenta, cyan, yellow and black dots represent the temperature at 0.3 m above snow surface with offset $-6°$C, 6.5m with offset $-3°$C, 9.5 m without offset, 12 m with offset $+3°$C, and 15.8 m with offset $+6°$C, respectively. The red and the green dots in (iii) are the results at height of 6.1 m and 14.4 m. The error bar for the meteorological data indicates the standard deviation of each one-hour interval.

Figure 3 (left) is a scatter-gram showing the seeing with atmospheric temperature near the snow surface. The correlation coefficient between the seeing and the temperature is $-0.91 \sim -0.74$. This correlation suggests that the seeing is sensitive to the temperature variation near the snow surface, or the surface boundary layer. The scatter-gram of the seeing and the wind shear, which are measured at heights of 6.1 m and 14.4 m, are plotted in Figure 3 (right). The correlation coefficient is 0.81. The friction inside the atmosphere is caused by the wind shear, resulting in turbulence. Hence the seeing would change due to variations in the turbulence strength near the snow surface.

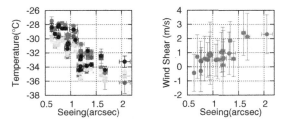

Figure 3. (Left) Correlation between seeing and surface temperature. The abscissa is seeing in arcsec, and the ordinate is the temperature averaged over each one-hour observation. (Right) Correlation between seeing and wind shear.

The general characteristics of Dome Fuji's seeing could not be evaluated from our results because the data were obtained over only a short duration in summer. As there is no sunrise in the Antarctic winter (i.e. polar night), the conditions of the atmosphere at Dome Fuji will be not then be the same as in summer.

3. Future seeing measurements

We aim to determine the winter-time seeing at Dome Fuji. In the 2012–2013 campaign there was no winter crew at Dome Fuji station, so we have to operate any instruments automatically. PLATO-F (Hengst *et al.* 2010; Ashley *et al.* 2010), which was installed at Dome Fuji in January 2011, will be the provide for this unmanned operation. It supplies a maximum of 2 kW of electric power and Iridium internet communications.

We developed a stand-alone DIMM telescope. We call this new telescope the "Dome Fuji Differential Image Motion Monitor (DF–DIMM)". We used a Meade LX200-8" ACF for the telescope and SBIG ST-i CCD cameras for detector and wide-field finder. We replaced the grease, bearings and cables. We placed some heaters inside for $-80°C$ operation. The C language, awk and bash scripts were used for DF–DIMM software. Pointing, focusing, seeing measurements and data transfer are carried out automatically. We plant to set up the DF–DIMM this austral summer, and then will start annual seeing measurements at Dome Fuji from 2013.

Acknowledgements

We acknowledge the National Institute of Polar Research, Japan for funding and for the logistics support. Kentaro Motohara gave us his user-friendly DIMM software and some technical support. This work has been supported in part by a Grant-in-Aid for Scientific Research 18340050 and 21244012 of the Ministry of Education, Culture, Sports, Science and Technology in Japan. Hirofumi Okita thanks the scholarships and grant-in-aid of Tohoku University International Advanced Research and Education Organization.

References

Aristidi, E., Agabi, A., Vernin, J., *et al.* 2003, *A&A*, 406, L19
Aristidi, E., Agabi, A., Fossat, E., *et al.* 2005, *A&A*, 444, 651
Aristidi, E., Fossat, E., Agabi, A., *et al.* 2009, *A&A*, 499, 955
Ashley, M. C. B., Bonner, C. S., Everett, J. R., *et al.* 2010, *Proc. SPIE*, 7735, 133
Bonner, C. S., Ashley, M. C. B., Cui, X., *et al.* 2010, *PASP*, 122, 1122
Hengst, S., Luong-Van, D. M., Everett, J. R., *et al.* 2010, *International Journal of Energy Research*, 34, 827
Ishii, S., Seta, M., Nakai, N., *et al.* 2010, *Polar Science*, 3, 213
Lawrence, J. S., Ashley, M. C. B., Tokovinin, A., & Travouillon, T. 2004, *Nature*, 431, 278
Loewenstein, R. F., Bero, C., Lloyd, J. P., *et al.* 1998, *ASP Conf. Ser.*, 141, 296
Peterson, J. B., Radford, S. J. E., Ade, P. A. R., *et al.* 2003. *PASP*, 115, 383
Saunders, W., Lawrence, J. S., Storey, J. W. V., *et al.* 2009, *PASP*, 121, 976
Sims, G, Ashley, M. C. B., Cui, X., *et al.* 2012, *PASP*, 124, 74
Swain, M. & Gallée, H. 2006, *PASP*, 118, 1190
Tomasi, C., Petkov, B., Benedetti, E., *et al.* 2008 *J Atmos Ocean Technol*, 25, 213
Travouillon, T., Ashley, M. C. B., Burton, M. G., Storey, J. W. V., & Loewenstein, R. F. 2003, *A&A*, 400, 1163
Travouillon, T., Ashley, M. C. B., Burton, *et al.* 2003, *A&A*, 409, 1169

Astrophysics from Antarctica
Proceedings IAU Symposium No. 288, 2012
M. G. Burton, X. Cui & N. F. H. Tothill, eds.

© International Astronomical Union 2013
doi:10.1017/S1743921312016638

A worldwide comparison of the best sites for submillimetre astronomy

P. Tremblin[1], N. Schneider[1,2], V. Minier[1], G. Al. Durand[1] and J. Urban[3]

[1] Laboratoire AIM Paris-Saclay (CEA/Irfu - Uni. Paris Diderot - CNRS/INSU), Centre d'études de Saclay, 91191 Gif-Sur-Yvette, France
email: `pascal.tremblin@cea.fr`

[2] Université de Bordeaux, LAB, UMR 5804, CNRS, 33270, Floirac, France

[3] Chalmers University of Technology, Department of Earth and Space Sciences, 41296 Göteborg, Sweden

Abstract. Over the past few years a major effort has been put into the exploration of potential sites for the deployment of submillimetre (submm) astronomical facilities. Amongst the most important sites are Dome C and Dome A on the Antarctic Plateau, and the Chajnantor area in Chile. In this context, we report on measurements of the sky opacity at 200 μm over a period of three years at the French-Italian station, Concordia, at Dome C, Antarctica. Based on satellite data, we present a comparison of the atmospheric transmission at 200, 350 μm between the best potential/known sites for submillimetre astronomy all around the world.

The precipitable water vapour (PWV) was extracted from satellite measurements of the Infrared Atmospheric Sounding Interferometer (IASI) on the METOP-A satellite, between 2008 and 2010. We computed the atmospheric transmission at 200 μm and 350 μm using the forward atmospheric model MOLIERE (Microwave Observation LIne Estimation and REtrieval). This method allows us to compare known sites all around the world without the calibration biases of multiple in-situ instruments, and to explore the potential of new sites.

1. Introduction

A major obstacle to ground-based observations in the submm range (and specifically at wavelengths shorter than 500 μm) is the atmosphere. This part of the electromagnetic spectrum is normally the preserve of space telescopes such as the Herschel space observatory (Pilbratt *et al.* 2010) although large submm facilities such as ALMA will be able to operate down to 420 μm and possibly below in the future (Hills *et al.* 2010). However, submm observations in the 200-μm window with ground-based instruments will always require exceptional conditions (see Marrone *et al.* 2005; Oberst *et al.* 2006).

For ground-based sites, previous studies (e.g. Schneider *et al.* 2009; Tremblin *et al.* 2011; Matsushita *et al.* 1999; Peterson *et al.* 2003) already showed that a few sites are well-suited for submm, mid-IR, and FIR astronomy and their transmission properties are rather well determined (for example by Fourier Transform Spectrometer obervations in the 0.5-1.6 THz range at Mauna Kea/Hawaii (Pardo *et al.* 2001)). The high-altitude ($\geqslant 5000$ m) Chilean sites are known for dry conditions (see Matsushita *et al.* 1999; Peterson *et al.* 2003), and site testing is now carried out at the driest place on Earth, Antarctica (see Chamberlin *et al.* 1997; Yang *et al.* 2010; Tremblin *et al.* 2011). Comparisons between Antarctic and Chilean sites are difficult and uncertain since they rely on ground-based instruments that use different methods and calibration techniques (see Peterson *et al.* 2003, for example). The working conditions are also an important issue, a single instrument moved from Chile to Antarctica will have a different behavior in the harsh polar environment ($-70°C$ in winter at Dome C). A meaningful comparison is possible if several independent instruments are used at each place. An example of such a study is the one of Tremblin *et al.* (2011) that obtained transmission data at Dome C thanks to radio-soundings and the radiometers HAMSTRAD (Ricaud *et al.* 2010) and SUMMIT08. However, it is rare to have many instruments at one site. The best solution is to use satellite data, which also enables to investigate any location on Earth. Thanks to the IASI (Infrared Atmospheric Sounding Interferometer) on the Metop-A satellite, it is now possible to conduct such a comparison over several years with no instrument bias and with the same working conditions for the detectors. We present here a 3-year study of the PWV of a selection of existing and upcoming submm-sites in Antarctica, Chile, Tibet, and Argentina, as well as for two SOFIA stations, Palmdale/California and Christchurch/New Zealand. A direct comparison is enabled by comparing the individual and cumulated quartiles of PWV and the transmission is given, using these PWV values and the atmospheric model MOLIERE-5 (Urban *et al.* 2004).

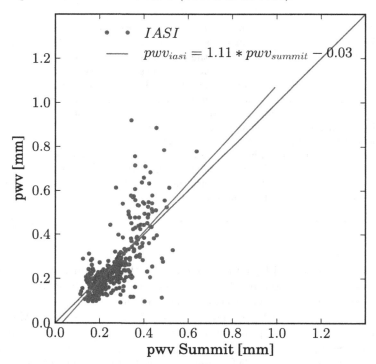

Figure 1. Correlation between in-situ measurements (SUMMIT08) and satellite measurments (IASI) between 2008 and 2010 at the Concordia station in Antarctica.

2. In-situ and satellite PWV measurements

IASI (Infrared Atmospheric Sounding Interferometer) is an atmospheric interferometer working in the infrared, launched in 2006 on the METOP-A satellite (Phulpin *et al.* 2007; Pougatchev *et al.* 2008; Herbin *et al.* 2009; Clerbaux *et al.* 2009). Vertical profiles of tropospheric humidity at ninety altitude levels (resolution 1 km) are retrieved with a typically 10% accuracy (Pougatchev *et al.* 2008). The amount of precipitable water vapour (PWV) is given by the integral of these vertical profiles. All the measurements in a zone of 110 km^2 around the site of interest are averaged. In-situ measurements at the Concordia station in Antarctica were performed between 2008 and 2010 thanks to a radiometer SUMMIT08. These data were compared with the satellite data (see Tremblin *et al.* 2011) and the correlation between the two is given in Fig. 1. There is a negative bias for SUMMIT08 (SUMMIT08 values are lower than the ones for IASI) at high PWV (more than 0.4 mm) that was identified as an instrumental effect of SUMMIT08. Nevertheless, the correlation at low PWV is very good and validates the use of the measurements from IASI to compare the PWV statistics between different sites for astronomical purposes.

The use of vertical satellite profiles is slightly trickier for a mountain site. Since we take all measurements in a zone of 110 km^2, we sometimes get profiles that do not contain mountain altitude but include lower ones. This would bias the retrieved PWV to high values. To overcome this difficulty, we generally truncated the profiles at the altitude of the site of interest. This method was already used by Ricaud *et al.* (2010) to compare IASI measurements with the HAMSTRAD radiometer, over the Pyrenees mountains. They showed a very good correlation for the integrated PWV. We also determined in this way the PWV content at high altitudes ($\geqslant 11$ km) over Palmdale, USA and Christchurch, New Zealand for the on-going and future flights of SOFIA. Table 1 shows the comparison between all the sites with the first decile and the quartiles of the PWV statistics between 2008 and 2010. These results clearly show that Antarctic sites are the driest sites followed by South-American sites and then northern-hemisphere sites. Our long-term satellite statistic of PWV shows that the site of Summit in Greenland offers comparable observing conditions (PWV and altitude) to Mauna Kea, which opens a new perspective for submm astronomy in the northern hemisphere.

Table 1. First decile and quartiles of the PWV for all the studied sites.

Time fract. 2008-2010	SOFIA Palm./Christ.	Dome A	Dome C	South Pole	Cerro Chaj.	Chaj. Plat.	Sum- mit	Cerro Macon	Mauna Kea	Yang- bajing
0.10	0.006/0.004	0.11	0.17	0.15	0.27	0.39	0.36	0.47	0.62	1.21
0.25	0.006/0.005	0.16	0.22	0.21	0.37	0.53	0.51	0.66	0.91	2.47
0.50	0.007/0.006	0.21	0.28	0.30	0.61	0.86	1.94	0.02	1.44	inf
0.75	0.009/0.007	0.26	0.39	0.49	1.11	1.63	1.96	1.66	2.57	inf

3. Transmission at 200 μm and 350 μm

For the determination of the tropospheric transmission corresponding to the PWVs of the various deciles and quartiles for each site, we use MOLIERE-5.7 (Microwave Observation and LIne Estimation and REtrieval), a forward and inversion atmospheric model (Urban *et al.* 2004), developed for atmospheric science applications. It has previously been used to calculate the atmospheric transmission up to 2000 GHz (≈ 150 μm) for a large number of astronomical sites (see Schneider *et al.* (2009) and http://submm.eu). The first decile and the quartiles of the transmission at 200 μm and 350 μm are given in Table 2. For the first quartile, only the transmission at Cerro Chajnantor catches up Antarctica thanks to the high altitude of the site. However for the second and third quartiles Antarctic sites have a better transmissions based on the long term statistics, especially at 350 μm. For the three sites, Chajnantor Plateau, Cerro Macon and Mauna Kea, the transmission window at 350 μm opens significantly while it is only rarely possible to observe at 200 μm.

The stability of the transmission can be compared thanks to the site photometric quality ratio (SPQR) introduced by De Gregori *et al.* (2012). It consists in the ratio of the monthly averaged transmission to its monthly standard deviation, on a daily time-scale. The averaged value of the SPQR ratio for the transmission at 200 μm between 2008 and 2010 are given in Table 2. A first look at the values shows that all temperate sites have a SPQR ratio lower than 1 while Antarctic sites have a ratio greater than 1. The averaged transmission is lower than its fluctuations on temperate sites i.e., the transmission is highly variable. Note that the Arctic site on the Summit mountain is also highly variable, hence Antarctica is really unique even among polar environments. On the Antarctic plateau, Dome A has the best SPQR ratio with a monthly-averaged transmission that is typically 3-4 times higher than the fluctuations. Dome C achieves also very good conditions with a ratio of the order of 2-3 while the South Pole has a monthly-averaged transmission of the same order of the fluctuations comparable to the conditions reached at Cerro Chajnantor.

Table 2. First decile and quartiles of the 350-μm (top) and 200-μm (middle) transmissions for all the studied sites. Bottom: averaged value of the SPQR ratio at 200 μm between 2008 and 2010.

Time fraction 2008-2010	Dome C	Dome A	South Pole	Cerro Chaj.	Chaj. Plat.	Cerro Macon	Mauna Kea	Summit	Yangbajing
0.10	0.62	0.72	0.61	0.65	0.56	0.49	0.47	0.42	0.19
0.25	0.57	0.67	0.56	0.58	0.46	0.41	0.36	0.31	0.02
0.50	0.51	0.62	0.47	0.44	0.31	0.29	0.21	0.15	0.00
0.75	0.41	0.57	0.34	0.24	0.12	0.15	0.07	0.02	0.00
Time fraction 2008-2010	Dome C	Dome A	South Pole	Cerro Chaj.	Chaj. Plat.	Cerro Macon	Mauna Kea	Summit	Yangbajing
0.10	0.17	0.32	0.16	0.20	0.09	0.05	0.05	0.03	0.00
0.25	0.11	0.22	0.11	0.12	0.04	0.02	0.02	0.01	0.00
0.50	0.07	0.16	0.05	0.04	0.01	0.00	0.00	0.00	0.00
0.75	0.03	0.11	0.01	0.00	0.00	0.00	0.00	0.00	0.00
SPQR ratio	2.7	3.6	1.3	1.1	0.7	0.7	0.7	0.6	0.3

4. Conclusions

Among all the sites studied, Cerro Chajnantor and the Antarctic Plateau present the best conditions for submm astronomy. However only Dome A and Dome C have a stable transmission that will allow unique science such as time-series and large surveys to be performed there. The method used to compare the different sites is robust and based on only one instrument, IASI, and the atmospheric model MOLIERE. A calculator to show the PWV statistics and to compute the corresponding transmission at any given wavelength is available to the community at http://irfu.cea.fr/submm and http://submm.eu for all the sites presented here and for the three years 2008, 2009 and 2010.

References

Chamberlin, R. A., Lane, A. P., & Stark, A. A. 1997, *ApJ*, 476, 428
Clerbaux, C., Boynard, A., Clarisse, L., *et al.* 2009, *ACP*, 9, 6041
De Gregori, S., de Petris, M., Decina, B., *et al.* 2012, *MNRAS*, 425, 222
Herbin, H., Hurtmans, D., Clerbaux, C., Clarisse, L., & Coheur, P. F. 2009, *ACP*, 9, 9433
Hills, R. E., Kurz, R. J., & Peck, A. B. 2010, *SPIE*, 7733, 773317
Marrone, D., Blundell, R., Tong, E., *et al.* 2005, *STT*

Matsushita, S., Matsuo, H., Pardo, J. R., & Radford, S. J. E. 1999, *PASJ*, 51, 603

Oberst, T. E., Parshley, S. C., Stacey, G. J., *et al.* 2006, *ApJ*, 652, L125

Pardo, J. R., Cernicharo, J., & Serabyn, E. 2001, *IEEE*, 49, 1683

Peterson, J. B., Radford, S. J. E., Ade, P., *et al.* 2003, *PASP*, 115, 383

Phulpin, T., Blumstein, D., Prel, F., *et al.* 2007, *AERSDPU III*, 6684, 12

Pilbratt, G. L., Riedinger, J. R., Passvogel, T., *et al.* 2010, *A&A*, 518, L1

Pougatchev, N., August, T., Calbet, X., *et al.* 2008, *EOS XIII*, 7081, 18

Ricaud, P., Gabard, B., Derrien, S., *et al.* 2010, *ITGRS*, 48, 2189

Schneider, N., Urban, J., & Baron, P. 2009, *PSS*, 57, 1419

Tremblin, P., Minier, V., Schneider, N., *et al.* 2011, *A&A*, 535, 112

Urban, J., Baron, P., Lautié, N., *et al.* 2004, *JQSRT*, 83, 529

Yang, H., Kulesa, C. A., Walker, C. K., *et al.* 2010, *PASP*, 122, 490

Astrophysics from Antarctica
Proceedings IAU Symposium No. 288, 2012
M. G. Burton, X. Cui & N. F. H. Tothill, eds.

© International Astronomical Union 2013
doi:10.1017/S174392131201664X

Winter sky brightness and cloud cover at Dome A, Antarctica

Anna M. Moore[1], Yi Yang[2,3], Jianning Fu[2], Michael C. B. Ashley[4], Xiangqun Cui[5], LongLong Feng[6,7], Xuefei Gong[5,7], Zhongwen Hu[5,7], Jon S. Lawrence[8,9], Daniel M. Luong-Van[4], Reed Riddle[1], Zhaohui Shang[7,10], Geoff Sims[4], John W. V. Storey[4], Nicholas F. H. Tothill[11], Tony Travouillon[12], Lifan Wang[3,6,7], Huigen Yang[7,13], Ji Yang[6], Xu Zhou[7,14], Zhenxi Zhu[6,7]

[1] Caltech Optical Observatories, California Institute of Technology, 1200 E. California Blvd., Pasadena, CA 91107, USA
email: amoore@astro.caltech.edu

[2] Department of Astronomy, Beijing Normal University, Beijing, China

[3] Department of Physics and Astronomy, Texas A&M University, College Station 77843, USA

[4] School of Physics, University of New South Wales, Sydney NSW 2052, Australia

[5] Nanjing Institute of Astronomical Optics & Technology, Nanjing 210042, China

[6] Purple Mountain Observatory, Nanjing 210008, China

[7] Chinese Center for Antarctic Astronomy, China

[8] Department of Physics and Astronomy, Macquarie University, Sydney NSW 2109, Australia

[9] Australian Astronomical Observatory, Sydney NSW 1710, Australia

[10] Tianjin Normal University, Tianjin 300074, China

[11] School of Computing, Engineering & Mathematics, University of Western Sydney, Locked Bag 1797, Penrith South DC, NSW 1797, Australia

[12] California Institute of Technology, 1200 E. California Blvd., Pasadena, CA 91107, USA

[13] Polar Research Institute of China, Shanghai 200136, China

[14] National Astronomical Observatories, Chinese Academy of Science, Beijing 100012, China

Abstract. At the summit of the Antarctic plateau, Dome A offers an intriguing location for future large scale optical astronomical observatories. The Gattini Dome A project was created to measure the optical sky brightness and large area cloud cover of the winter-time sky above this high altitude Antarctic site. The wide field camera and multi-filter system was installed on the PLATO instrument module as part of the Chinese-led traverse to Dome A in January 2008. This automated wide field camera consists of an Apogee U4000 interline CCD coupled to a Nikon fisheye lens enclosed in a heated container with glass window. The system contains a filter mechanism providing a suite of standard astronomical photometric filters (Bessell B, V, R) and a long-pass red filter for the detection and monitoring of airglow emission. The system operated continuously throughout the 2009, and 2011 winter seasons and part-way through the 2010 season, recording long exposure images sequentially for each filter. We have in hand one complete winter-time dataset (2009) returned via a manned traverse. We present here the first measurements of sky brightness in the photometric V band, cloud cover statistics measured so far and an estimate of the extinction.

Keywords. site testing, surveys

1. Introduction

Dome A is the highest point on the Antarctic plateau and a site of recent and extensive astronomical site testing (Yang *et al.* (2009)). The primary goals of the Gattini Dome A project are to (i) obtain the sky brightness in the B, V, and R photometric bands above the Dome A site, (ii) to measure the cloud cover extent during the 2009 and 2010 winter seasons (later extended to 2011 by the National Science Foundation Office of Polar Programs) and (iii) to provide aurora and airglow statistics. Though the experiment was not necessarily designed with high precision photometry in mind an additional goal became (iv) to perform photometry of bright target stars in the field monopolizing on the unprecedented window function available from such a site. The camera hardware, performance and ambitious journey to the highest point on the Antarctic plateau is described in literature (Moore *et al.* (2010)).

1.1. *Data and reduction*

The camera operated throughout the austral winter seasons of 2009 and 2011, and part way through 2010. During operation the filter wheel was cycled continuously with two exposures taken for each filter, of length 30s and 100s. The results presented here concentrate on images taken with the Bessel V band filter of exposure time 100s taken during the 2009 season. The images were bias and dark subtracted and calibrated using 100 or so of the brightest stars across the field. The IRAF 'apphot' routine was used to perform aperture photometry. Approximately 1000 images per day were collected by the system. A limitation on the data reduction was the absence of an adequate flat field image given the difficulty of obtaining such an image on-sky with such a wide field and the stationary nature of the camera. This will be remedied with the acquisition of multi-filter flat field images given the return of the camera to Caltech in mid-2012.

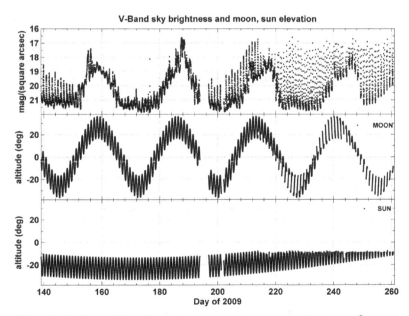

Figure 1. (Upper) Median V band sky brightness in magnitudes/arcsec^2 measured during the 2009 winter season. Recording starts in mid-May 2009. (Lower) Corresponding solar and (middle) lunar elevation are plotted for reference.

1.2. *V-band Sky brightness*

Fig. 1 shows the median sky brightness in V band magnitudes/arcsec^2 for the 2009 austral winter season. The figure shows the corresponding solar and lunar elevation angles for reference. The data has no pre-selection of any form. No attempt as yet has been made to identify areas of low star contamination, and it is acknowledged this could be a contributor, in places, to the median sky brightness displayed in Fig. 1 given the pixel size on the sky is approximately 150 arcsec^2. The general form of the median sky brightness is influenced, as expected, by the lunar monthly cycle and the solar diurnal cycle. However, during periods of low solar elevation and moonless conditions, we see evidence of a minimum of just above 22 magnitude/arcsec^2.

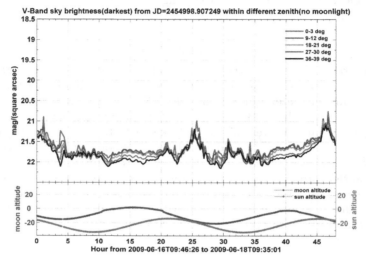

Figure 2. (Top) 48 hour time series of median V band sky brightness under moonless conditions for a range of zenith distances. (Lower) Corresponding lunar and solar elevation. Times shown are UT. For color figures, see on-line version.

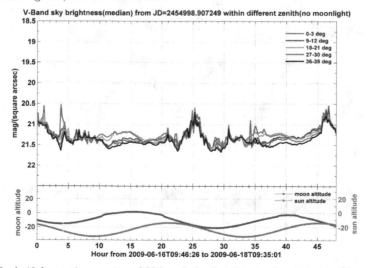

Figure 3. (Top) 48 hour time series of V band sky brightness for darkest 2% under moonless conditions for a range of zenith distances. (Lower) Corresponding lunar and solar elevation. Times shown are UT. For color figures, see on-line version.

For clarity, Fig. 2 shows the equivalent plot for a range of zenith distances over a 48 hour period close to mid-winter. The plot shows a range of values from 21 to 21.5 magnitudes/arcsec^2 with an approximate diurnal peak. The origin of the peak is unknown at this time but it is not caused by direct solar illumination of the upper atmosphere. More likely is an increased aurora activity and/or an airglow related increase. More interestingly, shown in Fig. 3 is the same plot but pre-selected for the darkest 2% of sky brightness values. Here we see a range of sky brightness values between 21.5 and 22.2 magnitudes/arcsec^2. We believe this reflects the sky brightness with the absence of strong aurora lines, given the brightest aurora line corresponding to [OI] emits at 557.7nm that falls within the large passband of the V filter. Work is ongoing in collaboration with the Nigel spectrometer team (Sims *et al.* (2012)) to classify aurora events in the Gattini dataset so that a realistic estimate of sky brightness with and without strong aurora lines can be determined. This is warranted as in practice, strong atmospheric emission lines such as [OI] can be avoided when designing large scale future experiments such as the use of a slightly shortened red cutoff SDSS g' filter in imaging surveys (Zhou *et al.* (2010), Moore *et al.* (2010)).

1.3. *Cloud cover*

A preliminary analysis of cloud cover was performed. Cloud cover was estimated by creating a reference 'pseudo' star from a weighted average of many stars across the field of view. An increase in the pseudo star magnitude it is assumed equates to the presence of some form of extinction whether this is cloud cover or local snow lifted above the height of instrument. Using this method, a total of 75% of the data collected during the 2009 winter season shows a dimming of the pseudo star of less than 0.5 magnitudes only, that corresponds to a cloud cover of less than 37%. This analysis is on-going.

1.4. *Atmospheric extinction*

Atmospheric extinction has been approximated for the 2009 V band data. The lack of adequate flat fields representing the response of uniform illumination across the 90° field of view of the instrument results in large errors for the approximated extinction values. Extinction values of 0.1 to 0.2 per airmass were calculated. However, accurate extinction will be calculated within the next few months given the arrival of the camera system to Caltech in August 2012 and with this the ability to measure accurate multi-filter flat fields in the laboratory.

1.5. *Acknowledgements*

The research is supported by the Chinese PANDA International Polar Year project and the Polar Research Institute of China. The project was funded by the following awards from the National Science Foundation Office of Polar Programs: ANT 0836571, ANT 0909664 and ANT 1043282. We thank Shri Kulkarni and Caltech Optical Observatories, Gerard Van Belle and Chas Beichman for financial contribution to this project. The operation of PLATO at Dome A is supported by the Australian Research Council, the Australian Antarctic Division, and the University of New South Wales. The authors wish to thank all the members of the 2008/2009/2010 PRIC Dome A heroic expeditions.

References

Yang, H., *et al.* 2009, *Publications of the Astronomical Society of the Pacific*, 121, 876
Moore, A. M., *et al.* 2010, *Proceedings of the SPIE*, 7733
Sims, G., *et al.* 2012, *Publications of the Astronomical Society of the Pacific*, 124, 916
Zhou, X., *et al.* 2010, *Publications of the Astronomical Society of the Pacific*, 122, 889

Astrophysics from Antarctica
Proceedings IAU Symposium No. 288, 2012
M. G. Burton, X. Cui & N. F. H. Tothill, eds.
© International Astronomical Union 2013
doi:10.1017/S1743921312016651

First look at HRCAM images from Dome A, Antarctica

Geoff Sims[1], Michael C. B. Ashley[1], Xiangqun Cui[2],
LongLong Feng[3,4], Xuefei Gong[2,4], Zhongwen Hu[2,4],
Jon S. Lawrence[5,6], Daniel M. Luong-Van[1], Zhaohui Shang[4,7],
John W. V. Storey[1], Nick Tothill[8], Lifan Wang[3,4,9], Huigen Yang[4,10],
Ji Yang[3], Xu Zhou[4,11] and Zhenxi Zhu[3,4]

[1] School of Physics, University of New South Wales, Sydney NSW 2052, Australia
email: g.sims@unsw.edu.au

[2] Nanjing Institute of Astronomical Optics & Technology, Nanjing 210042, China

[3] Purple Mountain Observatory, Nanjing 210008, China

[4] Chinese Center for Antarctic Astronomy, China

[5] Department of Physics and Astronomy, Macquarie University, Sydney NSW 2109, Australia

[6] Australian Astronomical Observatory, Sydney NSW 1710, Australia

[7] Tianjin Normal University, Tianjin 300074, China

[8] University of Western Sydney, Sydney NSW, Australia

[9] Department of Physics and Astronomy, Texas A&M University, College Station 77843, USA

[10] Polar Research Institute of China, Shanghai 200136, China

[11] National Astronomical Observatories, Chinese Academy of Science, Beijing 100012, China

Abstract. HRCAM (High Resolution CAMera) is a Canon 50D 15-megapixel digital SLR camera equipped with a Sigma 4.5 mm f/2.8 fish-eye lens. It was installed at Dome A on the Antarctic plateau in January 2010 and photographs the sky every 15 minutes. Primarily functioning as a site-testing instrument, data obtained from HRCAM provide valuable statistics on cloud cover, sky transparency and the distribution and frequency of auroral activity. We present a first look at data from HRCAM during 2010, including an overview of how we intend to reduce the images. We also demonstrate the potential of stellar photometry by using linear combinations of the in-built Canon RGB filters to convert instrumental magnitudes into the photometric BVR bands.

Keywords. Dome A, site testing, aurora, sky brightness, sky transparency

1. Introduction

The highest point on the Antarctic plateau, Dome A, has been home to a host of site testing and scientific instruments since the first Plateau Observatory (PLATO) was installed in 2008. HRCAM (High Resolution CAMera) is one such instrument, whose purpose is twofold: primarily it is a site testing camera to provide statistics on cloud cover, auroral distribution and sky transparency; secondly it serves as a visual reference to check any epochs at which other instruments record unusual results. In the case of HRCAM-3 at Ridge A (see Section 2), in the absence of a meteorological tower it was able to provide an indication of wind speed and direction by the direct observation of a flag. In many ways HRCAM is complementary to various instruments such as Nigel (Sims *et al.* 2010), Gattini (Moore *et al.* 2010) and CSTAR (Yuan *et al.* 2008), since it monitors similar phenomena but with the added advantage of all-sky coverage.

2. Instrumentation & Control

Instrumentation: HRCAM is composed of a Canon 50D digital SLR (DSLR) camera and a Sigma 4.5 mm f/2.8 fish-eye lens. This combination provides full sky coverage on the camera's 15.1 megapixel and 1.6x crop† sensor. The camera and lens assembly are contained in an enclosure which is designed to operate down to −80°C. The lens is flush with the top and exposed directly to the sky. A reflective metal cover surrounds the lens and is designed to shed snow; it has a low emissivity so that it can be heated efficiently. Figure 1 shows a photograph of HRCAM, shortly after being installed at Dome A. The holes on the bottom right in the Figure are for a dehumidifying system. The shutter is rated for 100,000 actuations, limiting the cadence of observations (based on the anticipated duration of the experiment).

Figure 1. HRCAM installed on the roof of the PLATO instrument module at Dome A in 2010.

Control: A single Milspec connector provides 24 VDC power and a 100 Mbps LAN connection. The enclosure includes an ARM-based computer running Linux, and the camera is controlled via a USB connection using gphoto2. The raw images are stored on an array of four 500 GB spinning hard-disks inside PLATO. Exposures are taken every 15 minutes and vary from 1/2000th of a second to 120 seconds depending on the sky brightness, and are automatically set by the software. Images are shot in 14-bit raw (.CR2) format, and thumbnail representations (80×80 px JPEG images; ~ 5 kB in size) are sent back via PLATO's Iridium satellite link. Despite the low resolution of the thumbnails, they are large enough to visually detect cloud, "diamond dust" (including 22° halos) and aurorae.

HRCAM clones: In addition to the original HRCAM at Dome A (which ran throughout 2010), there are two other exact clones on other sites on the high plateau: HRCAM-F at Dome F (2011), and HRCAM-3 at Ridge A (2012). As a result of the various deployment dates, there is so far virtually no overlap in data between any of the various HRCAMs.

3. Reduction

As with most consumer DSLRs, the Canon 50D uses a Bayer colour filter array (CFA). The structure of this array is different from conventional astronomical CCD imagers and thus requires a different reduction process. We first convert the raw images into FITS files using dcraw (Coffin 2012) and cr2fits.py (Jordahl 2012). The 14-bit data are preserved with no scaling or white balance adjustments. The resulting image is then split into four colour channels: R, G_1, G_2 and B, each channel being 25% of the original resolution. A schematic of the pipeline is shown in Figure 2.

---

† Most consumer-grade digital cameras, including the Canon 50D, have CCD sensors smaller than the standard 36×24 mm (35 mm) film format, which results in the sensor capturing only a subset of the 'full frame' image.

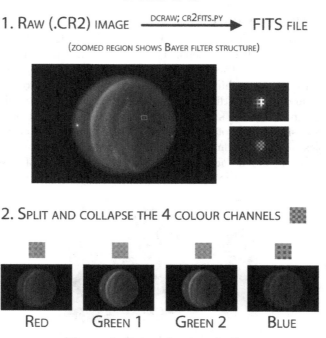

Figure 2. Data reduction pipeline.

4. Results & sample images

Photometry: As shown in Figure 3, the transmission profiles of the Bayer CFA are different to the standard BVR photometric filters. We investigated the possibility of using linear combinations of the colours to more closely match the standard filters. Aperture photometry was performed on a sample green image, using 14 stars, each at a similar airmass and with $0 < m_v < 6$. The rms error associated with the simple photometry was 0.29 mag. With a colour correction term $(B - G)$ added (determined using a least-squares method), this error dropped to 0.23 mag; an improvement of $\sim 20\%$.

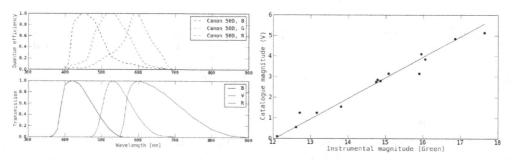

Figure 3. *Left*: comparison of Canon RGB and photometric BVR filter profiles; *Right*: Preliminary photometry test.

Sample images: Four sample images are given in Figure 4.

5. Acknowledgements

The author acknowledges the Astronomical Society of Australia (ASA) for providing travel support to attend this meeting.

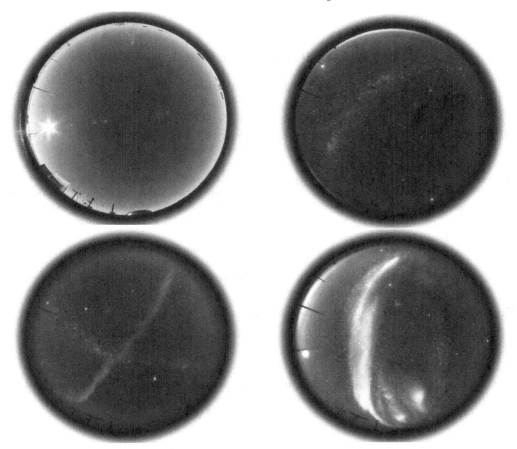

Figure 4. Sample HRCAM images. *Upper left*: blue sky during summer; *Upper right:* clear sky during winter, green aurorae visible on the horizon; *Lower left:* auroral arc with bright meteor; *Lower right:* bright auroral display.

References

Coffin, D. 2012, *WWW*, http://www.cybercom.net/~dcoffin/dcraw/
Jordahl, K. 2012, *WWW*, https://github.com/kjordahl/cr2fits
Moore, A. *et al.* 2010, *Proc. SPIE*, 7733, 77331S
Sims, G. *et al.* 2010, *Proc. SPIE*, 7733, 77334M
Yuan, X. *et al.* 2008, *Proc. SPIE*,7012, 70124G

Astrophysics from Antarctica
Proceedings IAU Symposium No. 288, 2012
M. G. Burton, X. Cui & N. F. H. Tothill, eds.

© International Astronomical Union 2013
doi:10.1017/S1743921312016663

CMB anisotropy science: a review

Anthony Challinor

Institute of Astronomy and Kavli Institute for Cosmology Cambridge, Madingley Road,
Cambridge, CB3 0HA, U.K.

DAMTP, Centre for Mathematical Sciences, Wilberforce Road, Cambridge, CB3 0WA, U.K.
email: a.d.challinor@ast.cam.ac.uk

Abstract. The cosmic microwave background (CMB) provides us with our most direct observational window to the early universe. Observations of the temperature and polarization anisotropies in the CMB have played a critical role in defining the now-standard cosmological model. In this contribution we review some of the basics of CMB science, highlighting the role of observations made with ground-based and balloon-borne Antarctic telescopes. Most of the ingredients of the standard cosmological model are poorly understood in terms of fundamental physics. We discuss how current and future CMB observations can address some of these issues, focusing on two directly relevant for Antarctic programmes: searching for gravitational waves from inflation via B-mode polarization, and mapping dark matter through CMB lensing.

Keywords. cosmology: cosmic microwave background

1. Introduction

It is now twenty years since the landmark discovery of fluctuations in the temperature of the cosmic microwave background radiation by the COBE satellite (Smoot *et al.* 1992). Over the intervening period, a now-standard cosmological model has emerged. The CMB fluctuations have been pivotal in putting this model on a firm observational footing (though many of its key ingredients continue to defy explanation in fundamental physics), and in measuring its parameters to a level of precision that is unprecedented in cosmology. While the game-changer in this field has undoubtedly been the full-sky measurements from the WMAP satellite, observations of the CMB from Antarctica have played an important role in this development and have achieved a number of significant 'firsts'. These include precision measurements of spatial flatness and the detection and, now, characterisation of linear polarization of the CMB.

This symposium covers a broad range of astrophysics so the purpose of this review is to set the scence for the other more specialised CMB contributions that follow. We begin by reviewing some of the basics of CMB science and the remarkable achievements made through observations of the CMB temperature and polarization fluctuations, highlighting the role of measurements made from Antarctica. Experiments in Antarctica are also very much at the cutting edge of future programmes seeking to address some of the questions raised by the standard cosmological model. Space limits us to discuss in detail only two of the main science goals of these experiments: the quest for gravitational waves and CMB lensing. For more complete recent reviews of CMB science, see Challinor & Peiris (2009) and Hu (2008).

2. The CMB and the standard cosmological model

In the standard cosmological model, named ΛCDM, the universe is well described on large scales by a spatially-flat, homogeneous and isotropic background metric with small fluctuations at the 10^{-5} level. The universe has evolved from a hot, dense phase during which matter and radiation were in thermal equilibrium at sufficiently early time. The CMB is the thermal relic radiation from this early phase and its existence is a cornerstone of the hot big bang model. The CMB radiation has now cooled to a temperature of 2.725 K but retains an almost perfect blackbody spectrum. Baryons and leptons make up 4.5% of the current energy density and cold dark matter (CDM; hypothesised matter with essentially only gravitational interactions and negligible thermal velocities) 22%. The remaining 73% is in the form of dark energy and drives the current accelerated expansion. Dark energy is not understood at all at a physical level but phenomenologically behaves like a smoothly distributed fluid with equation of state close to $p = -\rho$, as for a cosmological constant Λ.

The flatness and large-scale smoothness of the universe are neatly explained by a hypothesised period of quasi-exponential expansion – cosmic inflation – in the early universe. During a period of only 10^{-32} s, the universe expanded in size by at least 60 e-folds. Inflation is not understood at a fundamental level, but it can be realised in simple models by a scalar field ϕ evolving slowly over a flat part of its self-interaction potential $V(\phi)$. A compelling feature of inflation is that it naturally provides a causal mechanism for generating primordial curvature perturbations and gravitational waves with nearly scale-free power spectra. Small-scale quantum fluctuations in light scalar fields, and the spacetime metric, are stretched beyond the Hubble radius during inflation to appear later as classical, long-wavelength cosmological perturbations that seed the growth of large-scale structure. In simple models (e.g. $V(\phi) \propto \phi^2$) inflation at energies $E_{\text{inf}} \sim 10^{16}$ GeV reproduces the observed level of perturbations.

2.1. *Temperature anisotropies*

The CMB carries an imprint of the primordial perturbations via small temperature anisotropies at the $O(10^{-5})$ level. The universe became transparent to CMB photons around the time of recombination, when atomic hydrogen (and helium) first formed. This defines a last-scattering surface centred on our current location, and spatial fluctuations in the CMB energy density, bulk velocity and gravitational potential over this surface project to give temperature anisotropies in the CMB. In more detail, for curvature perturbations, the fractional anisotropy $\Theta(\hat{\mathbf{n}})$ along a direction $\hat{\mathbf{n}}$ at time t_0 is given approximately by

$$\Theta(\hat{\mathbf{n}}) = \Theta_0 + \psi - \hat{\mathbf{n}} \cdot \mathbf{v}_b + \int_{t_*}^{t_0} (\dot{\psi} + \dot{\phi})\, dt\,. \tag{2.1}$$

Here, Θ_0 is the fractional fluctuation in the CMB temperature on the last-scattering surface (time t_*), $\mathbf{v}_b$ is the baryon peculiar velocity, and ψ and ϕ are the gravitational potentials ($\phi = \psi$ in general relativity when non-relativistic matter is dominant). Each term has a simple physical interpretation: we see the intrinsic temperature fluctuation Θ_0, modified by the gravitational-redshifting effect of the potential ψ and the Doppler shift from scattering off moving matter. The final *integrated Sachs-Wolfe* term in Eq. (2.1) involves the integral of the time derivatives of ψ and ϕ; if a potential well is getting shallower in time (as happens during dark-energy domination), photons receive a net blueshift in crossing the well and the CMB appears hotter. In practice, around 10% of photons were re-scattered after the universe reionized which, on all but the largest scales, reduces the *primary anisotropies* sourced around recombination by a factor $e^{-\tau}$, where $\tau \approx 0.1$ is the Thomson optical depth.

The small amplitude of the temperature anisotropies means they can be calculated very accurately with linear perturbation theory. The fluctuations on the last-scattering

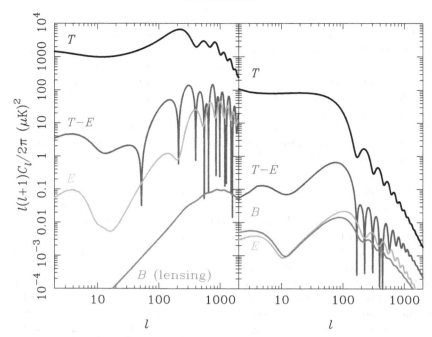

Figure 1. Temperature (black), E-mode (green), B-mode (blue) and TE cross-correlation (red) CMB power spectra from curvature perturbations (left) and gravitational waves (right) for a tensor-to-scalar ratio $r = 0.24$. The B-mode spectrum induced by weak gravitational lensing is also shown in the left-hand panel (blue).

surface are therefore a linearly-processed version of the nearly scale-free primordial perturbation. On scales large compared to the Hubble radius at last-scattering, only gravity is important but on smaller scales the acoustic physics of the primordial plasma and photon diffusion dominate. Gravity-driven infall will tend to enhance a positive density perturbation, but this is resisted by photon pressure setting up acoustic oscillations in the plasma. The sine and cosine-like modes of oscillation extrapolate back to decaying and constant modes at early times. Inflation is democratic, putting equal power into each mode at generation, but any decaying mode is totally negligible by the time the acoustic oscillations begin. This process leaves only cosine-like oscillations in the plasma, so that oscillations on all scales start off in phase. However, different scales oscillate at different frequencies and scales which have reached extrema of their oscillations by last-scattering have enhanced power in the anisotropies on the corresponding angular scales. In this way, the sound horizon $r_s(t_*)$, i.e. the (comoving) distance a sound wave can have propagated by time t_*, introduces a preferred length scale to the fluctuations. It is a fortunate coincidence that the corresponding angular scale r_s/d_A (where d_A is the angular-diameter distance back to last-scattering) is around $1°$ and so straightforward to observe at frequencies around $100\,\mathrm{GHz}$ where the CMB is brightest.

Figure 1 shows the predicted angular power spectrum, C_l^T, from inflationary curvature perturbations. The power spectrum is the variance of the multipoles Θ_{lm} in a spherical-harmonic expansion $\Theta(\hat{\mathbf{n}}) = \sum_{lm} \Theta_{lm} Y_{lm}(\hat{\mathbf{n}})$. The multipole index l corresponds roughly to anisotropies at scale $180°/l$. The plateau in C_l on large scales is from the combination of primary anisotropies on scales large enough to be unaffected by acoustic processing, and from the integrated-Sachs-Wolfe effect from late-time decay of the gravitational potentials. On intermediate scales, we have acoustic peaks. Finally, on smaller scales

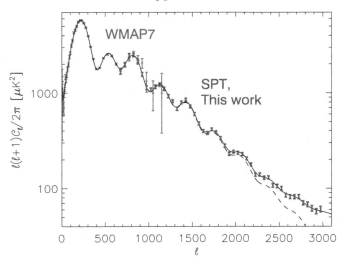

Figure 2. Measurements of the temperature power spectrum from WMAP (orange; Larson *et al.* 2011) and 790 deg² of the SPT 150-GHz survey (blue; Keisler *et al.* 2011). Also shown are the CMB spectrum (dashed) and the total spectrum (CMB and extragalactic foregrounds; solid) in the best-fitting ΛCDM model. Reproduced with permission from Keisler *et al.* (2011).

the power decays rapidly. This *damping tail* is sourced by perturbations on scales small enough that photons had time to diffuse out of overdensities by last-scattering, thus damping out the acoustic oscillations. This process imprints another scale, the diffusion scale, into the CMB.

The picture outlined above is spectacularly confirmed by measurements of the temperature anisotropy. Theory only allows us to predict the statistical properties of the primordial perturbation. In simple models of inflation, the statistics are Gaussian, and so fully characterised by their power spectrum. Primordial Gaussianity is borne out by careful measurements of the statistics of the CMB anisotropies. For this reason, the main focus of observational CMB research for the past 20 years has been to obtain precise estimates of the CMB angular power spectrum and to confront these against theoretical models. An example of such measurements from the WMAP satellite (Larson *et al.* 2011) and the South Pole Telescope (SPT; Keisler *et al.* 2011) is shown in Fig. 2. Nine acoustic peaks have now been measured and the SPT measurements thoroughly characterise the damping tail. The error bars on the power spectrum include the effects of instrument noise and cosmic/sample variance – at each l we estimate the power spectrum from the empirical variance from a sample of only $(2l+1)f_{sky}$ independent quantities (where f_{sky} is the sky fraction covered by the survey). The Planck survey (The Planck Collaboration 2006), which is expected to report its first CMB results in early 2013, will improve considerably the statistical power of the CMB power spectrum measurements between $l = 500$ where WMAP becomes noise limited, and $l = 2000$ where Planck's poorer resolution and sensitivity loses out to SPT despite the greatly extended sky coverage. Beyond $l \sim 2000$, measurements start to be contaminated by the unresolved background of extra-Galactic sources at all frequencies.

Given that the physics of the CMB is so well understood, the standard cosmological model can be tested very precisely and its parameters determined to high precision (see e.g. Komatsu *et al.* 2011; Keisler *et al.* 2011). Here, for brevity, we can highlight only three examples.

Primordial power spectrum: This affects the overall morphology of the CMB power spectrum. Parameterising the spectrum of primordial curvature perturbations as a power-law, $\mathcal{P}_{\mathcal{R}}(k) \propto k^{n_s-1}$, WMAP constrains $n_s = 0.963 \pm 0.014$ (assuming no contribution to the CMB from gravitational waves; Larson *et al.* 2011). This is beautifully consistent with the inflationary prediction of a nearly scale-free spectrum ($n_s = 1$). A departure from scale-invariance is detected at almost the 3σ level and provides important constraints on the dynamics of inflation [i.e. the slope and curvature of $V(\phi)$]. The combination of the n_s constraint and upper limits on the gravitational wave power spectrum (see Section 3) already rule out several simple inflation models.

Matter densities: The relative heights of the acoustic peaks are influenced by the physical densities of baryons and CDM. For example, increasing the baryon fraction adds inertia but not pressure support to the plasma, reducing the bulk modulus. This increases the overdensity at the midpoint of the acoustic oscillations boosting the compressional peaks (1st, 3rd etc.). Precise measurements of the physical baryon density, $\Omega_b h^2 = 0.02258 \pm 0.00056$ (Larson *et al.* 2011), have been made via this route, nicely consistent with constraints from big-bang nucleosynthesis. This 3% precision should improve to around 1% with Planck data. Similarly, the CDM density is measured to be $\Omega_c h^2 = 0.1109 \pm 0.0056$. This provides inescapable evidence of the need for non-baryonic dark matter independently of other lines of reasoning such as the clustering and internal kinematics of galaxies.

Curvature: The angular scale of the acoustic peaks r_s/d_A is now very precisely measured. In standard models, the matter densities determined from the relative peak heights fully determine r_s, thus allowing an accurate measurement of the angular-diameter distance to last scattering. This distance is very sensitive to spatial curvature through its geometrical focusing effect, but this can always be compensated by altering the radial distance to last-scattering (through the Hubble constant H_0 or, equivalently, the dark energy density). This leads to a *geometric degeneracy* whereby models with the same physical densities at high redshift, the same primordial power spectrum, and the same angular-diameter distance to last-scattering give almost identical angular power spectra. (An example is given later in Fig. 4.) The degeneracy can be broken by adding other astrophysical distance measures such as the Hubble constant, the angular-diameter distance at lower redshift (inferred from the relic of the baryon acoustic oscillations – BAO – in the clustering of galaxies; Eisenstein *et al.* 2005) or the luminosity distance inferred from supernovae. An early, important example of determining curvature via this route was from the 1998 flight of BOOMERanG (de Bernardis *et al.* 2000). By precisely characterising the first acoustic peak, the team were able to establish that space was flat at the 10% level. More recent measurements are consistent with flatness at the 0.5% level (Komatsu *et al.* 2011), strongly supporting one of the main predictions of inflationary cosmology.

2.2. *Polarization*

The other key CMB observable is polarization. Thomson scattering of unpolarized radiation with a quadrupole ($l = 2$) anisotropy in its total intensity generates linear polarization. The relevant epoch for the generation of polarization in the CMB is around recombination since at early times scattering is too efficient to allow a significant quadrupole to grow, while after recombination scatterings are very rare (until the universe reionizes). The expected linear polarization from curvature perturbations has an r.m.s. of $5\,\mu K$.

Linear polarization can be described by two Stokes parameters Q and U. These depend on a choice of basis and measure the difference in intensity transmitted by linear polarizers aligned with the basis directions (Q) or at $45°$ to them (U). While Stokes parameters provide a local, operational definition of polarization, their coordinate dependence makes

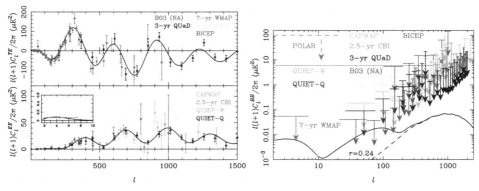

Figure 3. *Left:* current measurements of the polarization power spectra TE (top) and EE (bottom) from WMAP7 (magenta; Larson *et al.* 2011), QUaD (black; Brown *et al.* 2009), BOOMERanG (blue; Piacentini *et al.* 2006; Montroy *et al.* 2006), DASI (cyan; Leitch *et al.* 2005), CAPMAP (green; Bischoff *et al.* 2008), CBI (orange; Sievers *et al.* 2007), BICEP (red; Chiang *et al.* 2010) and QUIET W-band (light grey; QUIET Collaboration 2012) and QUIET Q-band (dark grey; QUIET Collaboration 2011). The lines are ΛCDM fits to temperature and polarization data. *Right:* current 95% upper limits on the BB power spectrum including the constraint from POLAR (dashed cyan; Keating *et al.* 2001). The dashed line is the contribution from gravitational waves for $r = 0.24$, the 95% upper limit from fits to the temperature and E-mode polarization data from WMAP7 combined with BAO and H_0 measurements (Komatsu *et al.* 2011), and the solid line includes the contribution from gravitational lensing.

them rather inconvenient for cosmological interpretation. Instead, linear polarization can be described in terms of two scalar fields, E and B (Seljak & Zaldarriaga 1997; Kamionkowski *et al.* 1997). The Stokes parameters are properly the components in an orthonormal basis of a rank-2 symmetric, trace-free tensor which can be expressed in term of second derivatives of E and B (neglecting sky curvature for simplicity, and using Cartesian coordinates):

$$\begin{pmatrix} Q & U \\ U & -Q \end{pmatrix} \propto \left(\partial_i \partial_j - \frac{1}{2} \delta_{ij} \nabla^2 \right) E + \epsilon_{k(i} \partial_{j)} \partial_k B \,. \tag{2.2}$$

This is analogous to decomposing a vector field into a gradient part (E) and a divergence-free curl part (B). Note that E and B are non-local in Q and U.

The E-modes are scalars under parity but B-modes are pseudo-scalar. In the absence of parity-violating physics, the two fields must be uncorrelated. This leaves three non-zero polarization power spectra: the E- and B-mode auto-correlations C_l^E and C_l^B, and the cross-correlation C_l^{TE} between E and the temperature anisotropies. The predicted angular power spectra for inflationary curvature perturbations are shown in the left-hand panel of Fig. 1. The main points to note are as follows: (i) polarization is a small signal; (ii) E-mode polarization peaks on smaller scales than the temperature, since it relies on diffusion in small-scale modes for its generation; (iii) the acoustic peaks in C_l^E are at the troughs of C_l^T since the temperature quadrupole derives mostly from the plasma bulk velocity which vanishes when the density is at an extremum; (iv) there is a 'bump' in the polarization on large scales generated by re-scattering once the universe reionizes; and (v) by symmetry, curvature perturbations cannot generate B-mode polarization except through second-order processes such as gravitational lensing (see Section 4). This last point makes B-modes a potentially powerful probe of gravitational waves; see Section 3.

Observations of CMB polarization are not yet as advanced as for the temperature anisotropies. Current power spectrum measurements are shown in Fig. 3. Antarctic experiments have played a very significant role, including the first detection of CMB polarization by the DASI interferometer in 2002 (Kovac *et al.* 2002), and the current best characterisation of the spectra by QUaD (Brown *et al.* 2009) and BICEP (Chiang *et al.* 2010). The measurements are in excellent agreement with expectations based on the temperature power spectrum, providing an important consistency test. Moreover, through large-angle E-modes, WMAP measures the optical depth to reionization to be $\tau = 0.088 \pm 0.015$ (Larson *et al.* 2011), providing an important integral constraint on astrophysical models of reionization. Future E-mode polarization measurements will tighten parameter constraints over those from the temperature anisotropies, particularly in non-standard models, and extend the angular range that can be reliably probed before foregrounds dominate. However, the real excitement over polarization is the prospect of detecting the signature of gravitational waves via B-modes and exploiting B-modes induced by weak lensing.

We end this section by emphasising that, despite the triumph of the standard cosmological model in fitting essentially all cosmological data (with just six parameters), the model raises several big questions. Did inflation happen? What is the nature of dark matter? Why is the universe accelerating? In the following sections, we review how ongoing and future CMB observations will help answer some of these questions.

3. Gravitational waves and B-mode polarization

Inflation naturally predicts the production of a stochastic background of primordial gravitational waves accompanying the primordial density perturbation (Starobinskiĭ 1979). The spectrum of gravitational waves depends only on the expansion rate during inflation. Since this is nearly constant during slow-roll inflation, with only a slow decrease, the primordial spectrum $\mathcal{P}_h(k)$ should be well approximated by a power-law with a slightly red spectrum. As the Friedmann equation relates the expansion rate directly to the energy density during inflation, a measurement of the gravitational wave power gives directly the energy density and hence the *energy scale* E_{inf} during inflation. It is conventional to express the amplitude of $\mathcal{P}_h(k)$ in terms of its ratio to the power spectrum of curvature perturbations $\mathcal{P}_{\mathcal{R}}(k)$ at a cosmologically-relevant scale (often $k_0 = 0.002\,\text{Mpc}^{-1}$). This *tensor-to-scalar* ratio r is related to E_{inf} by

$$r = 8 \times 10^{-3}(E_{\text{inf}}/10^{16}\,\text{GeV})^4, \qquad (3.1)$$

where we have taken the scalar amplitude $\mathcal{P}_{\mathcal{R}}(k_0) = 2.36 \times 10^{-9}$. Note that $r \sim 10^{-2}$ for inflation occurring around the GUT scale, $E_{\text{inf}} \sim 10^{16}\,\text{GeV}$.

Gravitational waves damp away due to the expansion of the universe when their wavelength is smaller than the Hubble radius. The best prospect for detection is therefore via the CMB which is sensitive to early times (after last-scattering) and large scales. Gravitational waves generate CMB temperature anisotropies due to the integrated effect of the anisotropic expansion they induce along the line of sight; see Fig. 1. However, the signal is limited to large angular scales, $l < 60$, corresponding to gravitational waves with wavelengths larger than the Hubble radius at last-scattering. On such scales, chance upwards fluctuations in the temperature anisotropies from curvature perturbations due to cosmic variance limit our ability to measure r. In the optimistic scenario that all other cosmological parameters are know, cosmic variance gives a 1σ error on r of 0.07 from the temperature anisotropies alone. In practice, degeneracies make the CMB-only limit a little worse: e.g. $r < 0.21$ (at 95% confidence) from WMAP7+SPT (Keisler *et al.* 2011), improving on $r < 0.36$ from WMAP7 alone (Komatsu *et al.* 2011). Gravitational waves also leave an imprint in the linear polarization of the CMB. Significantly, they generate E- *and* B-modes with roughly equal power, unlike curvature perturbations which only generate B-modes at second order through gravitational lensing. In principle, B-mode measurements of r can do much better than inferences from the temperature or E-mode polarization since the former is only limited by the cosmic variance of the lens-induced B-modes†. The problem is that the B-mode signal is very small (see Fig. 1); the limit $r < 0.24$ implies that the r.m.s. from gravitational waves is less than $200\,\text{nK}$. The measurement therefore requires exquisite sensitivity and control of systematic effects to maintain polarization purity, and careful rejection of polarized emission from our Galaxy.

Current upper limits on the B-mode power spectrum are shown in Fig. 3. The best constraints over nearly the full range of scales come from BICEP (Chiang *et al.* 2010) at degree scales and QUaD (Brown *et al.* 2009) on smaller scales. The BICEP constraint of $r < 0.73$ (95% confidence) is not yet competitive with that from the temperature anisotropies, although it is rather less model dependent. There are two main scales to attempt detection of B-modes from gravitational waves: $l < 10$ where the signal is generated by scattering at reionization, and $l \sim 100$ where the signal from scattering around recombination peaks. The reionization signal needs a nearly full-sky survey and so broad frequency coverage to remove Galactic emission which is dominant over most of the sky. The best near-term constraints on these scales will come from Planck with forecasts indicating $r < 0.05$ may be achievable (Efstathiou & Gratton 2009). The signal from recombination can be constrained by targeting clean, connected regions of the sky (typically around $1000\,\text{deg}^2$) in areas of low Galactic emission. By good fortune, one of the cleanest such regions is accessible from Antarctica. A series of results from BICEP's successors, BICEP2, Keck and POLAR (see contributions from Pryke and Kuo in this volume), as well as the balloon-borne SPIDER (Filippini *et al.* 2010) and several experiments based in Atacama, are eagerly anticipated over the next five years. These should push down the errors on r to around 0.01. This is an interesting target for inflationary physics since the signal from a large class of simple models – "large-field" such as monomial potentials – would be detectable. Looking further ahead, the constraint on r could plausibly be improved to the 10^{-4}–10^{-3} level with a future polarization satellite.

4. Weak gravitational lensing of the CMB

The fluctuations in the temperature of the CMB are mostly imprinted at the epoch of last scattering. However, CMB photons undergo small gravitational deflections due to the clumpy distribution of matter (weak gravitational lensing) as they propagate from last-scattering to the present epoch. The r.m.s. deflection is only $2.7\,\text{arcmin}$ but is coherent over several degrees. The lensing effect is similar to seeing the CMB fluctuations from the last-scattering surface through patterned glass, and subtly distorts their statistics. With telescope resolution of a few arcminutes or better, these distortions can be detected and used to reconstruct the lensing deflection. This opens up a new cosmological probe of structure formation at epochs and scales that are difficult to access with more direct probes (such as galaxy clustering). Lensing is an emerging field in observational CMB research and results from the SPT are at the forefront of this.

Weak lensing has several important effects on the CMB; see Lewis & Challinor (2006) for a detailed review. Magnification and demagnification of the acoustic-scale features leads to a smoothing of the acoustic peaks, reaching the 10% level at $l > 2000$ in

---

† In principle, the cosmic variance from lensing can even be removed by "delensing" the observed Q and U maps with a reconstruction of the lensing deflection field (Seljak & Hirata 2004). The latter can be obtained from the CMB itself with high-resolution polarization observations; see Section 4.

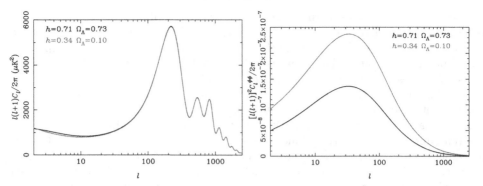

Figure 4. Breaking the angular-diameter distance degeneracy with CMB lensing. The unlensed (and unobservable!) temperature power spectra (left) for the standard ΛCDM model (black) and a closed model with low H_0 to match the angular scale of the acoustic peaks (red) are very nearly degenerate. The degeneracy is broken in the power spectrum of the lensing deflection angle (right) since matter is more clustered at late times in the model with low H_0.

temperature, and rather larger in E-mode polarization. On smaller scales, for which the unlensed CMB is very smooth, lensing generates small-scale power that dominates the primary anisotropies for $l > 4000$. The lens remapping moves around the polarization amplitude while preserving the direction, generating B-modes from the primary E-modes with an almost white spectrum for $l \ll 1000$; see Fig. 1. As noted in Section 3, this will become an important source of confusion for CMB searches for gravitational waves. Finally, lensing introduces four-point non-Gaussianity with a very specific and predictable shape from which the full angular power spectrum $l(l+1)C_l^{\phi\phi}$ of the lensing deflections can be reconstructed‡. Through these lensing effects, the CMB is sensitive to parameters that have degenerate effects in the primary anisotropies. For example, Fig. 4 compares the unlensed temperature power spectra and the deflection power spectra for the standard ΛCDM model and a closed model with low H_0. These models lie along the geometric degeneracy of the unlensed CMB power spectra. However, the deflection spectra are quite different since matter is more clustered at late times in the low-H_0 model. Other parameters that benefit similarly from lensing information include sub-eV neutrino masses and early dark energy.

The first measurements of the deflection power spectrum from the four-point function of the temperature anisotropies have recently been reported by the Atacama Cosmology Telescope (ACT; Das *et al.* 2011) and SPT (van Engelen *et al.* 2012). These are in excellent agreement with expectations for the standard ΛCDM model; see Fig. 5. The current SPT measurements constitute a 6.3σ detection but they are from only $590\,\mathrm{deg}^2$ of sky. The significance can be expected to increase several-fold with the analysis of the full $2500\,\mathrm{deg}^2$ survey, similar to what should be achieved with Planck. Already, combining the SPT lens reconstruction with the temperature power spectrum from WMAP breaks the geometric degeneracy in ΛCDM models with curvature giving a significant detection of dark energy from the CMB alone; see the right-hand panel of Fig. 5. The degeneracy is also broken by the effect of lensing on the high-l temperature power spectrum itself, as measured, for example, by SPT (Keisler *et al.* 2011).

Lens reconstruction from the CMB temperature suffers from statistical noise due to chance correlations in the unlensed CMB that mimic the effect of lensing. This is such that temperature reconstructions will never give cosmic-variance-limited measurements of the

‡ In linear theory, the deflection is the gradient of the *lensing potential* ϕ. The lensing potential is an integrated measure of the gravitational potential along the line of sight.

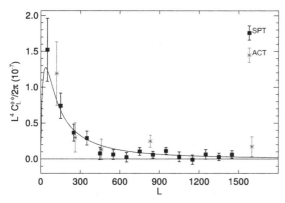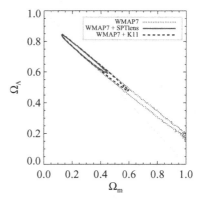

Figure 5. *Left:* Current measurements of the lensing deflection power spectrum from SPT (black; van Engelen *et al.* 2012) and ACT (red; Das *et al.* 2011). The solid line is a ΛCDM fit to CMB temperature and polarization data, but not to the lensing data. *Right:* 95% confidence regions in the Ω_Λ–Ω_m plane for ΛCDM models with curvature. The geometric degeneracy is evident in the WMAP7-alone constraints (red), but the tail of low-H_0 closed models is cut-off by the higher-resolution SPT data which is sensitive to the lensing effect in the temperature power spectrum (green; Keisler *et al.* 2011). Even tighter constraints are obtained by combining the SPT lens reconstruction from the left-hand figure with the WMAP7 data (blue). Figure reproduced with permission from van Engelen *et al.* (2012).

deflection power spectrum for multipoles $l > 100$. Polarization measurements are very helpful here (Hu & Okamoto 2002), since they intrinsically have more small-scale power and the B-mode of polarization is not confused by primary anisotropies. In principle, polarization can provide cosmic-variance limited reconstructions to multipoles $l \approx 500$, i.e. on all scales where linear theory applies. For this reason, lens reconstruction from polarization is an important science goal for the polarization upgrades to the SPT (McMahon *et al.* 2009) and ACT (Niemack *et al.* 2010), as well as proposed successors to the Planck satellite (Bock *et al.* 2008, 2009; The COrE Collaboration 2011).

5. Outlook

The future of CMB observations lies on several fronts. Precise polarization measurements on large scales will greatly improve limits on the stochastic background of gravitational waves predicted from inflation in the early universe. Wide-area, high-resolution temperature and polarization measurements will allow precise reconstruction of the gravitational-lensing effect in the CMB and provide a new window to the large-scale clustering of matter around redshift two. In addition, arcminute-scale observations will provide catalogues of thousands of galaxy clusters over a broad redshift range and with well-understood selection functions, and measure the Doppler signatures from the bulk flows of matter in the post-reionization universe. The cluster catalogues will be used to probe the growth of structure and evolution of the volume element to high redshift. These programmes address directly many of the outstanding issues raised by the standard cosmological model, such as the physics of inflation and the cause of the current accelerated expansion.

References

Bischoff C. *et al.*, 2008, *ApJ*, 684, 771
Bock J. *et al.*, 2009, ArXiv e-prints 0906.1188

Bock J. *et al.*, 2008, ArXiv e-prints 0805.4207

Brown M. L. *et al.*, 2009, *ApJ*, 705, 978

Challinor A. & Peiris H., 2009, in *American Institute of Physics Conference Series*, Vol. 1132, Novello M., Perez S., eds., pp. 86–140

Chiang H. C. *et al.*, 2010, *ApJ*, 711, 1123

Das S. *et al.*, 2011, *Phys. Rev. Lett.*, 107, 021301

de Bernardis P. *et al.*, 2000, *Nature*, 404, 955

Efstathiou G. & Gratton S., 2009, *JCAP*, 6, 11

Eisenstein D. J. *et al.*, 2005, *ApJ*, 633, 560

Filippini J. P. *et al.*, 2010, in *SPIE Conference Series*, Vol. 7741, Holland W. S., Zmuidzinas J., eds., pp. 77411N–77411N-12

Hu W., 2008, ArXiv e-prints 0802.3688

Hu W. & Okamoto T., 2002, *ApJ*, 574, 566

Kamionkowski M., Kosowsky A., & Stebbins A., 1997, *Phys. Rev. Lett.*, 78, 2058

Keating B. G., O'Dell C. W., de Oliveira-Costa A., Klawikowski S., Stebor N., Piccirillo L., Tegmark M., & Timbie P. T., 2001, *ApJ. Lett.*, 560, L1

Keisler R. *et al.*, 2011, *ApJ*, 743, 28

Komatsu E. *et al.*, 2011, *ApJS*, 192, 18

Kovac J. M., Leitch E. M., Pryke C., Carlstrom J. E., Halverson N. W., & Holzapfel W. L., 2002, *Nature*, 420, 772

Larson D. *et al.*, 2011, *ApJS*, 192, 16

Leitch E. M., Kovac J. M., Halverson N. W., Carlstrom J. E., Pryke C., Smith M. W. E., 2005, *ApJ*, 624, 10

Lewis A. & Challinor A., 2006, *Phys. Rep.*, 429, 1

McMahon J. J. *et al.*, 2009, in *American Institute of Physics Conference Series*, Vol. 1185, Young B., Cabrera B., Miller A., eds., pp. 511–514

Montroy T. E. *et al.*, 2006, *ApJ*, 647, 813

Niemack M. D. *et al.*, 2010, in *SPIE Conference Series*, Vol. 7741, Holland W. S., Zmuidzinas J., eds., pp. 77411S–77411S-21

Piacentini F. *et al.*, 2006, *ApJ*, 647, 833

QUIET Collaboration, 2012, ArXiv e-prints 1207.5034

QUIET Collaboration, 2011, *ApJ*, 741, 111

Seljak U. & Hirata C. M., 2004, *Phys. Rev. D*, 69, 043005

Seljak U. & Zaldarriaga M., 1997, *Phys. Rev. Lett.*, 78, 2054

Sievers J. L. *et al.*, 2007, *ApJ*, 660, 976

Smoot G. F. *et al.*, 1992, *ApJ. Lett.*, 396, L1

Starobinskiĭ A. A., 1979, JETP Lett., 30, 682

The COrE Collaboration, 2011, ArXiv e-prints 1102.2181

The Planck Collaboration, 2006, ArXiv e-prints astro-ph/0604069

van Engelen A. *et al.*, 2012, *ApJ*, 756, 142

Astrophysics from Antarctica
Proceedings IAU Symposium No. 288, 2012
M. G. Burton, X. Cui & N. F. H. Tothill, eds.

© International Astronomical Union 2013
doi:10.1017/S1743921312016675

Precision CMB Measurements from Long Duration Stratospheric Balloons:
Towards B-modes and Inflation

William C. Jones

Princeton University, Department of Physics,
323 Jadwin Hall, Washington Road, Princeton, NJ 08544, USA
email: wcjones@princeton.edu

Abstract. Observations of the Cosmic Microwave Background (CMB) have played a leading role in establishing an understanding of the structure and evolution of the Universe on the largest scales. This achievement has been enabled by a series of extremely successful experiments, coupled with the simplicity of the relationship between the cosmological theory and data. Antarctic experiments, including both balloon-borne telescopes and instruments at the South Pole, have played a key role in realizing the scientific potential of the CMB, from the characterization of the temperature anisotropies to the detection and study of the polarized component. Current and planned Antarctic long duration balloon experiments will extend this heritage of discovery to test theories of cosmic genesis through sensitive polarized surveys of the millimeter-wavelength sky. In this paper we will review the pivotal role that Antarctic balloon borne experiments have played in transforming our understanding of the Universe, and describe the scientific goals and technical approach of current and future missions.

Keywords. Cosmology, Cosmic Microwave Background, Ballooning, Polarimetry, Cryogenic Detectors

---

The cosmic microwave background radiation represents a uniquely powerful laboratory for probing topics in fundamental physics ranging from cosmology to particle physics. The spectrum and the statistical properties of the anisotropy in the temperature and polarization of the CMB rely almost exclusively on the primordial initial conditions and well understood physics in the linear regime (Kamionkowski 1999, Kamionkowski & Kosowsky 1999, Scott 1999). As a result, modeling uncertainties have played a negligible role in the interpretation of the increasingly powerful observational data, allowing for unambiguous tests of competing cosmological theories and imposing tight constraints on the parameters that describe them (Samtleben *et al.* 2007, Komatsu *et al.* 2011, Larson *et al.* 2011, Challinor 2012).

Following the discovery of CMB temperature anisotropies by the DMR instrument on the COBE satellite, studies of the CMB focused on the characterization of the angular power spectrum of these fluctuations (Wright *et al.* 1992, Smoot *et al.* 1992). The statistical properties of these extremely faint fluctuations, having amplitudes of a few hundred μK, encode a rich set of information about the contents and the evolutionary history of the Universe; cosmological theories make specific predictions regarding the shape of the power spectrum of the intensity and polarization fluctuations on the sky. Figure 1 shows a few example power spectra corresponding to qualitatively different cosmological models, all of which were consistent with observations prior to 1999 (see Figure 2).

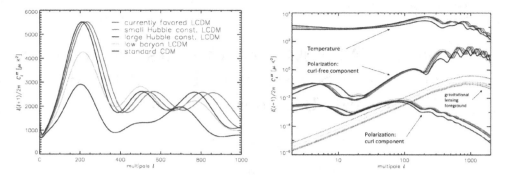

Figure 1. *At left*, the power spectra of temperature fluctuations corresponding to a range of cosmological models, all of which were consistent with the data available circa 1999. *At right*, the corresponding temperature (above, at several thousand μK^2) and polarization (both E-mode and B-mode) spectra, displayed on a log plot to illustrate the amplitude of the polarized component relative to the already faint temperature anisotropies.

CMB fluctuations on the sky represent a single realization of a Gaussian random process; observational data are used to infer the underlying power spectrum from which our realization is drawn. Cosmological models are tested and model parameters determined from these estimates through Bayesian inference. The experimental challenge of extracting this information is framed by the required instrumental sensitivity, sky coverage, control of systematics and the ability to disentangle the cosmological signal from that of the astrophysical foregrounds.

CMB temperature anisotropies

Boomerang was a pioneering balloon borne instrument designed to image the CMB temperature fluctuations, with fidelity to angular scales ranging from 5° to 20′ (Lange *et al.* 1996). The experiment employed two innovations that enabled a dramatically improved millimeter-wavelength imaging capability relative to contemporary instruments. The first of these were the cryogenic bolometric detectors at the heart of the receiver. These devices, cooled to a third of a degree above absolute zero, provided an instantaneous sensitivity limited only by the thermal radiation of the backgrounds. These "spider web" bolometers, named after their web-like silicon nitride structure, were designed to minimize the suspended heat capacity and the cross section to cosmic ray particles (Lange *et al.* 1995), allowing the devices to take full advantage of the capabilities afforded by the second of these innovations: the Antarctic long duration balloon platform.

From the point of view of atmospheric opacity and stability, Antarctica is arguably the best site in the world to perform observations at millimeter and far-infrared wavelengths with terrestrial telescopes (Burton *et al.* 1994, Storey *et al.* 1998, Battistelli *et al.* 2012); a review of South Pole based CMB experiments can be found elsewhere in these proceedings (Halverson 2012). In addition to the favorable conditions of the lower atmosphere, the stability of the stratospheric circumpolar winds during the Austral summer offers a unique capability for long duration balloon (LDB) flights (Gregory 2006). Stratospheric balloons launched from the ice shelf near McMurdo Station now routinely provide in excess of a week of observation in a near-space environment (Gregory & Stepp 2004).

Boomerang launched on December 29, 1998 from Williams Field, on the ice shelf approximately 10 km from McMurdo Station. Prior to 1998, many experiments had made statistically significant detections of anisotropy over a range of angular scales (e.g.

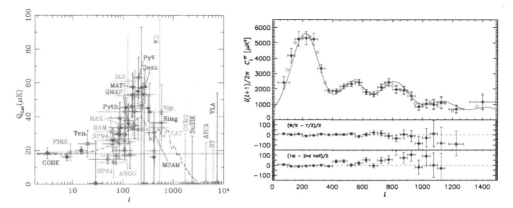

Figure 2. *At left*, the status of measurements of the angular power spectrum of the CMB, circa 1999 (Scott 1999). A number of experiments had detected significant levels of anisotropy, including hints of a peak near a multipole, $\ell \sim 200$, corresponding to angular scales of one degree. *At right*, the final angular power spectrum (upper panel), and corresponding null tests (lower two panels), derived from the Boomerang data (de Bernardis *et al.* 2000, Ruhl *et al.* 2003, Jones *et al.* 2006a). The characteristic scale of the features evident in Figure 3 result in a corresponding series of harmonic peaks in the power spectrum.

Netterfield *et al.* 1995, Torbet *et al.* 1999), but no single instrument had imaged the CMB with high signal-to-noise over a large area of the sky, providing a definitive characterization of the shape of the CMB power spectrum. The state of the measurement of the angular power spectrum circa 1999 is shown in Figure 2, reproduced with permission from Scott (1999).

The raw sensitivity and high fidelity of the Boomerang data provided an unambiguous measure of the CMB temperature anisotropy from super horizon scales, sampling primordial density fluctuations, to the harmonic peaks resulting from oscillations in the photon-baryon fluid prior to decoupling, while also clearly showing the expected suppression of power on small scales (above a multipole $\ell \gtrsim 900$) resulting from the combination of projection effects and Silk damping (photon diffusion) on small scales. This was the first single experiment to clearly measure all three of these distinct features in the CMB temperature anisotropy.

The scientific impact of the data from the 1998 Antarctic flight was extraordinary. For the first time, when combined with measurements of the Hubble constant and the expansion history via observations of supernovae, there was a clear determination that the geometry of the Universe was Euclidean, and that it consists of a surprising mix of Dark Energy ($\sim$ 70%), Dark Matter ($\sim$ 30%) and only trace amounts of familiar baryonic matter (de Bernardis *et al.* 2000). Within weeks of the publication of these data, the results from North American flight of the Maxima experiment were released, broadly confirming the Boomerang result (Hanany *et al.* 2000). While the value of these parameters have since been determined with greater precision, the ΛCDM model established by Boomerang's first Antarctic LDB flight remains the "Standard Model" of Cosmology (Lange *et al.* 2001).

CMB polarization

Following the successful characterization of the bulk features of the CMB temperature power spectrum, experimental groups directed their efforts toward the detection and characterization of the extremely faint polarized component of CMB. Polarization in the

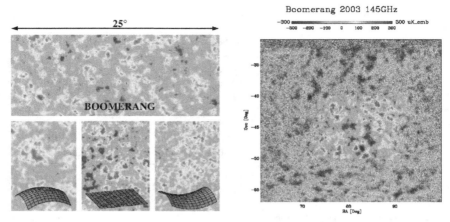

Figure 3. *At left*, the characteristic angular size of the CMB temperature fluctuations encodes the angular diameter distance to the surface of last scattering. A measure of this scale, together with a measure of the Hubble constant, constrains the topology of space-time to be closed (bottom left panel), Euclidean (bottom middle panel) or open (bottom right panel). The top panel shows the data from Antarctic long duration balloon flight of Boomerang in 1998 (Netterfield *et al.* 2002, Ruhl *et al.* 2003). Together with the data from the Maxima experiment, which imaged a much smaller portion of sky (Hanany *et al.* 2000), the data from the 1998 LDB flight provided the first resolved images of the CMB, and with it an unambiguous measure of the apparent angular size of the sound horizon corresponding to the epoch of last scattering. *At right*, the Boomerang temperature map derived from the data obtained in the 2003 long duration balloon (LDB) flight (Jones *et al.* 2006a). The integration time was distributed between a central deep field, and a shallow field covering roughly 10x more area, resulting in the apparent non-uniformity of the noise. The central deep field has approximately the same resolution and signal to noise as that achieved by the *Planck* HFI. Unlike the data in the left panel, this image has had no spatial filtering applied to highlight the small scale structure in the CMB.

CMB is generated by several processes. Thomson scattering in the optically thin plasma present during the epoch of recombination is generated by quadrupole anisotropies in the ambient radiation field. Quadrupoles induced from scalar perturbations, as are expected from primordial density fluctuations, produce a curl-free vector field (the "E-mode") with an amplitude a few percent of temperature anisotropy. Quadrupoles produced from tensor perturbations, such as gravitational waves from inflation, contribute a non-zero curl (the "B-mode").

In the context of the most simple inflationary theories, the amplitude of this component is simply related to the energy scale of the theory underlying the strong-electroweak phase transition and is not expected to exceed an rms level of a few hundred nano-Kelvin. On angular scales smaller than a degree, gravitational lensing of the CMB E-mode induces a B-mode signature even in the absence of a primordial component (Rees 1968, Kosowsky *et al.* 1999).

Based on mature technologies and insensitive to atmospheric emission, coherent interferometric receivers represented the most sensitive and robust technology for microwave polarimetry available at the time. The first statistical detection of a diffuse E-mode signal, later confirmed to have the angular spectrum expected from the CMB, was achieved at the South Pole in with DASI, a 30 GHz interferometer (Leitch *et al.* 2002, Kovac *et al.* 2002).

Coherent receivers, including both interferometric (DASI, CBI) and quasi-total power detectors (CAPMAP, WMAP) remained the technology of choice for the first generation of CMB polarization experiments. However, moving beyond detection to the precise

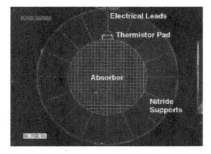

Figure 4. *At left*, a polarization sensitive bolometer, as employed in the 2003 flight of Boomerang, Bicep, QUaD and the *Planck* HFI instrument (Jones *et al.* 2003). *At right*, one of *Spider*'s six large format arrays of 512 antenna-coupled detectors (Bonetti *et al.* 2012).

characterization of CMB polarization required a dramatic increase in system sensitivity over the range of frequencies relevant to the CMB and Galactic foregrounds, 30 to 300 GHz.

At frequencies above 60 GHz, the noise performance of receivers based on coherent amplification are less competitive with incoherent (bolometric) techniques due to the quantum limit associated with phase sensitive amplification. This is particularly true for the low background cryogenic bolometers flown in space and on stratospheric balloons, which are limited in sensitivity only by the photon noise resulting from the 3 K cosmic background radiation (Lange *et al.* 2002).

Despite the advantages in raw sensitivity, and unlike their coherent analogues, traditional bolometric detectors are intrinsically insensitive to polarization.† However, the extreme sensitivity requirements of the B-mode science motivated a concerted effort to develop massively scalable bolometric receivers that would preserve the favorable polarimetric capabilities of the coherent systems.

Antarctic ballooning again played a leading role in both the technical development and scientific discovery of the next generation of CMB experiments. Indeed, just three months after the end of the first Antarctic Boomerang flight, the Caltech group proposed an ambitious program of technology development that would enable new discoveries from the South Pole, Antarctic LDB flights and ultimately on space missions.

The plan of work was to develop and fly two new technologies; polarization sensitive bolometers (PSBs) and large-format arrays of antenna-coupled bolometers (Lange *et al.* 1999). Having reached fundamental limits to the sensitivity of a given detector, massively scaling the number of detectors in a focal plane was the only way to realize the necessary increase in system sensitivity.

The first of these, polarization sensitive bolometers, saw first light on Boomerang in the 2003 Antarctic LDB campaign. They were later duplicated in larger numbers in the Bicep and QUaD experiments, and ultimately flew in the High Frequency Instrument (HFI) on the *Planck* spacecraft (Jones *et al.* 2003, 2007, 2006b; Takahashi *et al.* 2010). In addition to unprecedentedly sensitive images of the unpolarized CMB anisotropies, the data from the 2003 flight of Boomerang provided the first measurements of the CMB EE and TE power spectra made with a bolometer as well as at frequencies above 30 GHz (see Figure 5; Jones *et al.* 2006a, Piacentini *et al.* 2006, Montroy *et al.* 2006, MacTavish *et al.* 2006).

---

† Prior to the 2003 flight of Boomerang, no attempt had been made to perform CMB polarimetry with bolometers since the pioneering effort of Caderni *et al.* (1978).

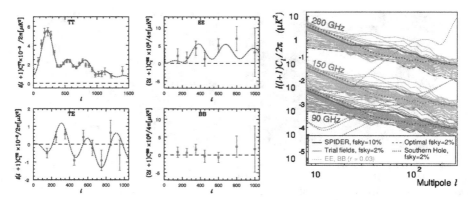

Figure 5. *At left*, the power spectrum of the CMB temperature and polarization anisotropies as measured by Boomerang during the January 2003 Antarctic LDB flight (MacTavish *et al.* 2006). *At right*, an illustration of the expected *minimum* amplitude of polarized Galactic emission (at these frequencies, primarily thermal dust emission) relative to the expected levels of the cosmological B-mode (shown in dotted lines for both the E- and B-modes). On scales larger than a few degrees, the Galactic foreground emission is likely to dominate the cosmological signal, requiring exquisite measurements over a range of frequencies to discriminate the two. For more discussion, see Fraisse *et al.* (2011).

The second of these, large format arrays of antenna-coupled detectors, are being developed for the Spider Antarctic LDB mission as well as South Pole based telescopes, and have only recently seen first light (Kuo *et al.* 2006, Fraisse *et al.* 2011, Brevik *et al.* 2010). Since the antenna, filter and detector can all be lithographed on a single wafer, antenna-coupled detectors are massively scalable. Whereas the number of detectors grew from eight PSBs in Boomerang, to dozens of PSBs in Bicep/QUaD and HFI, over two thousand background-limited detectors will fly on Spider.

Several groups, including UC Berkeley, NIST, ANL, GSFC and Caltech/JPL are now fielding innovative and complementary array technologies that provide the sensitivity, frequency coverage and control of systematics required to realize the scientific potential of CMB polarization (Hubmayr *et al.* 2012, McMahon *et al.* 2012, George *et al.* 2012, Westbrook *et al.* 2012, Aubin *et al.* 2010). EBEX and Spider represent the near future of Antarctic LDB CMB experiments. The two experiments are highly complementary, employing different technologies while probing complementary angular scales and electromagnetic frequencies. EBEX will field 1400 TES detectors spanning 150–400 GHz, covering 1% of the sky with 8′ resolution. Spider will fly 2400 detectors between 95 and 220 GHz, covering 10% of the sky with 30′ resolution. The EBEX Antarctic LDB experiment, which has deployed as of this writing, is the first highly multiplexed transition edge sensor (TES) array to have flown on a balloon (Reichborn-Kjennerud *et al.* 2010). Together, these Antarctic LDB experiments will constrain the B-mode CMB polarization at angular scales ranging from 30° to 10′, providing tight constraints on the shape of the B-mode power spectrum.

The future

Antarctic long duration balloon flights have played a pivotal role in the technological and scientific achievements in the CMB community during the last fifteen years, and remain at the leading edge of the search for the signature of inflation in the CMB. The benefits of the near-space environment provided by the stratospheric balloon platform include:

- Dramatically increased raw sensitivity – two to three times the best acheived from terrestrial telescopes at 90–150 GHz.
- The ability to extend frequency coverage above 150 GHz without significant degradation in performance due to atmospheric continuum emission
- Sensitivity to angular scales roughly five times larger than possible from the ground
- Access to 10% of the full sky near the Southern Galactic Pole that is relatively free from Galactic emission

The extremely low signal-to-background of the B-mode science requires not only massive scaling of background-limited detector arrays, but also new levels of systematic control. The community has developed a variety of creative solutions for both, but continued support of the development of detectors *and* polarimetric systems will be required as the next generation of experiments push orders of magnitude beyond the sensitivity of the first generation of CMB polarization experiments.

Galactic foregrounds will pose a significant, if not dominant, challenge in observational efforts to detect and characterize the cosmological B-mode signal. Disambiguation of the Galactic and cosmological signals will require sensitive surveys in closely spaced frequency bands spanning 60 to 400 GHz.

In addition to serving as a proving ground for space technologies, the balloon program provides an unparalleled training ground for the young scientists who will facilitate future space missions. Having been central to the successes of the past fifteen years, the Antarctic LDB program is poised to continue that tradition through the next decade.

Acknowledgements

The dramatic achievements of the past fifteen years, both technical and scientific, are the result of the combined efforts of an extraordinarily vibrant, talented and competitive community of researchers. This summary has been narrowly focused on the role played by Antarctic Long Duration Balloons in the development of the field, and in no way presumes to provide a comprehensive survey of contribution of the broader CMB community to the field. The author would like to acknowledge the support of the David and Lucille Packard Foundation, NASA grant NNX12AE95G, NSF award ANT-1043515 and the CSBF, and to thank the organizers of the IAU General Assembly for their efforts and hospitality.

References

Aubin, F., *et al.* 2010, in *SPIE Conference Series*, 7741, 77411T
Battistelli, E. S., *et al.* 2012, *MNRAS* 423, 1293
Bonetti, J., *et al.* 2012, *J. Low Temp. Phys.* 167, 146
Brevik, J. A., *et al.* 2010, in *SPIE Conference Series*, 7741, 77411H
Burton, M., *et al.* 1994, *PASA*, 11, 127
Caderni, N., Fabbri, R., Melchiorri, B., Melchiorri, F., & Natale, V. 1978, *Phys. Rev. D* 17, 1908
Challinor, A. D. 2012, *these proceedings*
D. D. Gregory & W. E. Stepp. 2004, *Advances in Space Research* 33, 1608
D. D. Gregory. 2006, *Advances in Space Research* 37, 2021
de Bernardis, P., *et al.* 2000, *Nature*, 404, 955
Fraisse, A. A., *et al.* 2011, ArXiv 1106.3087
George, E. M., *et al.* 2012, ArXiv 1210.4971
Halverson, N. 2012, *these proceedings*
Hanany, S., *et al.* 2000, ApJ 545, L5
Hubmayr, J., *et al.* 2012, *J. Low Temp. Phys.* 167, 904

Jones, W. C., Bhatia, R., Bock, J. J., & Lange, A. E. 2003, in *SPIE Conference Series* 4855, ed. T. G. Phillips & J. Zmuidzinas, 227

Jones, W. C., *et al.* 2006a, ApJ 647, 823

Jones, W. C., *et al.* 2006b, *New Ast. Rev.* 50, 945

Jones, W. C., *et al.* 2007, *A&A* 470, 771

Kamionkowski, M. 1999, *Nuc. Phys. B Proceedings Supplements* 70, 529

Kamionkowski, M. & Kosowsky, A. 1999, *Ann. Rev. Nuclear and Particle Science* 49, 77

Komatsu, E., *et al.* 2011, *ApJS*, 192, 18

Kosowsky, A. 1999, *New Ast. Rev.* 43, 157

Kovac, J. M., *et al.* 2002, *Nature* 420, 772

Kuo, C. L., *et al.* 2006, *Nuclear Instruments and Methods in Physics Research A* 559, 608

Lange, A., *et al.* 1995, *Space Sci. Rev.* 74, 145

Lange, A. E. 2002, in *Proc. Far-IR, Sub-mm & mm Detector Technology Workshop*, ed. J. Wolf, J. Farhoomand, & C. R. McCreight, NASA/CP-211408

Lange, A. E., Bock, J. J., & Zmuidzinas, J. 1999, in *NASA Space Astrophysics Detector Development*

Lange, A. E., *et al.* 1996, in *ESA Special Publication 388*, Submillimetre and Far-Infrared Space Instrumentation, ed. E. J. Rolfe & G. Pilbratt, 105

Lange, A. E., *et al.* 2001, *Phys. Rev. D* 63, 042001

Larson, D., *et al.* 2011, *ApJS* 192, 16

Leitch, E. M., *et al.* 2002, *Nature* 420, 763

MacTavish, C. J., *et al.* 2006, *ApJ* 647, 799

McMahon, J., *et al.* 2012, *J. Low Temp. Phys.* 167, 879

Montroy, T. E., *et al.* 2006, *ApJ* 647, 813

Netterfield, C. B., Jarosik, N., Page, L., Wilkinson, D., & Wollack, E. 1995, *ApJ* 445, L69

Netterfield, C. B., *et al.* 2002, *ApJ* 571, 604

Piacentini, F., *et al.* 2006, *ApJ* 647, 833

Rees, M. J. 1968, *ApJ*, 153, L1

Reichborn-Kjennerud, B., *et al.* 2010, in *SPIE Conference Series* 7741, 77411C

Ruhl, J. E., *et al.* 2003, *ApJ*, 599, 786

Samtleben, D., Staggs, S., & Winstein, B. 2007, *Ann. Rev. Nuclear and Particle Science* 57, 245

Scott, D. 1999, ArXiv astro-ph/9911325

Smoot, G. F., *et al.* 1992, *ApJ*, 396, L1

Storey, J. W. V, Ashley, M. C. B., Burton, M. G., & Phillips, M. A. 1998, in *SPIE Conference Series* 3354, ed. A. M. Fowler, 1158

Takahashi, Y. D., *et al.* 2010, *ApJ* 711, 1141

Torbet, E., *et al.* 1999, *ApJ* 521, L79

Westbrook, B., *et al.* 2012, *J. Low Temp. Phys.* 167, 885

Wright, E. L., *et al.* 1992, *ApJ*, 396, L13

Astrophysics from Antarctica
Proceedings IAU Symposium No. 288, 2012
M. G. Burton, X. Cui & N. F. H. Tothill, eds.

© International Astronomical Union 2013
doi:10.1017/S1743921312016687

A CMB B-mode Search with Three Years of BICEP Observations

Colin Bischoff[1] for the BICEP Collaboration

[1]Harvard-Smithsonian Center for Astrophysics
60 Garden St. MS 42, Cambridge, MA 02138, USA
email: **cbischoff@cfa.harvard.edu**

Abstract. The search for *B*-mode, or curl-type, polarization in the Cosmic Microwave Background is the most promising technique to constrain or detect primordial gravitational waves predicted by the theory of inflation. The BICEP telescope, which observed from the South Pole for three years from 2006 through 2008, is the first experiment specifically designed to target this signal. We review the observational motivations for inflation, the advantages of *B*-mode observations as a technique for detecting the gravitational wave background, and the design features of BICEP that optimize it for this search. The final analysis of all three seasons of BICEP data is in progress, representing a 50% increase in integration time compared to the result from Chiang *et al.* (2010). A preview of the three year result includes *E*-mode and *B*-mode maps, as well as the projected constraint on *r*, the tensor-to-scalar ratio.

Keywords. cosmology: observations, cosmic microwave background, polarization, instrumentation: polarimeters

---

1. The Cosmic Microwave Background and Inflation

The Cosmic Microwave Background (CMB) serves as a unique and invaluable window to our early universe. First discovered by Penzias and Wilson in 1964, the CMB is a relic thermal radiation that was originally emitted during the period of recombination, about 300,000 years after the Big Bang, when the universe had cooled enough that protons and electrons could form neutral Hydrogen. Recombination occurred at a temperature of ∼3,000 K; since then, the size of the universe has increased by a factor of 1,100, so the CMB is observed to have a temperature of 3 K today.

The temperature of the CMB is very nearly uniform across the entire sky, with small anisotropies at the 10^{-5} level. This uniformity poses a theoretical problem, because the causal horizon at the time of recombination corresponds to an angular scale of only three degrees. The question of how different regions of the universe that are causally disconnected achieved such similar temperatures is known as the horizon problem.

A solution to the horizon problem is offered by the theory of inflation. The theory proposes that the universe underwent a period of accelerated expansion for a tiny fraction of a second after the Big Bang (Guth, 1981). The expansion increased the scale factor of the space-time metric by 60 *e*-foldings, or 10^{26}, transforming sub-atomic scales to astronomical ones. The observed temperature uniformity of the universe is explained because the regions that were causally disconnected at recombination were in close, causal proximity before inflation.

In addition to the horizon problem, there are several other observed features of our universe that are neatly explained by the theory of inflation.

- Observations of the CMB have demonstrated that the spatial geometry of the universe is flat on cosmological scales (de Bernardis *et al.*, 2000). Inflation naturally explains this flatness, since our entire observable universe inflated from a small locally flat region of a pre-inflation space with arbitrary curvature.

• From temperature anisotropies in the CMB, we know that the initial density perturbations were nearly scale-invariant, with a small red tilt. Inflation naturally generates such a spectrum of perturbations, as quantum fluctuations are blown up to astronomical scales. After inflation ends, these initial perturbations will evolve linearly, starting from the time when their wavelength enters the causal horizon. The CMB encodes the state of each mode at the time of recombination.

• Related to the horizon problem is the presence of super-horizon modes in the CMB. The process described above to generate density perturbations will create modes with wavelength larger than the causal horizon. After inflation, these modes are frozen and unable to evolve, so we observe their primordial amplitudes.

• Guth's initial motivation to invent inflation did not involve the CMB at all; he proposed the theory to explain the observed lack of magnetic monopoles, which could have been diluted to unobservable density by the expansion of space.

While inflation does an impressive job of setting up initial conditions that match our universe, it remains a highly speculative theory. Proposed mechanisms for inflation all involve physics beyond the Standard Model. The energy scale involved, typically the Grand Unified Theory (GUT) scale (10^{16} GeV), is so enormous that direct experimental tests are impossible. However, an additional prediction of the theory of inflation is that it would create a cosmic background of gravitational waves, which would in turn leave an imprint on the Cosmic Microwave Background. Detecting the gravitational wave signal would provide a "smoking gun" for inflation, and is one of the most important goals in cosmology today.

2. Searching for Inflation with CMB Polarization

The amplitude of the gravitational wave background that is generated by inflation is parametrized by r, the tensor-to-scalar ratio, which specifies the power in gravitational waves (tensor perturbations) as a fraction of the power in density (scalar) perturbations (for details, see Baumann et al., 2009). The parameter r depends on the energy scale of inflation, V_ϕ, as shown in equation (2.1); inflation occurring at the GUT scale corresponds to a tensor-to-scalar ratio in the range 0.01 to 0.1.

$$V_\phi \simeq \left(3 \times 10^{16} \text{ GeV}\right) r^{1/4} \tag{2.1}$$

The gravitational wave background does affect the power spectrum of the CMB temperature anisotropies. Experiments that map the CMB temperature have placed upper limits on r, with the best limit to date set at $r < 0.17$ at 95% confidence by Keisler et al. (2011) using data from the South Pole Telescope in combination with full sky CMB maps from WMAP, measurements of baryon acoustic oscillations, and the Hubble parameter. This technique is limited, however, by the large amplitude, and correspondingly large sample variance, of the temperature power spectrum, which overwhelms the signal of interest.

The polarization of the CMB provides a way to search for the signal of inflation against a much lower background. An arbitrary polarization field can be decomposed into gradient-type and curl-type components (Kamionkowski, Kosowsky, and Stebbins, 1997, Hu and White, 1997), known respectively as E-modes and B-modes in analogy to electromagnetism. As illustrated in Figure 1, a plane-wave polarization fluctuation is a pure E-mode if the polarization is aligned parallel or perpendicular to the wave vector, $\vec{k}$; for a pure B-mode, the polarization is aligned at $\pm 45°$ to the wave vector.

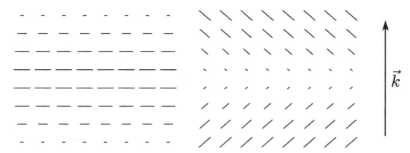

Figure 1. Pure E-mode (left) and B-mode (right) polarization patterns. For both cases, the wave vector of the modulation points vertically. E-mode polarization is parallel or perpendicular to $\vec{k}$; B-mode polarization is oriented at $\pm 45°$ to $\vec{k}$.

The density perturbations, which dominate the CMB temperature anisotropy, contribute only E-modes to the polarization anisotropy, while the gravitational wave background sources both E and B-modes. This can be understood intuitively based on the fact that the odd-parity B-modes are characterized by a handedness in addition to their amplitude and wave vector; a scalar density perturbation has only amplitude and wave vector while tensor gravity waves also feature a polarization degree of freedom that can produce odd parity. With no contribution from density perturbations, a search for B-modes in the CMB polarization is a clean and powerful method to detect or limit the amplitude of gravitational waves sourced by inflation.

Figure 2 shows the current state of measurements of the E-mode and B-mode power spectra. The E-mode polarization was first detected by DASI, observing from the South Pole, as reported in Kovac *et al.* (2002); this detection has since been confirmed by many other experiments. There have been no detections of B-mode polarization to date. The tightest upper limit on B-modes from gravitational waves comes from two seasons of BICEP data (Chiang *et al.*, 2010), corresponding to $r < 0.72$ at 95% confidence. This limit is weaker than the upper limit from CMB temperature measurements, but future polarization experiments will have the sensitivity to fully exploit the potential of B-modes in the search for inflation.

3. The BICEP Telescope

Background Imaging of Cosmic Extragalactic Polarization, or BICEP, is the first CMB telescope designed specifically to search for the B-mode signal of inflation. BICEP was initially deployed to the Amundsen-Scott South Pole Station in the austral summer of 2005–2006 and made observations during three winter seasons from 2006 through 2008. Results from the analysis of the first two observing seasons have been published in Chiang *et al.* (2010) along with a detailed instrument characterization in a companion paper, Takahashi *et al.* (2010). The E-mode power spectrum and B-mode upper limits from Chiang *et al.* are included in Figure 2.

A discussion of the design of the BICEP telescope highlights the challenges that must be addressed by any experiment targeting the B-mode signal.

- First, and most important, the signal is very small, so instrumental sensitivity is critical. The BICEP focal plane consists of 49 pairs of feedhorn-coupled polarization-sensitive bolometers (PSB), as shown in Figure 3. Orthogonal absorbers are each coupled to a neutron transmutation doped (NTD) Germanium bolometer and the two temperature

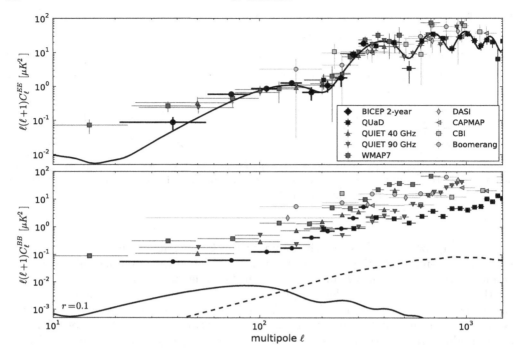

Figure 2. Measurements to date of the E-mode (top) and B-mode (bottom) power spectra, from Chiang *et al.* (2010), Brown *et al.* (2009), QUIET Collaboration (2011), QUIET Collaboration (2012), Larson *et al.* (2011), Leitch *et al.* (2005), Bischoff *et al.* (2008), Sievers *et al.* (2007), and Montroy *et al.* (2006). The E-mode spectrum has been detected at high signal-to-noise by many experiments, but only upper limits exist for the B-mode spectrum. The solid line shown for the E-mode spectrum is the theory curve for ΛCDM cosmology. The solid line shown for the B-mode spectrum is the expected signal from gravitational waves with $r = 0.1$, peaking at $\ell \sim 80$, while the signal due to gravitational lensing of the E-modes, which peaks at $\ell \sim 10^3$, is shown as a dashed line.

measurements are differenced to obtain a measurement of linear polarization. The sensitivity of an individual PSB pair is background limited, so improved instrument sensitivity can only be achieved by increasing the number of detectors, or moving to a lower background environment, such as a stratospheric balloon or satellite.

• While B-modes are geometrically distinct, instrumental systematics can cause leakage of E-mode polarization or temperature anisotropies into the B-modes. The temperature power spectrum is five orders of magnitude larger than the B-mode spectrum corresponding to $r = 0.1$, so these systematics must be very tightly controlled. The key to achieving this goal for BICEP lies in the clean design of the optical system—a cryogenic on-axis refractor with two plastic lenses cooled to 4K. A three axis mount allows BICEP to observe the same region of the sky at different boresight orientations; in combination with a highly redundant scan strategy, this provides additional suppression of systematics. The success of this optical design led to its adoption for successor projects, BICEP2 and the Keck Array.

• The B-mode signal from gravitational waves peaks at angular scales of 2–3° (multipoles of 70–100), so requirements for angular resolution are modest. The aperture of the BICEP telescope is only 25 cm, which yields beams with FWHM of 0.93° and 0.60° at 100 and 150 GHz, respectively. The small aperture means that the far-field of the optics is easily accessible for calibration. BICEP used a flat mirror to redirect the field of view

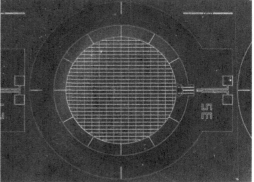

Figure 3. (left) The BICEP focal plane, consisting of 49 feedhorns – 25 operating at 100 GHz, 22 operating at 150 GHz, and 2 operating at 220 GHz. (right) A pair of polarization-sensitive bolometers, consisting of orthogonal absorbers (vertical and horizontal lines) each coupled to separate bolometers.

towards a calibration source mounted on an adjacent building. This technique yields high quality beam maps that are used to set upper limits on optical systematics.

• Galactic foreground signals, dominated by thermal dust at high frequencies and synchrotron emission at low frequencies, can be minimized by focusing observations on the cleanest regions of the sky. The South Pole observing site gives BICEP continuous access to the Southern Hole, two percent of the sky with uniquely low foregrounds. By observing at two frequencies, BICEP has a lever to distinguish between foreground and CMB, if it proves necessary.

• Other advantages of observing from the South Pole include cold, dry atmospheric conditions, which provide the best microwave observing on Earth, and excellent logistics support from the United States Antarctic Program.

4. Data Analysis for the BICEP Three-Year Result

The result published in Chiang *et al.* used only the first two observing seasons (2006 and 2007); work is underway on a final analysis that will incorporate all three seasons of BICEP data. A change in the data selection will include several channels that were cut from the two-year result due to irregular bolometer time constants. Between the third season and the additional channels, the total integration time is increased by 50%.

The new analysis will also incorporate a novel technique for removing temperature to polarization leakage due to instrumental systematics. This technique uses high signal-to-noise CMB temperature maps provided by WMAP (Jarosik *et al.*, 2011) to deproject modes corresponding to certain classes of systematics. For example, a relative error in the calibrated responsivity between detectors in a PSB pair leads to direct coupling of the temperature map to polarization, known as "monopole leakage". A pointing offset between the two detectors in a pair results in false polarization signal proportional to the first spatial derivative of the temperature map, known as "dipole leakage". The deprojection procedure, described in detail in Aikin (2012, in preparation), applies a linear regression to determine the coefficient for each leakage mode over each observing phase (nine hours). The derived coefficients can be compared to other estimates of the instrumental systematics to make sure that the effect is well understood. While the deprojection procedure has been formulated for monopole, dipole, and quadrupole leakage (the last is due to differences in beam ellipticity between paired detectors), the instrument

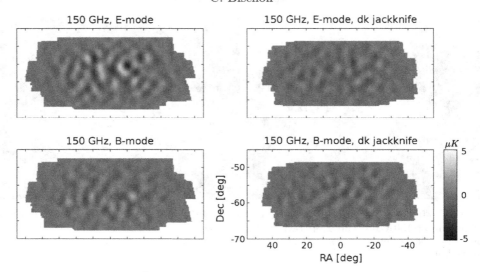

Figure 4. Maps of E-mode (top left) and B-mode (bottom left) polarization from three years of BICEP. Jackknife maps constructed by differencing data at alternate telescope boresight orientations are shown for both E-modes (top right) and B-modes (bottom right). All maps are consistent with noise, except for the non-jackknife E-modes.

characterization in Takahashi *et al.* (2010) demonstrated that only the monopole leakage exceeds (slightly) the benchmark corresponding to $r = 0.1$. For this reason, the BICEP three-year analysis will deproject monopole leakage only. However, the well-characterized dataset has been invaluable for testing out the higher-order corrections that may prove necessary for more sensitive experiments.

Maps of E-mode and B-mode polarization made from three years of BICEP data are shown in Figure 4. An aggressive bandpass filter has been applied to show only modes in the multipole range of interest for the search for inflation, $50 < \ell < 120$. In addition to the actual E-mode and B-mode maps, two jackknife maps are shown, with E-modes and B-modes calculated by differencing maps made at alternate telescope boresight orientations. The sky signal subtracts out of the jackknife maps and residuals are found to be consistent with noise. Examination of Figure 4 confirms that the E-mode map shows significantly higher power, due to the real CMB E-mode signal; the B-mode map is consistent with the jackknife maps, so no B-modes are detected.

While some work is still necessary before publication of the three-year CMB power spectra, simulations with realistic noise and correct data selection can be used to project the power of the result for constraining r. Figure 5 shows a histogram of 95% confidence upper limits derived from 500 independent simulations of the entire BICEP experiment. The solid grey bars include three years of data while the hollow bars include only the two years of data used for Chiang *et al.*. These histograms show that large variation in upper limits that can be obtained even for a non-detection, depending on the observed level of B-mode power. However, we do have the Chiang *et al.* result, indicated by the dashed vertical line, which is a strong prior on our expectation for the upper limit from the three-year analysis.

In conclusion, a final result for three years of BICEP will be published soon, with 50% additional integration time and improved systematics mitigation in comparison to Chiang *et al.*. However, this result will soon be eclipsed by vastly more sensitive data from BICEP2, which deployed during the austral summer of 2009–2010 and is currently in its third and final observing season, and the Keck Array, which deployed one year

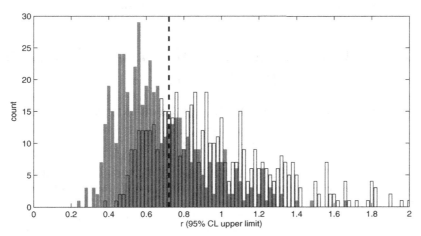

Figure 5. Histograms of 95% confidence upper limits on r, derived from simulations of the full three-year (solid grey bars) or two-year (hollow bars) BICEP data. The vertical dashed line indicates the upper limit actually obtained for the two year data, as published in Chiang *et al.* (2010).

after BICEP2 and will be operating through 2016. For more about BICEP2 and the Keck Array, see Clem Pryke's article in these proceedings.

References

Aikin, R., in preparation
Baumann, D. *et al.* 2009, *AIP-CP*, 1141, 10
Bischoff, C. *et al.* 2008, *ApJ*, 684, 771
Brown, M. L. *et al.* 2009, *ApJ*, 705, 978
Chiang, H. C. *et al.* 2010, *ApJ*, 711, 1123
de Bernardis, P. *et al.* 2000, *Nature*, 404, 955
Guth, Alan H. 1981, *Phys. Rev. D*, 23, 347
Hu, W. & White, M. 1997, *Phys. Rev. D*, 56, 596
Jarosik, N. *et al.* 2011, *ApJS*, 192, 14
Kamionkowski, M., Kosowsky, A., & Stebbins, A. 1997, *Phys. Rev. D*, 55, 7368
Keisler, R. *et al.* 2011, *ApJ*, 743, 28
Kovac, J. M. *et al.* 2002, *Nature*, 420, 772
Larson, D. *et al.* 2011, *ApJS*, 192, 16
Leitch, E. M. *et al.* 2005, *ApJ*, 624, 10
Montroy, T. E. *et al.* 2006, *ApJ*, 647, 813
Penzias, A. A. & Wilson, R. W. 1964, *ApJ*, 142, 419
QUIET Collaboration 2011, *ApJ*, 741, 111
QUIET Collaboration 2012, arXiv:1207.5034
Sievers, J. L. *et al.* 2007, *ApJ*, 660, 976
Takahashi, Y. D. *et al.* 2010, *ApJ*, 711, 1141

Astrophysics from Antarctica
Proceedings IAU Symposium No. 288, 2012   © International Astronomical Union 2013
M. G. Burton, X. Cui & N. F. H. Tothill, eds.   doi:10.1017/S1743921312016699

CMB Polarization with BICEP2 and Keck-Array

Clement Pryke for the BICEP2 and Keck-Array Collaborations

University of Minnesota Physics, 116 Church Street S.E.,
Minneapolis, MN, 55455, USA

Abstract. BICEP2 is an evolution from the highly successful BICEP CMB polarization experiment. In turn Keck-Array is an array of BICEP2 like receivers to achieve an additional increase in sensitivity. All these experiments are located at the South Pole in Antarctica and target the CMB *B*-mode polarization signal which is predicted to exist in many simpler models of Inflation at angular scales of several degrees. The design and performance of BICEP2 and Keck-Array is described and some preliminary polarization maps are presented.

Keywords. CMB, Polarization, Antarctica

1. Introduction

The theory known as inflation posits that the Universe underwent a tremendous burst of hyper expansion (factor $\sim 10^{60}$) at a tiny fraction of a second after the initial singularity ($\sim 10^{-35}$ s). During this expansion quantum fluctuations of the metric of all kinds are "forced real" injecting — amongst other things — gravity waves into the fabric of space which then propagate through it until the present time. At $\approx 380,000$ years after the beginning the Universe made the transition from plasma to neutral gas (recombination) and the light we see today as the Cosmic Microwave Background (CMB) was released. If these Inflationary gravity waves were present at recombination then they will have produced quadrupolar variations in the intensity of the CMB photons as they were last scattered from the plasma electrons. When electrons are exposed to unpolarized light which has a quadrupolar anisotropy the re-scattered light is partially polarized. The inhomogeneities which give rise to the temperature anisotropy of the CMB drive flows of material which, through Dopler shifting, generate quadrupolar anisotropy and hence polarization. However the resulting polarization pattern is naturally aligned with its own gradient and, by analogy with electrodynamics, is known as *E*-mode. The gravity waves, on the other hand, respect no such rule and can produce patterns with a curl, or *B*-mode, component. A number of experiments are targeting this exciting, although perhaps elusive, signal. They are in essence remote sensing gravity wave experiments looking for the tell-tale signature of Inflation written on the last scattering surface.

The polarization of the CMB was first detected by the DASI experiment which was sited at South Pole (Kovac *et al.* 2009). DASI was followed at Pole by the twin experiments QUaD (Brown *et al.* 2009) and BICEP (Chiang *et al.* 2009) which together still hold the record for best sensitivity at smaller and larger angular scales respectively. BICEP has now been replaced by BICEP2 and QUaD with the Keck-Array. The science goal of all these experiments is the same — to push down to the sensitivity level required to detect the inflationary *B*-modes. The signal is conventionally described in terms of the tensor to scalar ratio *r*. From temperature anisotropy and other observations we know that $r \leqslant 0.2$. This paper will describe BICEP2 and Keck-Array and their preliminary results.

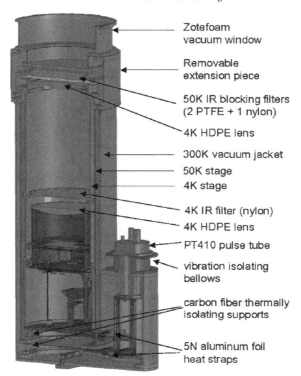

Zotefoam
vacuum window

Removable
extension piece

50K IR blocking filters
(2 PTFE + 1 nylon)

4K HDPE lens

300K vacuum jacket

50K stage

4K stage

4K IR filter (nylon)

4K HDPE lens

PT410 pulse tube

vibration isolating
bellows

carbon fiber thermally
isolating supports

5N aluminum foil
heat straps

Figure 1. Cross section of a Keck-Array cryostat. (Reproduced from Sheehy *et al.* (2010).)

2. The BICEP/Keck Strategy

The BICEP/Keck strategy for the detection of inflationary B-modes is to go as deep as possible, as quickly as possible, at 150 GHz in the "Southern Hole" region (centered at RA= 0^h, Dec=$-57°$). At this frequency the CMB is bright, but also maximally distinct from galactic foreground emission in such areas of high galactic latitude sky. The ~ 600 square degree patch size is selected to allow adequate separation of E and B-modes at the angular scale where the cosmological signal is predicted to peak ($\ell \sim 80$). The "Southern Hole" is also at high (and constant) elevation angle as viewed from the Pole which helps to reduce (and stabilize) possible ground contamination. The Sun is also a potential source of contamination but is always relatively far from the observing field and, of course, below the horizon for six months of the year.

The South Pole is an excellent site for these type of observations as is detailed elsewhere in these Proceedings. At 150 GHz water vapor is highly opaque. Due to the high altitude and extreme cold the median precipitable water vapor is lower on the Antarctic plateau than at any other site †. Importantly what moisture there is is well mixed in the atmosphere and thus presents a stable baseline to the observations.

BICEP2 and Keck-Array use near identical optics to BICEP1. The light enters through a 30 cm vacuum window, passes through IR blocking filters, and then through an all-cold telecentric refracting telescope (see Figure 1). This system delivers only modest angular resolution (0.6 deg FWHM at 150 GHz) but is highly stable, compact, and low cost. In addition, due to the small size, it can be fitted with a long absorptive fore-baffle which

† Some sites have lower PWV for a small fraction of the time but what matters for CMB observing is the "average" conditions.

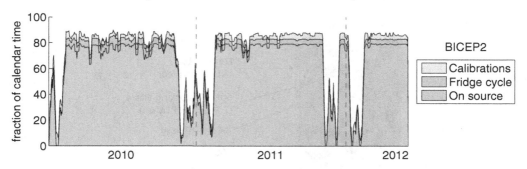

Figure 2. Observing efficiency of BICEP2 through 2.5 observing seasons. (Reproduced from Ogburn *et al.* (2012).)

results in extremely low sidelobe response. A further advantage of big beam telescopes is that the far field is only ∼ 100 m away enabling calibration sources to be placed on masts of modest height. The focal planes are cooled to 250 mK using closed cycle helium refrigerators.

The basic operating principle of the telescopes is similar to many others. The mount scans in azimuth sweeping the pencil beams of the detectors across the sky while the detector readout signals are recorded. After a number of scans the mount steps in elevation and the process repeats forming a raster pattern. Each pixel contains bolometers sensitive to orthogonal linear polarizations. In offline analysis relative gain calibration is applied and the pair difference taken to measure polarization. The data is binned into maps with appropriate accounting for the angle of the pair as projected onto the sky. Atmospheric emission is unpolarized and so cancels in the differencing operation.

Working at Pole enforces a strong discipline on the experimental effort. For nine months of the year the station is "closed" — no planes can arrive or leave. Thus all equipment installation, maintenance and upgrades must be conducted during the brief summer season. Once the last plane departs the telescopes settle into relentless homogeneous observation watched over by their dedicated winter-over scientists.

3. BICEP2

While BICEP and QUaD had focal planes assembled from individually packaged NTD detectors (diced wafers) BICEP2 made the transition to monolithic tiles of detectors (intact wafers). These new generation focal planes consist of four tiles each containing 64 dual polarization pixels (i.e. 256 pixels and 512 total TES detectors). Instead of using feedhorns the beam forming is accomplished "on wafer" by making each pixel a phased array of planar crossed dipole antennas.

BICEP2 was deployed in late 2009 and has observed through the 2010, 2011 and 2012 Austral winters. Observing efficiency has been excellent as is seen in Figure 2. The per detector sensitivity is better than BICEP1 and the yield is also good — Figure 3 compares the sensitivity distributions — BICEP2 has more than ten times the mapping speed of BICEP1.

Although the analysis does not primarily rely on it, as a "by-product" it is possible to make maps of *E*-modes and *B*-modes. Figure 4 shows such maps made with six months of BICEP2 data — note the dramatic reduction in *B*-mode noise. Ground based polarization experiments benefit tremendously from performing an instantaneous pair difference of orthogonal detectors — low frequency noise from both atmosphere and instrument is almost completely common mode and the pair difference noise is empirically found to be

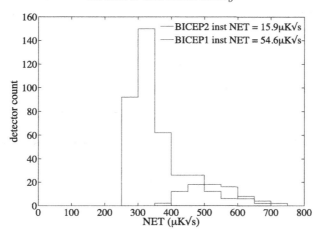

Figure 3. Distribution of sensitivity of BICEP2 detectors versus that of BICEP1. Note that not only does BICEP2 have many more pixels than BICEP1 but that each of them is also more sensitive. (Figure courtesy of Justus Brevik.)

almost white. (The alternative is fast modulation of the polarization sensitivity angle — BICEP/Keck has not chosen to pursue this option.)

Pair differencing demands excellent spatial coincidence of the beams on the sky, and this requirement is especially demanding when the angular size of the beams is only a few times smaller than the effect under study. If there is a small offset in the A/B beam centroids then this produces a leakage of the gradient of the much brighter temperature anisotropy into the polarization signal. However under rotation of the array by 180 degrees the leakage is inverted and hence cancels in the polarization map. The BICEP2 detectors exhibit substantial A/B centroid offset but by designing the observation pattern appropriately the leakage is demonstrably suppressed by a large factor. This is visually demonstrated in Figure 4 where the B-mode map noise is reduced by a large factor versus BICEP1 — these maps effectively are a B-mode limit many times better than the published BICEP1 result (Chiang *et al.* 2009). Further suppression of centroid offset and other beam systematic effects is achieved in analysis using a technique dubbed "deprojection". A full analysis of the first two years of BICEP2 data is in an advanced state and an r-limit will be released very soon.

4. Keck-Array

The focal plane of BICEP2 is already packed with the maximum number of detectors which it will support with full efficiency and acceptable beam quality. Good though the sensitivity of BICEP2 is, pursuit of the gravity wave signal demands further increase in sensitivity. The only option then is to increase the numbers of receivers and this is exactly what Keck-Array does — deploys a set of five BICEP2 like receivers on the telescope mount originally built for DASI (and then re-used for QUaD). This was made possible by the transition from liquid helium cooling to pulse tube cooler technology, and in addition required a very compact cryostat design — see Figure 1. Note the extremely small (few centimeter) distance from the 300 K cryostat shell to the 4 K telescope assembly. While the pulse tube coolers use ~ 10 kW each presenting a non negligible load on the station power plant, this is still logistically far preferable to the resources which would be necessary to provide vast amounts of liquid helium through each winter season.

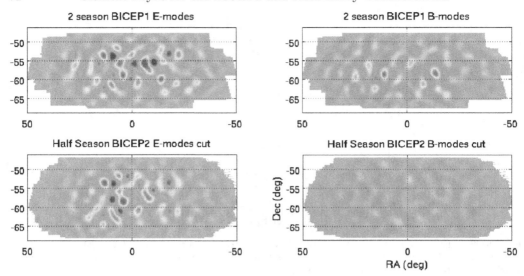

Figure 4. Maps of the CMB polarization E-mode and B-mode filtered to the angular scale range where the inflationary gravity wave signal is expected to be most prominent ($50 < \ell < 120$). The upper maps are the published BICEP1 results which hold the current world record in terms of sensitivity ($r < 0.72$, Chiang *et al.* (2010)). The lower maps are new preliminary BICEP2 results using half a season of data. The E-mode maps were already signal dominated for BICEP1 and so show little change going to BICEP2. However the BICEP2 B-mode maps show a large decrease in noise due to the much higher sensitivity of the BICEP2 instrument.

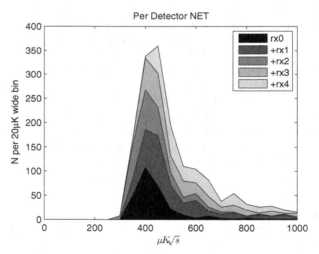

Figure 5. Histogram showing the sensitivity of Keck-Array detectors with a per receiver breakdown. Further improvements are planned — especially to the worst performing receiver. (Reproduced from Kernasovskiy *et al.* (2012).)

The sensitivity of the Keck-Array as measured during the 2012 Austral winter season is shown in Figure 5. One of the receivers badly underperforms and upgrades to this and others will be performed during the 2012/13 summer season.

The pulse tube coolers cease to function when inclined at angles > 45 degrees from the vertical. (The line of sight rotation axis of the telescope mount precludes angling the cooler from the optical axis as is done for SPT.) Therefore a 45 degree flat mirror is

Figure 6. Calibration of Keck-Array. Left: Keck-Array is located in the bowl shaped shield in the foreground, while the calibration source is on the building in the background. Center: The flat mirror mounted on Keck-Array. Right: A calibration source transmits from its mast. (Reproduced from Vieregg *et al.* (2012).)

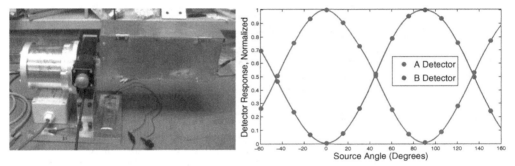

Figure 7. A rotating polarized calibration source and test results. Note the excellent polarization purity of the new antenna coupled detectors. (Reproduced from Vieregg *et al.* (2012).)

required to redirect the beams parallel to the horizon to calibration sources mounted on an adjacent building. Because of the large bowl shaped shield within which the mount is located this mirror has to be extremely large and high above the telescopes. In addition there is a strong desire to be able to mount and unmount this calibration flat without the use of a crane. An elegant solution has been implemented using honeycomb panels and large carbon fiber rods — see Figure 6.

Figure 7 shows a rotating polarized source and test data. The source transmits near 100% polarized broadband radiation while the entire assembly is rotated. The stationary detectors record data which modulates sinusoidally as the source aligns and anti-aligns with the polarization sensitivity direction of each given detector. The polarization purity of the new antenna coupled detectors is excellent — when anti-aligned the signal is < 1% of that when aligned. (The older horn fed detectors had cross polar response of ~ 3%.)

Figure 8 shows test data taken with a chopped thermal (unpolarized) source. In this case the telescope scans out a raster pattern passing the beam of each detector across the source direction, thus producing a beam map. The illumination of the aperture plane is a truncated Gaussian leading to a far field beam pattern with an approximately Gaussian main lobe surrounded by ring sidelobes. The figure shows the beam of a single detector with an elliptical Gaussian fit and the residual. The first sidelobe is apparent and stacking over detectors multiple sidelobes are detected in excellent agreement with optical

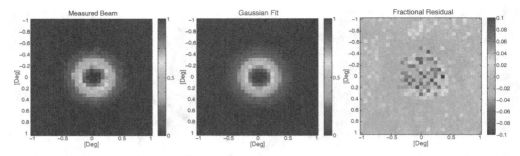

Figure 8. Left: far field beam pattern of a single Keck-Array detector as measured with an unpolarized thermal source. Center: the best fitting elliptical Gaussian. Right: the residual between data and model. (Reproduced from Vieregg *et al.* (2012).)

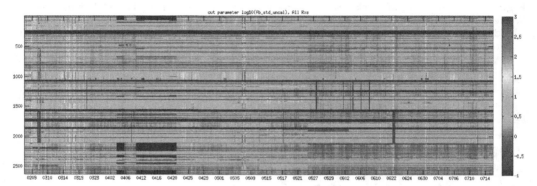

Figure 9. One of the Keck-Array cut parameters which monitors the raw data quality. The color scale shows the log of the standard deviation of the data as measured across each forward or backward "half-scan". Red/yellow vertical bars correspond to bad weather while horizontal dark blue bands are non functional pixels. The dark blue vertical bar around 0622 corresponds to a period when one of the five receivers was offline. Note that the horizontal scale is approaching six months.

modeling. Further beam mapping is planned although this data is mainly a cross check and is not explicitly needed for CMB analysis. The deprojection algorithm effectively fits for the beam parameters from the CMB data itself. The A/B centroid offsets of the Keck-Array detectors are somewhat smaller than that of BICEP2 and further major improvements have been achieved in newer test detectors.

Keck-Array is an extremely complicated system with ~ 2500 channels of TES detectors being read out through several layers of SQUID's. There are many parameters to tune and much scope for optimization. During winter running the data is transferred back to the US daily via satellite (~ 25 GB/day) and must be constantly monitored for quality assurance. To aid in this process, and as an integral part of the data reduction, sophisticated automatic data cutting machinery has been built. Figure 9 shows one of the lowest level cut parameters.

5. Conclusion

The BICEP2 and Keck-Array experiments have been described. As of the time of writing BICEP2 is completing its third and final season of observations. Meanwhile Keck-Array is completing its first season with the full complement of five receivers. Four more seasons of observations are planned with further sensitivity upgrades and a switch to additional observing frequencies (90 and/or 220 GHz). Meanwhile a published result from BICEP2 is imminent. With the massive sensitivity which Keck-Array brings to bear detection of B-mode polarization is essentially guaranteed. Our projections indicate that the B-mode from galactic foreground in our field will be at a level equivalent to $r \sim 0.02$ with an uncertainty of perhaps factor two. If a signal is detected then the angular power spectrum will give hints as to whether the origin is galactic or cosmological. Ultimately however confirmation at other frequencies and by other experiments will be required.

References

J. Kovac, E. Leitch, C. Pryke, J. Carlstrom, & W. Holzapfel 2001, *Nature*, 442, 772. astro-ph/0209478

M. Brown *et al.* 2009, *ApJ*, 705, 978. arxiv/0906.1003

C. Chiang *et al.* 2010, *ApJ*, 711, 1123. arxiv/0906.1181

R. W. Ogburn *et al.* 2012, *Proceedings of SPIE*, 8452. arxiv/1208.0638

C. Sheehy *et al.* 2010, *Proceedings of SPIE*, 7741. arxiv/1104.5516

S. Kernasovskiy *et al.* 2012, *Proceedings of SPIE*, 8452. arxiv/1208.0857

A. Vieregg *et al.* 2012, *Proceedings of SPIE*, 8452. arxiv/1208.0844

Astrophysics from Antarctica
Proceedings IAU Symposium No. 288, 2012
M. G. Burton, X. Cui & N. F. H. Tothill, eds.

© International Astronomical Union 2013
doi:10.1017/S1743921312016705

The South Pole Telescope:
Latest Results and Future Prospects

Bradford Benson[1] & the SPT Collaboration (pole.uchicago.edu)

[1] Kavli Institute for Cosmological Physics, University of Chicago
5640 South Ellis Avenue, Chicago, IL 60637
email: bbenson@kicp.uchicago.edu

Abstract. The South Pole Telescope is a 10 meter telescope optimized for sensitive, high-resolution measurements of the cosmic microwave background (CMB) anisotropy and millimeter-wavelength sky. In November 2011, the SPT completed the 2500 deg^2 SPT-SZ survey. The survey has led to several major cosmological results, derived from measurements of the fine angular scale primary and secondary CMB anisotropies, the discovery of galaxy clusters via the Sunyaev-Zel'dovich (SZ) effect and the resulting mass-limited cluster catalog, and the discovery of a population of distant, dusty star forming galaxies (DSFGs). In January 2012, the SPT was equipped with a new polarization sensitive camera, SPTpol, which will enable detection of the contribution to the CMB polarization power spectrum from lensing by large scale structure (the so-called "lensing B-modes") and, on larger angular scales, a detection or improved upper limit on the primordial inflationary signal ("gravitational-wave B-modes"), thereby constraining the energy scale of Inflation. Development is underway for SPT-3G, the third-generation camera for SPT. The SPT-3G survey will cross the threshold from statistical detection of B-mode CMB lensing to imaging the fluctuations at high signal-to-noise; enabling the separation of lensing and inflationary B-modes and improving the constraint on the sum of the neutrino masses Σm_ν to a level relevant for exploring the neutrino mass hierarchy.

Keywords. cosmology, cosmic microwave background, clusters of galaxies, polarization

---

1. Introduction

The South Pole Telescope (SPT, Fig. 1) is a 10 meter telescope optimized for low-noise, high-resolution imaging surveys of the sky at millimeter (mm) and submillimeter (submm) wavelengths. In particular, all aspects of the SPT—the site, the telescope, the RF shielding and the cryogenic receivers—have been optimized for making ultra-sensitive measurements of the cosmic microwave background (CMB) anisotropy from degree to arcminute angular scales over thousands of square degrees of the sky (Carlstrom *et al.* 2011). The telescope is located at the NSF Amundsen-Scott South Pole station, the best developed site on Earth for mm-wave observations, with 30 times less atmospheric fluctuation power than found at the ALMA site in the Atacama desert (Bussman *et al.* 2005, Radford 2011). The telescope is an off-axis, classical Gregorian design which gives a wide diffraction-limited field of view, low scattering and high efficiency with no blockage of the primary aperture. The current telescope optics produce a 1′ FWHM beam at 150 GHz with a conservative illumination of the inner 8 meters of the telescope and a ∼1 deg^2 diffraction-limited field of view (Padin *et al.* 2008). The SPT is designed to modulate the beams on the sky by slewing the entire telescope at up to 4 deg s^{-1} and eliminating the need for a chopping mirror. The SPT observing program consists of three planned, underway, or proposed surveys: 1) SPT-SZ (2007-2011), 2) SPTpol (2012-2015) and 3) SPT-3G (2016-2019).

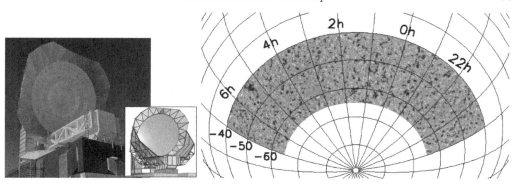

Figure 1. *(Left)* Photograph of the SPT with a new RF shield extending past the 10 meter primary. The inset shows the mechanical design of the second component of new shielding to be installed in November 2012. *(Right)* The 2500 deg^2 SPT-SZ survey map at 95 GHz.

2. SPT-SZ

The SPT-SZ survey was completed in November 2011 and covers 2500 deg^2 of the sky (see Fig. 1) at 95, 150 and 220 GHz with unprecedented depth and angular resolution. The SPT-SZ observations have led to significant results and new discoveries in three main areas: using the SZ effect to discover new galaxy clusters (particularly at high redshift), measurements of fine-scale CMB anisotropy, and the systematic discovery of strongly lensed high-redshift star forming galaxies.

The SPT-SZ cluster survey produced the first discovery of galaxy clusters via the SZ effect (Staniszewski *et al.* 2009), the first cosmological constraints from an SZ cluster survey (Vanderlinde *et al.* 2010), and constraints on primordial non-Gaussianity from the most massive clusters in the survey (Williamson *et al.* 2011). Benson *et al.* (2012) was the first result to use an SZ cluster survey to demonstrate significant improvements on the dark energy equation of state, w, and the sum of the neutrino masses, Σm_ν, measuring $w = -0.973\pm0.063$ and $\Sigma m_\nu < 0.28$ eV at 95% confidence, a factor of 1.25 and 1.4 improvement, respectively, over the constraints without SPT cluster data. Reichardt *et al.* (2012a) released a catalog of 158 SZ-selected optically confirmed clusters (see Fig. 2) from the first 720 deg^2 of the SPT-SZ survey, more than doubling the number of comparably massive clusters discovered at redshift $z > 0.5$. The unique sample of massive, high-redshift clusters discovered in the SPT-SZ survey has motivated numerous follow-up observations spanning wavebands from radio to X-rays, with large programs on the *Herschel*, *Spitzer*, *Hubble*, *XMM-Newton* and *Chandra* space telescopes.

The high sensitivity and angular resolution of the SPT-SZ survey are also ideal for precision measurements of fine-scale CMB anisotropy, and have led to many significant results. The SPT-SZ survey made the first detection of secondary CMB anisotropy due to the background of lower mass SZ clusters (Lueker *et al.* 2010), the first CMB-based constraint on the evolution of the ionized fraction during the epoch of reionization (Reichardt *et al.* 2012b, Zahn *et al.* 2012), the most-significant detection of non-Gaussian power in the CMB from gravitational lensing (Van Engelen *et al.* 2012), and the first detection of galaxy bias from CMB-lensing (Bleem *et al.* 2012). The SPT-SZ survey has also led to the best current measurement of the primordial fine-scale CMB anisotropy power spectrum (Fig. 2, Keisler *et al.* 2011). These data are sensitive to the expansion rate during recombination and thus to the number of relativistic particle species present at that epoch. This has allowed SPT to place the tightest yet constraints on the number of light particle species beyond the standard three neutrinos (e.g., sterile neutrinos).

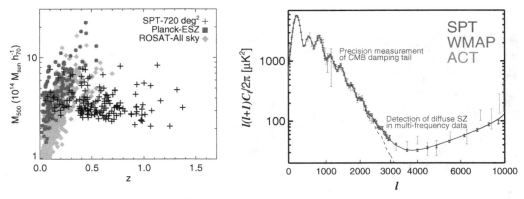

Figure 2. *(Left)* Mass vs. redshift for three cluster samples: (1) SZ-selected clusters from 720 deg^2 of the SPT-SZ survey, (2) SZ-selected clusters from *Planck*, and (3) X-ray selected clusters from the ROSAT all-sky survey. High-resolution SZ surveys, such as SPT-SZ, have a nearly redshift-independent selection. *(Right)* SPT measurement of the CMB power spectrum using 790 deg^2, or 1/3 of the full SPT-SZ survey, along with data from WMAP and ACT. A best-fit ΛCDM theory spectrum is shown with dashed (CMB) and solid (CMB+foregrounds) lines.

3. SPTpol

On January 27, 2012, first light on SPT was achieved for SPTpol, a new polarization sensitive camera with significantly higher mapping speed and, crucially, polarization sensitivity. The 768-pixel SPTpol focal plane contains detectors at two observing frequencies, 90 and 150 GHz, composed of two different detector architectures fabricated at Argonne National Lab and NIST, respectively, in collaboration with the SPT team. At 150 GHz, SPTpol consists of seven arrays of corrugated feedhorn-coupled TES polarimeters providing 588 dual-polarization pixels, or 1176 total bolometers (Hubmayr *et al.* 2011). At 95 GHz, SPTpol uses 180 individually packaged dual-polarization absorber-coupled polarimeters (a total of 360 bolometers) that are coupled to the telescope through machined contoured feedhorns (Chang *et al.* 2012). Fig. 3 shows a photo of the focal plane.

Projected constraints on the B-mode power spectrum from three-years of SPTpol data are shown in Fig. 3. With *Planck* priors, this data will lead to a constraint on the tensor-to-scalar ratio of $\sigma(r) = 0.028$ and a 1σ uncertainty on the sum of the neutrino masses $\sigma(\Sigma m_\nu) = 0.096$ eV. This constraint on Σm_ν will be roughly four times better than the KATRIN beta decay experiment, which has a predicted sensitivity of $\sim$0.6 eV (90% C.L.) for Σm_ν (Gonzalez-Garcia *et al.* 2010), and compliments other cosmological probes.

4. SPT-3G

We are currently developing a third-generation camera for the SPT, SPT-3G, which will exploit two technological advances to achieve a significant leap in sensitivity: 1) an improved wide-field optical design that allows more than twice as many diffraction-limited optical elements, and 2) multi-chroic pixels, sensitive to multiple observing bands in a single detector element. The combination will deliver a factor-of-20 improvement in mapping speed over the already impressive SPTpol camera. In total, the focal plane will consist of 15,234 detectors in three observing bands (90, 150, and 220 GHz) and will represent a fundamental step forward in CMB polarization measurements.

The SPT-3G camera will enable the advance from statistical detection of B-mode polarization power to high signal-to-noise measurements of the individual modes, i.e., maps. This data set will enable a broad range of cosmological results on dark energy,

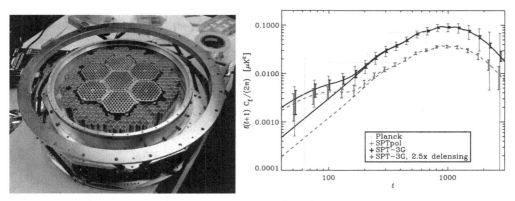

Figure 3. *(Left)*: Photo of the SPTpol focal plane. *(Right)* Projected B-mode constraints from four years of observing with the SPT-3G camera (black, thick-cross), with overplotted projected constraints from *Planck* (blue, dashed-cross) and SPTpol (purple, thin-cross). Model curves (solid lines) are for $\Sigma m_\nu = 0$, with $r = 0$ and $r = 0.04$. The red-medium-cross and dashed lines show the added sensitivity to primordial gravitational-wave B-modes from de-lensing, which causes a 2.5 reduction in lensed B-mode power.

neutrino mass, General Relativity, the epoch of reionization, and Inflation. For example, the 2500 deg^2 SPT-3G survey would enable an extremely tight constraint on r ($\sigma(r) <$ 0.01, see Fig. 3), and a constraint on Σm_ν to a level relevant for exploring the mass hierarchy ($\sigma(\Sigma m_\nu) < 0.06$ eV). It will also yield a cluster catalog 10 times larger than SPT-SZ, when combined with DES will yield a dark energy figure-of-merit > 100.

In a broader context, the high-sensitivity and high-resolution mm-wave temperature and polarization data are the first in a series of targeted surveys of the large region of the Southern sky with low Galactic foregrounds. They will inform and complement the upcoming LSST and CCAT surveys, as well as future planned spectroscopic surveys. The SPT-3G survey will uniquely contribute a measurement of the projected mass distribution over the full 2500 deg^2 region, enabling powerful new probes of large scale structure and adding value to all-sky surveys such as *Planck* and *WISE*.

References

Benson, B. A., *et al.* 2012, *ApJ submitted*, arXiv:astro-ph/1112.5435
Bleem, L. E., *et al.* 2012, *ApJ*, 753, 9
Bussman, R. S., Holzapfel, W. L., & Kuo, C. L. 2005, *ApJ*, 622, 1343
Carlstrom, J. E., *et al.* 2011, *PASP*, 123, 568
Chang, C. L., *et al.* 2012, *Journal of Low-Temperature Physics*, 183
Gonzalez-Garcia, M., Maltoni, M., & Salvado, J. 2010, *Journal of High Energy Physics*, 08, 117
Hubmayr, J., *et al.* 2012, *Journal of Low-Temperature Physics*, Jan, 904
Keisler, R., *et al.* 2011, *ApJ*, 743, 28
Lueker, M., *et al.* 2010, *ApJ*, 719, 1045
Padin, S., *et al.* 2008, *Applied Optics*, 47, 24, 4417
Radford, S. 2011, *Revista Mexicana de Astronomia*, arXiv:astro-ph/1107.5633
Reichardt, C. L., *et al.* 2012, *ApJ submitted*, arXiv:astro-ph/1203.5775
Reichardt, C. L., *et al.* 2012, *ApJ*, 755, 70
Van Engelen, A., *et al.* 2012, *ApJ*, 756, 142
Vanderlinde, K., *et al.* 2010, *ApJ*, 722, 1180
Vieira, J. D., *et al.* 2010, *ApJ*, 719, 763
Williamson, R., *et al.* 2011, *ApJ*, 738, 139
Zahn, O., *et al.* 2012, *ApJ*, 756, 65

Astrophysics from Antarctica
Proceedings IAU Symposium No. 288, 2012
M. G. Burton, X. Cui & N. F. H. Tothill, eds.

© International Astronomical Union 2013
doi:10.1017/S1743921312016717

Toward a 10,000-element B-Mode Experiment

Chao-Lin Kuo[1] for the BICEP3 and POLAR1 Collaborations

[1] Dept. of Physics, Stanford University
Kavli Institute for Particle Astrophysics and Cosmology
385 Via Pueblo Mall, Stanford, CA94305, USA
email: clkuo@stanford.edu

Abstract. In this paper, we introduce two compact, large-throughput CMB polarimeter designs (POLAR1 and BICEP3). These pathfinder experiments will pave the way for a comprehensive multi-frequency South Pole B-mode survey that, when jointly analyzed with arcminute-scale polarization data, can conclusively answer the question whether there is an appreciable fraction ($> 1\%$) of the primordial perturbations in the form of tensor modes (gravitational waves).

Keywords. Cosmology: cosmic microwave background, instrumentation: polarimeters

1. Introduction

Over the past year, several CMB polarization experiments targeting B-modes started observation with approximately a thousand detectors. Historically, the CMB community tends to be overly optimistic when it comes to predicting the performance of the experiments. However, based on the *achieved* sensitivity *improvements* over previous projects that had published limits on r, it is quite reasonable to expect the ongoing experiments to reach $r \equiv T/S = 0.02 \sim 0.05$ (Table 1.). This range corresponds to the expected level of Galactic foregrounds in the cleanest part of the sky at the optimal frequency. A few other experiments that will soon be operational should also be able to reach a similar sensitivity.

The next phase of CMB polarization research is about following up these measurements both in sensitivity and in frequency coverage. For the current generation of experiments, the combination of foregrounds, lensing, and potentially instrument systematics will likely limit the *detectable* r to > 0.05. This is especially true if we adopt the widely accepted detection threshold of 5-σ in elementary particle physics. Such a standard would be very appropriate for the discovery of primordial gravitational waves, an extraordinary claim in physics of fundamental interactions.

To make further progress, the main challenge is still the overall survey speed. The current CMB experiments are nearly background-limited and fully utilizing the available throughput in the optical systems. Following up these experiments would require expanding the frequency range for each focal plane element (multichroic) and/or increasing the total throughput. The former is a technical challenge that many groups are attempting. The latter strategy, which will be discussed in this article, is more straightforward and can also be combined with multichroic focal plane technologies when they become mature in the future.

Dealing with Lensing: Two Approaches

The lensing-induced B-mode signal is proportional to $\sim \ell^2$ at low multipoles, with a power equivalent to $r = 0.0255$ at $\ell = 85$. Based on Table 1, the ongoing experiments will start to hit lensing after a few years of observation. To make further progress on

Table 1. Past and Ongoing B-mode Experiments

Experiment	NET (μK-$\sqrt{s}$)	yrs collected	yrs expected	r-limit (95%)
BICEP	55	3	3	0.72 (2 yrs)
QUIET-Q	69	0.7	0.7	2.2
QUIET-W	87	1.3	1.3	2.8
BICEP2	15.9	2.5	3	?
Keck Array	11.5	1.5	4.5	?
SPTPOL	~ 16	0.5	4	?
POLARBEAR	~ 19	0.5	3	?

r, there are two complementary approaches. The first approach is to increase the sky coverage and remove lensing in the power spectrum in the same way instrumental noise is debiased in temperature power spectrum measurements. After Planck, the expected amount of lensing will be determined to better than 3%, simply due to the fact that all cosmological parameters, as well as the E-mode polarization, will be very well determined. The measured B-modes at $\ell \sim 100$ can be compared with this expected level and the significance for excess B-modes can be calculated.

Alternatively, the lensing deflection field can be reconstructed from arcminute-scale B-mode measurements. The expected lensing contribution to degree-scale B-modes can be derived by the mathematical "lensing" of the observed E-polarization using the reconstructed deflection field, which can then be subtracted from the observed B-map. This *delensing* procedure can be iterated to achieve a better removal fraction.

These two approaches are complementary because very different assumptions are made in the lensing removal process. It would be a powerful confirmation if the same conclusion on r is deduced using two methods. The first method especially puts a premium on overall system sensitivity. By increasing the sky coverage, the sensitivity to primordial B-modes continues to improve as $f_{sky}^{-1/2}$ until the clean part of the sky runs out. At South Pole the accessible clean sky is approximately 3000-4000 square degrees.

2. Compact Large-Throughput CMB Experiments

The primordial tensor-induced B-mode signal peaks at $\ell \sim 100$, corresponding to an angular scale of a few degrees. A 30 cm aperture telescope operating at 150 GHz has a FWHM of $30'$, which is more than sufficient to resolve the primordial B-polarization. This suggests that if systematic effects are under control, it is advantageous to pack a given number of detectors into optical systems that are *as compact as possible*, since the total cost of the experiment is a sharp function of the size and weight of the optics. Another very significant advantage for compact optical systems is that they allow *active rotation* of the entire instrument. As has been demonstrated in BICEP, QUaD and QUIET, redundant polarization measurements at multiple instrument angles (especially pairs separated by 180°) are crucial for distinguishing instrumental effects from polarization signals from the sky. This systematics mitigation technique cannot be overemphasized.

In this section, we discuss two compact optical designs with very large throughputs: the crossed-Dragone reflectors and the refractive telescopes.

Crossed-Dragone Reflectors

A crossed-Dragone telescope consists of two off-axis, lightly curved reflectors. In studies of CMB telescope optics, crossed-Dragone emerges as the front running design due to its superior field-of-view and low polarization systematics. Because of its overwhelming advantages, a similar optical design is being considered for the NASA-CMBPOL mission concept "EPIC-IM".

A crossed-Dragone has no intermediate stop and one must therefore control where the spillover goes. The simplest solution is to use random diffusive surfaces to scatter the spillover onto the cold sky (15-20K at the South Pole). Such surfaces can be used to cover the inside of an optical enclosure that completely surrounds both reflectors (Figure 1a). Full non-sequential ray tracing simulations indicate that on average it takes ~ 3 reflections for a photon to escape the enclosure and evenly distribute on the sky. Using low-emissivity surfaces the loading from the spillover will be negligible. A reflective Winston cone ground shield will limit the diffuse sidelobes to within 30° of the boresight direction and away from the Galaxy and the ground.

A 1.6-m crossed-Dragone telescope, known as **POLAR1**, is being developed to test these novel ideas. If the spillover termination scheme is demonstrated, such design would clearly be a very economical solution for the next generation ground-based lensing machine, dubbed POLAR Array, which aims to survey a significant fraction of the sky to a few μK depth. The lensless design also greatly simplifies broadband anti-reflection coating, making it an ideal platform for future multichroic CMB receivers.

Because of the resolution advantage compared to refractors, crossed-Dragone reflectors can also play an important role in synchrotron monitoring at lower frequencies, in the context of tensor-mode searches. At 44 GHz, POLAR1's 1.6-m optics provides a FWHM well-matched to those from compact (30-60cm) refractors at 100-150 GHz. With a similar spillover termination technique, the reflector can feed ~ 180 polarimeter pairs at Q-band, compared with QUIET's 19 polarimeters.

Refractors

The use of refractors for CMB started with BICEP (2005-2008), followed by BICEP2 (2010-) and Keck Array (2011-). These experiments are described in detail in Bischoff *et al.* and Pryke *et al.* in these proceedings. The optical elements are entirely contained in

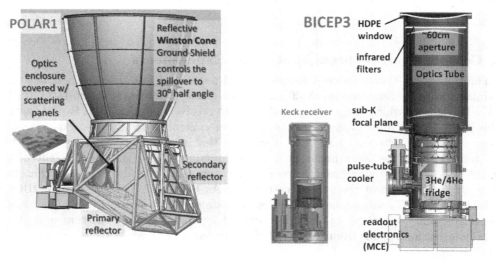

Figure 1. *Left:* POLAR1, a crossed-Dragone reflector with spillover scattered toward the sky. *Right:* BICEP3, a 60-cm cold refractor shown next to a Keck receiver for comparison.

receiver cryostats, which provide stable, low-loading conditions, and a convenient optical cold stop. Even a simple two-lens refractor can provide superb throughput. Keck Array now has over 1,200 polarimeter pairs fed by 5 separate cryogenic 30-cm refractors at the South Pole.

The Keck Array does require substantial logistic resources for a remote site in Antarctica. Each of the 5 receiver cryostats is cooled by its own water/glycol-chilled Pulse Tube cooler, consuming over 10kW per receiver. Managing a large number of compressor Helium lines through the azimuthal cable wrap also proved to be a complicated task. Finally, each cryostat needs its own closed-cycle Helium-3 refrigerator, now becoming increasingly expensive. While these difficulties have all been surmounted in Keck Array, it is desirable to increase the number of detectors in each cryogenic system for the future.

The most straightforward solution is to simply increase the aperture size of the optics from BICEP's 30-cm. Figure 1b shows a 60-cm B-mode refractor, known as **BICEP3**, next to a Keck receiver for scale. At 150 GHz, this single 60-cm, $f/1.6$ two-lens system can feed $\sim 2,000$ $2f\lambda$ polarimeter pairs on the 44cm-dia. focal plane. The cryogenic challenge associated with a larger window can now be solved with large-aperture reflective mesh filters that we have successfully developed. *The optical throughput of BICEP3 is as large as SPT-3G, yet it fits comfortably in the original BICEP mount.* A BICEP3 proposal will be submitted to NSF for 2013/2014 deployment. Adding BICEP3 would more than double the throughput of the BICEP/Keck program. In addition, lessons learned from BICEP3 will be valuable for future expansion plans.

3. A Focused Tensor-mode Program at the South Pole

CMB at the South Pole has a long and successful history. Exciting new data are expected from the ongoing SPT and Keck Array. Benson *et al.* (in these proceedings) describes an ambitious plan (SPT-3G) to upgrade the SPT polarization receiver. This will provide excellent arcminute-scale polarization measurements over 2,500 square degrees of sky. This data will supply deflection measurements, and push SZ-cluster science, reionization, and CMB lensing cosmology to a new regime.

Looking forward, continuing the search for B-mode polarization generated from primordial tensor modes remains the most compelling science goal. The program described in this paper emphasizes a very focused attack on the searches for the primordial B-modes using compact, large-throughput optical systems. A comprehensive South Pole-based B-mode initiative consisting of a combination of 1.6-m reflectors and 60-cm refractors spanning a wide frequency range of 40-220 GHz will greatly improve the expected measurements on r. By itself, this program will perform a deep search for tensor modes using the power spectrum-based lensing subtraction. When delensed by the deflection-field measured by SPT-3G, these measurements can provide a thorough, and likely conclusive search for inflationary primordial gravitational waves from the ground, reaching a 1-σ uncertainty on r well below 0.01, or, in a more optimistic scenario, a high signal-to-noise detection!

References

Bischoff, C. *et al.* 2012, *These Proceedings.*
Pryke, C. *et al.* 2012, *These Proceedings.*
Benson, B. *et al.* 2012, *These Proceedings.*

Astrophysics from Antarctica
Proceedings IAU Symposium No. 288, 2012
M. G. Burton, X. Cui & N. F. H. Tothill, eds.

Neutrino Astronomy: An Update

Francis Halzen

Department of Physics and Wisconsin IceCube Particle Astrophysics Center,
University of Wisconsin-Madison
email: halzen@wipac.wisc.edu

Abstract. Detecting neutrinos associated with the still enigmatic sources of cosmic rays has reached a new watershed with the completion of IceCube, the first detector with sensitivity to the anticipated fluxes. In this review, we will briefly revisit the rationale for constructing kilometer-scale neutrino detectors and summarize the status of the field.

Keywords. Neutrinos, cosmic rays, astrophysics

---

1. Introduction

Soon after the 1956 observation of the neutrino (Reines, 1956), the idea emerged that it represented the ideal astronomical messenger. Neutrinos reach us from the edge of the Universe without absorption and with no deflection by magnetic fields. Neutrinos have the potential to escape unscathed from the inner neighborhood of black holes, and, the subject of this update, from the cosmic accelerators where cosmic rays are born. Their weak interactions also make cosmic neutrinos very difficult to detect. Immense particle detectors are required to collect cosmic neutrinos in statistically significant numbers (Klein, 2008). Already by the 1970s, it had been understood that a kilometer-scale detector was needed to observe the "cosmogenic" neutrinos produced in the interactions of cosmic rays with background microwave photons (Roberts, 1992).

Today's estimates of the sensitivity for observing potential cosmic accelerators such as Galactic supernova remnants, active galactic nuclei (AGN), and gamma-ray bursts (GRB) unfortunately point to the same exigent requirement (Gaisser, 1995). Building a neutrino telescope has been a daunting technical challenge.

Given the detector's required size, early efforts concentrated on transforming large volumes of natural water into Cherenkov detectors that catch the light emitted by the secondary particles produced when neutrinos interact with nuclei in or near the detector (Markov, 1960). After a two-decade-long effort, building the Deep Underwater Muon and Neutrino Detector (DUMAND) in the sea off the main island of Hawaii unfortunately failed (Babson, 1990). However, DUMAND pioneered many of the detector technologies in use today and inspired the deployment of a smaller instrument in Lake Baikal (Balkanov, 2003) as well as efforts to commission neutrino telescopes in the Mediterranean (Aggouras, 2005; Aguilar, 2006; Migneco, 2008). These have paved the way toward the planned construction of KM3NeT.

The first telescope on the scale envisaged by the DUMAND collaboration was realized instead by transforming a large volume of transparent natural deep Antarctic ice into a particle detector, the Antarctic Muon and Neutrino Detector Array (AMANDA). In operation from 2000 to 2009, it represented the proof of concept for the kilometer-scale neutrino observatory, IceCube (ICPDD, 2001; Ahrens, 2004).

Neutrino astronomy has already achieved spectacular successes: neutrino detectors have "seen" the Sun and detected a supernova in the Large Magellanic Cloud in 1987. Both observations were of tremendous importance; the former showed that neutrinos have a tiny mass, opening the first crack in the Standard Model of particle physics, and the latter confirmed the basic nuclear physics of the death of stars. Fig. 1 illustrates the cosmic neutrino energy spectrum covering an enormous range, from microwave energies (10^{-12} eV) to 10^{20} eV (Becker, 2008). The figure is a mixture of observations and theoretical predictions. At low energy the neutrino sky is dominated by neutrinos produced in the Big Bang. At MeV energy, neutrinos are produced by the sun and by supernova explosions; the flux from the 1987 event is shown. At higher energies the neutrino sky is dominated by neutrinos produced by cosmic-ray interactions in the atmosphere, measured up to energies of 100 TeV by the AMANDA experiment (Achterberg, 2007). Atmospheric neutrinos are a key to our story, because they are the dominant background for extraterrestrial searches. The flux of atmospheric neutrinos falls dramatically with increasing energy; events above 100 TeV are rare, leaving a clear field of view of the sky for extraterrestrial sources.

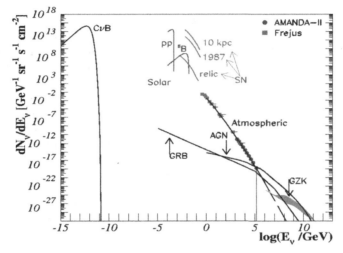

Figure 1. The cosmic-neutrino spectrum. Sources are the big bang ($C\nu B$), the Sun, supernovae (SN), atmospheric neutrinos, gamma-ray bursts, active galactic nuclei, and cosmogenic (GZK) neutrinos. The data points are from a detector at the Fréjus underground laboratory (Rhode, 1996) (red) and from AMANDA (Achterberg, 2007) (blue).

The highest energy neutrinos in Fig. 1 are the decay products of pions produced by the interactions of cosmic rays with microwave photons (Ahlers, 2010). Above a threshold of $\sim 4 \times 10^{19}$ eV, cosmic rays interact with the microwave background introducing an absorption feature in the cosmic-ray flux, the Greisen-Zatsepin-Kuzmin (GZK) cutoff. As a consequence, the mean free path of extragalactic cosmic rays propagating in the microwave background is limited to roughly 75 megaparsecs. Therefore, the secondary neutrinos are the only probe of the still-enigmatic sources at further distances. The calculation of the neutrino flux associated with the observed flux of extragalactic cosmic rays is straightforward, and yields on the order of one event per year in a kilometer-scale detector. The flux, labeled GZK in Fig. 1, shares the high-energy neutrino sky with neutrinos anticipated from gamma-ray bursts and active galactic nuclei (Gaisser, 1995).

2. The First Kilometer-Scale Neutrino Detector: IceCube

A series of first-generation experiments (Spiering, 2009; Katz, 2011) have demonstrated that high-energy neutrinos with ~ 10 GeV energy and above can be detected by observing

Cherenkov radiation from secondary particles produced in neutrino interactions inside large volumes of highly transparent ice or water instrumented with a lattice of photo-multiplier tubes. Construction of the first second-generation detector, IceCube, at the geographic South Pole was completed in December 2010 (Klein, 2010); see Fig. 2.

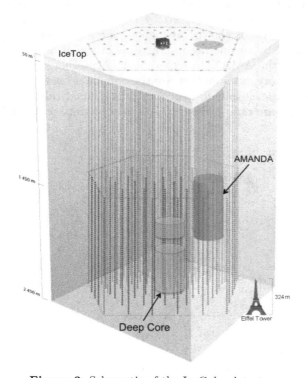

Figure 2. Schematic of the IceCube detector.

IceCube consists of 80 strings, each instrumented with 60 ten-inch photomultipliers spaced 17 m apart over a total length of one kilometer. The deepest module is located at a depth of 2.45 km so that the instrument is shielded from the large background of cosmic rays at the surface by approximately 1.5 km of ice. Strings are arranged at apexes of equilateral triangles that are 125 m on a side. The instrumented detector volume is a cubic kilometer of dark, highly transparent and sterile Antarctic ice. Radioactive background is dominated by the instrumentation deployed into this natural ice.

Each optical sensor consists of a glass sphere containing the photomultiplier and the electronics board that digitizes the signals locally using an onboard computer. The digitized signals are given a global time stamp with residuals accurate to less than 3 ns and are subsequently transmitted to the surface. Processors at the surface continuously collect the time-stamped signals from the optical modules, each of which functions independently. The digital messages are sent to a string processor and a global event builder. They are subsequently sorted into the Cherenkov patterns emitted by secondary muon tracks, or electron and tau showers, that reveal the direction of the parent neutrino (Halzen, 2006).

Based on data taken during construction, the actual effective area of the completed IceCube detector is increased by a factor 2(3) at PeV(EeV) energy over what had been expected (Ahrens, 2004). The neutrino collecting area is expected to increase with improved calibration and development of optimized software tools for the final detector, which has been operating stably in its final configuration since May 2011. Already reaching an angular resolution of better than 0.5 degree for muon tracks, this resolution can be reduced off-line to $\leqslant 0.2$ degree for individual events. The absolute pointing has been determined by measuring the shadowing of cosmic ray muons by the moon to 0.1 degree.

IceCube detects 10^{11} muons per year at a trigger rate of 2,700 Hz. Among these it filters 10^5 neutrinos, one every 6 minutes, above a threshold of ~ 100 GeV. The DeepCore infill array identifies a similar sample with energies as low as 10 GeV; see Fig. 2. These muons and neutrinos are overwhelmingly of atmospheric origin and are the decay products of pions and kaons produced by collisions of cosmic-ray particles with nitrogen and oxygen in the atmosphere. Atmospheric neutrinos are a background for cosmic neutrinos, at least at energies below 100 TeV where their flux cuts off sharply; see Fig. 3. At the highest energies, a small charm component is anticipated; its magnitude is uncertain and remains to be measured. As in conventional astronomy, IceCube must look through the atmosphere for cosmic neutrinos.

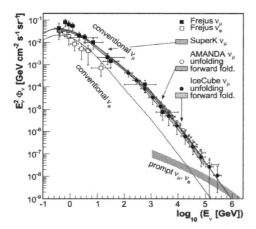

Figure 3. Measurements of the atmospheric neutrino energy spectrum; the Fréjus results (Daum, 1995), SuperK (Gonzalez-Garcia, 2006), AMANDA forward folding analysis (Abbasi, 2009) and unfolding analysis (Abbasi, 2010), and IceCube (40 strings) forward folding analysis (Abbasi, 2011b) and unfolding analysis (Abbasi, 2011a). All measurements include the sum of neutrinos and antineutrinos. The expectations for conventional ν_μ and ν_e flux are from Barr, 2004. The prompt flux is from Enberg, 2008.

3. Two Cosmic-Ray Puzzles

Despite their discovery potential touching a wide range of scientific issues, the construction of ground-based gamma-ray telescopes and kilometer-scale neutrino detectors has been largely motivated by the possibility of opening a new window on the Universe in the TeV energy region, and above in the case of neutrinos. In this review we will revisit the prospects for detecting gamma rays and neutrinos associated with cosmic rays, thus revealing their sources at a time when we are commemorating the 100th anniversary of their discovery by Victor Hess in 1912.

Cosmic accelerators produce particles with energies in excess of 10^8 TeV; we still do not know where or how (Sommers, 2009). The flux of cosmic rays observed at Earth is shown in Fig. 4. The energy spectrum follows a sequence of three power laws. The first two are separated by a feature dubbed the "knee" at an energy† of approximately 3 PeV. There is evidence that cosmic rays up to this energy are Galactic in origin. Any association with our Galaxy disappears in the vicinity of a second feature in the spectrum referred to as the "ankle"; see Fig. 4. Above the ankle, the gyroradius of a proton in the Galactic magnetic field exceeds the size of the Galaxy, and we are almost certainly witnessing the onset of an extragalactic component in the spectrum that extends to energies beyond 100 EeV. Support for this assumption now comes from three experiments (Abbasi, 2008) that have observed the telltale structure in the cosmic-ray spectrum resulting from the absorption of the particle flux by the microwave background, the so-called Greisen-Zatsepin-Kuzmin (GZK) cutoff. Neutrinos are produced in GZK interactions; it was already recognized in the 1970s that their observation requires kilometer-scale neutrino detectors. The origin of the cosmic-ray flux in the intermediate region covering PeV-to-EeV energies remains a mystery, although it is routinely assumed that its origin is some mechanism extending the reach of Galactic accelerators.

Acceleration of protons (or nuclei) to TeV energy and above requires massive bulk flows of relativistic charged particles. These are likely to originate from exceptional gravitational forces in the vicinity of black holes or neutron stars. The gravity of the collapsed

† We will use energy units TeV, PeV and EeV, increasing by factors of 1000 from GeV energy.

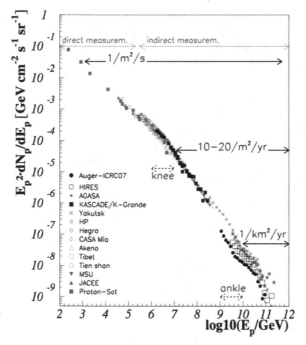

Figure 4. At the energies of interest here, the cosmic-ray spectrum follows a sequence of three power laws. The first two are separated by the "knee," the 2nd and 3rd by the "ankle." Cosmic rays beyond the ankle are a new population of particles produced in extragalactic sources.

objects powers large currents of charged particles that are the origin of high magnetic fields. These create the opportunity for particle acceleration by shocks. It is a fact that electrons are accelerated to high energy near black holes; astronomers detect them indirectly by their synchrotron radiation. Some cosmic sources must accelerate protons, because we observe them as cosmic rays.

The detailed blueprint for a cosmic-ray accelerator must meet two challenges: the highest energy particles in the beam must reach above 10^3 TeV for Galactic sources (10^8 TeV for extragalactic) and meet the total energy (luminosity) requirement to accommodate the observed cosmic-ray flux. Both represent severe constraints that have limited the imagination of theorists.

Supernova remnants were proposed as possible sources of Galactic cosmic rays as early as 1934 by Baade and Zwicky (Baade, 1934); their proposal is still a matter of debate after more than 70 years (Butt, 2009). Galactic cosmic rays reach energies of at least several PeV, the "knee" in the spectrum. Their interactions with Galactic hydrogen in the vicinity of the accelerator should generate gamma rays from the decay of secondary pions that reach energies of hundreds of TeV. Such sources should be identifiable by a relatively flat energy spectrum that extends to hundreds of TeV without attenuation; they have been dubbed PeVatrons. The search to pinpoint them has so far been unsuccessful.

Although there is no incontrovertible evidence that supernovae accelerate cosmic rays, the idea is generally accepted because of energetics: three Galactic supernova explosions per century converting a reasonable fraction of a solar mass into particle acceleration can accommodate the steady flux of cosmic rays in the Galaxy. Energetics also drives speculations on the origin of extragalactic cosmic rays.

By integrating the cosmic-ray spectrum in Fig. 4 above the ankle, we find that the energy density of the Universe in extragalactic cosmic rays is $\sim 3 \times 10^{-19}\,\mathrm{erg\ cm^{-3}}$ (Gaisser, 1997). This value is rather uncertain because of our ignorance of the precise energy where the transition from Galactic to extragalactic sources occurs. The power required for a population of sources to generate this energy density over the Hubble time of 10^{10} years is $2 \times 10^{37}\,\mathrm{erg\ s^{-1}}$ per $(\mathrm{Mpc})^3$. (In the astroparticle community, this flux is also known as $5 \times 10^{44}\,\mathrm{TeV\ Mpc^{-3}\ yr^{-1}}$).

A gamma-ray-burst fireball converts a fraction of a solar mass into the acceleration of electrons, seen as synchrotron photons. The observed energy in extragalactic cosmic rays can be accommodated with the reasonable assumption that shocks in the expanding GRB fireball convert roughly equal energy into the acceleration of electrons and cosmic rays (Waxman, 1995). It so happens that $2 \times 10^{51}\,\mathrm{erg}$ per GRB will yield the observed energy density in cosmic rays after 10^{10} years, given that their rate is on the order of 300 per Gpc^3 per year. Hundreds of bursts per year over the Hubble time produce the observed cosmic-ray density, just like three supernovae per century accommodate the steady flux in the Galaxy.

Problem solved? Not really: it turns out that the same result can be achieved assuming that active galactic nuclei convert, on average, $2 \times 10^{44}\,\mathrm{erg\ s^{-1}}$ each into particle acceleration. As is the case for GRBs, this is an amount that matches their output in electromagnetic radiation. Whether GRBs or AGN, the observation that these sources are required to radiate similar energies in photons and cosmic rays is unlikely to be an accident. We discuss the connection next; it will lead to a prediction of the neutrino flux.

4. Neutrinos (and Photons) Associated with Cosmic Rays

How many gamma rays and neutrinos are produced in association with the cosmic-ray beam? Generically, a cosmic-ray source should also be a neutrino beam dump. Cosmic rays accelerated in regions of high magnetic fields near black holes inevitably interact with radiation surrounding them. These may be photons radiated by the accretion disk in AGN and synchrotron photons that co-exist with protons in the exploding fireball producing a GRB. In these interactions, neutral and charged pion secondaries are produced by the processes

$$p + \gamma \to \Delta^+ \to \pi^0 + p \ \text{ and } \ p + \gamma \to \Delta^+ \to \pi^+ + n. \tag{4.1}$$

While secondary protons may remain trapped in the high magnetic fields, neutrons and the decay products of neutral and charged pions escape. The energy escaping the source is therefore distributed among cosmic rays, gamma rays and neutrinos produced by the decay of neutrons, neutral pions and charged pions, respectively.

In the case of Galactic supernova shocks, cosmic rays inevitably interact with the hydrogen in the Galactic disk, producing equal numbers of pions of all three charges in hadronic collisions $p + p \to n\,[\pi^0 + \pi^+ + \pi^-] + X$; n is the pion multiplicity. Their secondary fluxes should be boosted by the interaction of the cosmic rays with high-density molecular clouds that are ubiquitous in the star-forming regions where supernovae are more likely to explode.

In a generic cosmic beam dump, accelerated cosmic rays, assumed to be protons for simplicity, interact with a photon or proton target. In either case, accelerated cosmic rays produce charged and neutral pions. Subsequently, the pions decay into gamma rays and neutrinos that carry, on average, $1/2$ and $1/4$ of the energy of the parent pion. We here assume that the four leptons in the decay $\pi^+ \to \nu_\mu + \mu^+ \to \nu_\mu + (e^+ + \nu_e + \bar\nu_\mu)$ equally share the charged pion's energy. The energy of the pionic leptons relative to the proton is:

$$x_\nu = \frac{E_\nu}{E_p} = \frac{1}{4}\langle x_{p\to\pi}\rangle \simeq \frac{1}{20}, \tag{4.2}$$

and

$$x_\gamma = \frac{E_\gamma}{E_p} = \frac{1}{2}\langle x_{p\to\pi}\rangle \simeq \frac{1}{10}. \tag{4.3}$$

Here

$$\langle x_{p\to\pi}\rangle = \left\langle\frac{E_\pi}{E_p}\right\rangle \simeq 0.2 \tag{4.4}$$

is the average energy transferred from the proton to the pion.

5. Sources of the Extragalactic Cosmic Rays

Waxman and Bahcall (Waxman, 1998) have presented an interesting benchmark for the neutrino flux expected from extragalactic cosmic ray accelerators, whatever they may be. The cosmic-ray flux, assuming an injected E^{-2} spectrum before modification by absorption on the microwave background, can be parameterized as

$$\frac{dN_p}{dE_p} = \frac{5\times10^{-11}}{E_p^2}\ \mathrm{TeV}^{-1}\,\mathrm{cm}^{-2}\,\mathrm{s}^{-1}\,\mathrm{sr}^{-1}. \tag{5.1}$$

Integrating this flux, from the ankle to a maximal accelerator energy of 10^9 TeV, accommodates the total energy requirement of $\sim 3\times10^{-19}$ erg cm^{-3}. The secondary neutrino flux is given by

$$\frac{dN_\nu}{dE} = \frac{1}{3}\left[\frac{2}{3}\right]\frac{1}{x_\nu}\frac{dN_p}{dE_p}\left(\frac{E}{x_\nu}\right). \tag{5.2}$$

Here the coefficients correspond to photo- and hadro-production of the neutrinos, respectively. $N_\nu\,(=N_{\nu_\mu}=N_{\nu_e}=N_{\nu_\tau})$ represents the sum of the neutrino and antineutrino fluxes which are not distinguished by the experiments. Oscillations over cosmic baselines yield approximately equal fluxes for the three flavors. For the cosmic-ray flux introduced above, we obtain a neutrino flux

$$\frac{dN_\nu}{dE} \simeq \frac{2\times10^{-12}}{E^2}\ \mathrm{TeV}^{-1}\,\mathrm{cm}^{-2}\,\mathrm{s}^{-1}\,\mathrm{sr}^{-1}. \tag{5.3}$$

Notice that we sneaked in the assumption that each cosmic ray interacts once and only once in the target—if not, the flux is multiplied by the number of interactions n_{int}; see Halzen (2010) for a discussion. In fact, Waxman and Bahcall have argued that, if the density of the source were such that a high-energy cosmic ray interacted more than once, it would be opaque to TeV photons. So, the neutrino flux represents an upper limit for extragalactic sources that emit TeV gamma rays.

It is important to realize that the high-energy protons may be magnetically confined to the accelerator. In the case of GRBs, for instance, protons adiabatically lose energy, trapped inside the fireball that expands under radiation pressure until it becomes transparent and produces the display observed by astronomers. Secondary neutrons do escape with high energies and decay into protons that are the source of the observed extragalactic cosmic-ray flux. In this case, cosmic rays and pionic neutrinos are directly related by the fact that, for each secondary neutron decaying into a cosmic ray proton, there are three neutrinos produced by the associated π^+ (see Eq. 3.1):

$$E\frac{dN_\nu}{dE} = 3\,E_n\,\frac{dN_n}{dE_n}\,(E_n)\,, \tag{5.4}$$

and, after oscillations, per neutrino flavor

$$E^2\frac{dN_\nu}{dE} \simeq \left(\frac{x_\nu}{x_n}\right)E_n^2\,\frac{dN_n}{dE_n}\,(E_n)\,, \tag{5.5}$$

where $x_n \sim 1/2$ is the relative energy of the secondary neutron; the neutron flux is identified with the observed cosmic ray flux. This straightforward prediction has been ruled out by IceCube with data taken during construction (Abbasi, 2012). There are alternative scenarios that, fortunately, also yield predictions within reach of the completed detector.

The key feature is that the normalization of the generic neutrino flux of Eq. 5.2 is correct for GRBs because the fireball model generically predicts that $n_{int} \simeq 1$. The GRB phenomenology that successfully accommodates the astronomical observations, as well as the acceleration of cosmic rays, is that of the creation of a hot fireball of electrons, photons and protons that is initially opaque to radiation. The hot plasma therefore expands by radiation pressure, and particles are accelerated to a Lorentz factor Γ that grows until the plasma becomes optically thin and produces the GRB display. The rapid time structure of the burst is associated with successive shocks (shells), of width $\Delta R = c \times t_v$, that develop in the expanding fireball. The rapid temporal variation of the radiation, t_v, is on the order of milliseconds, and can be interpreted as the collision of internal shocks with different Lorentz factors. Electrons, accelerated by first-order Fermi acceleration, radiate synchrotron gamma rays in the strong internal magnetic field, and thus produce the spikes observed in the emission spectra. The number of interactions of protons with the synchrotron photons is simply determined by the optical depth of the fireball shells of width ΔR to $p\gamma$ interactions and is generically on the order of $n_{int} \simeq 1$.

Throughout the discussion of the neutrino flux associated with extragalactic cosmic rays we have neglected the fact that neutrinos, unlike cosmic rays, are not absorbed by microwave photons resulting in a neutrino flux increased by "a factor" that depends on the cosmological evolution of the sources. We have also assumed that the highest energy cosmic rays are protons. Experiments disagree on the composition but the cosmogenic flux is inevitably reduced in the presence of heavy primaries.

No compelling prediction is possible for AGN, complex systems with many possible sites for acceleration and interaction of the cosmic rays. Our discussion has, however, introduced the rationale that generic cosmic-ray sources produce a neutrino flux comparable to their flux of cosmic rays (Gaisser, 1997) and pionic TeV gamma rays (Alvarez-Muniz, 2002). In this context we introduce Fig. 5 which shows the present IceCube upper limits on the neutrino flux from nearby AGN as a function of their distance. Also shown is the TeV gamma-ray emission from the same sources; except for CenA and M87, the muon-neutrino limits have reached the level of the TeV photon flux. This is a notable fact because of the equipartition of the cosmic-ray, gamma-ray and neutrino fluxes from a cosmic ray accelerator. One can sum the sources shown in the figure into a diffuse flux; the result is, after dividing by $4\pi/c$, 3×10^{-12} TeV cm^{-2} s^{-1} sr^{-1}, or approximately 10^{-11} TeV cm^{-2} s^{-1} sr^{-1} for all neutrino flavors. This flux matches the "maximal" flux previously argued for; see Fig. 3.

IceCube's sensitivity is rapidly approaching this benchmark flux as shown in Fig. 6. In fact, the IceCube "limit" obtained with one year of data taken with 59 strings is a $1.8\,\sigma$ signal (Sullivan, 2012). Although not significant, it is of interest that the benchmark flux argued for above rises above the atmospheric neutrino background for energies exceeding

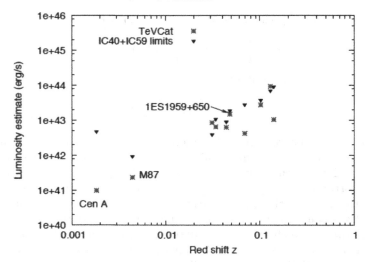

Figure 5. IceCube neutrino flux limits are compared with the TeV photon flux for nearby AGN. Figure courtesy of T. Gaisser.

100 TeV at a flux level already reached by the completed IceCube detector after one year of operation.

The improved performance of IceCube at EeV energy has created the opportunity to detect cosmogenic neutrinos. We anticipate 2.3 events in three years of running the completed detector assuming a flux derived from the "best fit" to the cosmic-ray data (Ahlers, 2010), and 4.8 events for the largest neutrino rate allowed by the constraint imposed on their accompanying photon flux by the observed flux of diffuse photons in the Universe (Ahlers, 2010).

Recently, in a dedicated search for cosmogenic neutrinos two events have been found (Ishihara, 2012) in the first year of data taken with the completed detector. They are showers of more than 10^5 photons, about 500 m in diameter, fully contained inside the detector. With no evidence of a muon track, they are initiated by electron or tau neutrinos; see Fig. 7. However, their energies, rather than super-EeV as expected for cosmogenic neutrinos, are in the PeV range: 1.1 and 1.3 PeV with a negligible statistical error and a 35% systematic error. The analysis of these events is ongoing and we expect this error to be significantly reduced in the near future. We have also determined the directions of the initial neutrinos exploiting the fact that the waveforms collected by the DOMs following and trailing the initial neutrino direction are identifiably different.

More importantly, we have designed a dedicated starting-event analysis to find more such events in the same data sample. Some of them should contain muon tracks whose arrival directions can be reconstructed with superior precision to that of the two shower events. The events are likely to represent new neutrino physics, or astrophysics, because their atmospheric origin is excluded, very conservatively, below the 10^{-2} level. Accommodating the events as the decay of charm particles produced in the atmosphere requires a flux that violates the IceCube diffuse limit obtained with data collected with 59 strings (Sullivan, 2012).

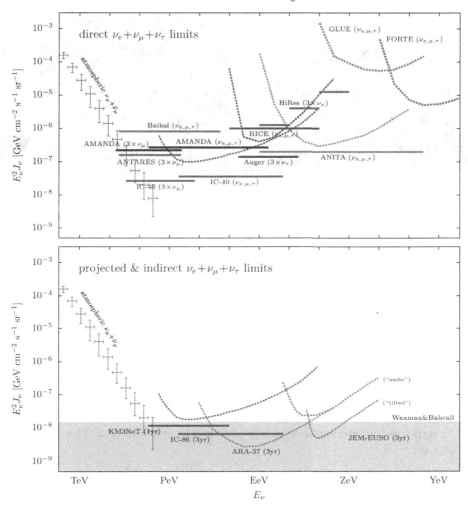

Figure 6. Limits on a diffuse neutrino flux from existing (top) and future (bottom) experiments; see Achterberg, 2007, *et al.* The shaded band indicates the anticipated neutrino fluxes associated with cosmic rays. Figure courtesy of M. Ahlers.

6. Sources of Galactic Cosmic Rays

Despite the commissioning of instruments with improved sensitivity, it has been impossible to conclusively pinpoint supernova remnants as the sources of the Galactic cosmic rays by identifying the accompanying gamma rays of pion origin. The position of the knee in the cosmic-ray spectrum indicates that some sources must accelerate cosmic rays to energies of several PeV. These so-called PeVatrons therefore produce pionic gamma rays whose spectrum extends to several hundred TeV without cutoff. In contrast, the widely studied supernova remnants RX J1713-3946 and RX J0852.0-4622 (Vela Junior) reach their maximum energy in the TeV region. In fact, recent data from Fermi LAT have directly challenged the hadronic interpretation of the GeV-TeV radiation from one of the best-studied candidates, RX J1713-3946 (Abdo, 2011).

It is difficult to hide a Galactic cosmic accelerator from view. A generic supernova remnant releasing an energy of $W \sim 10^{50}$ erg into the acceleration of cosmic rays will inevitably generate TeV gamma rays in the interaction of the accelerated nuclei with

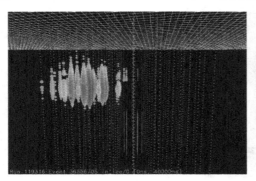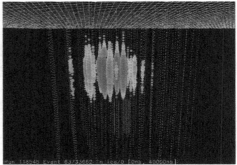

Figure 7. Displays of the two observed events. Each colored sphere represent a DOM that sent a time-stamped waveform to the event builder. Colors indicate the arrival timing of the photon (red=early, blue=late). The size of the sphere indicates the number of photons detected by each DOM.

the hydrogen in the Galactic disk. The emissivity in pionic gamma rays Q_γ is simply proportional to the density of cosmic rays n_{cr} and the density of the target $n \sim 1/\mathrm{cm}^3$ of protons in the disk. Here n_{cr} $(>1\,\mathrm{TeV}) \simeq 4 \times 10^{-14}\,\mathrm{cm}^{-3}$ is obtained by integrating the proton spectrum for energies in excess of 1 TeV. For an E^{-2} spectrum

$$Q_\gamma \simeq c \left\langle \frac{E_\pi}{E_p} \right\rangle \lambda_{pp}^{-1}\, n_{cr}\ (>1\,\mathrm{TeV}) \simeq 2cx_\gamma \sigma_{pp}\, n\, n_{cr} \tag{6.1}$$

or

$$Q_\gamma(> 1\,\mathrm{TeV}) \simeq 10^{-29}\,\mathrm{TeV\,cm^{-3}\,s^{-1}} \left(\frac{n}{1\,\mathrm{cm}^{-3}} \right). \tag{6.2}$$

The proportionality factor in Eq. (6.1) is determined by particle physics; $x_\gamma \simeq 0.1$ is the average energy of secondary photons relative to the cosmic ray protons and $\lambda_{pp} = (n\sigma_{pp})^{-1}$ is the proton interaction length ($\sigma_{pp} \simeq 40\,\mathrm{mb}$) in a density n. The corresponding luminosity is

$$L_\gamma(>1\,\mathrm{TeV}) \simeq Q_\gamma\, \frac{W}{\rho_E}, \tag{6.3}$$

where W/ρ_E is the volume occupied by the supernova remnant. Here we made the approximation that the volume of the young remnant is given by W/ρ_E, or that the density of particles in the remnant is not very different from the ambient energy density $\rho_E \sim 10^{-12}\,\mathrm{erg\,cm^{-3}}$ of Galactic cosmic rays.

We thus predict (Gonzalez-Garcia, 2009) a rate of TeV photons from a supernova at a nominal distance d on the order of 1 kpc of

$$\int_{E>1\,\mathrm{TeV}} \frac{dN_\gamma}{dE_\gamma}dE_\gamma = \frac{L_\gamma(> 1\mathrm{TeV})}{4\pi d^2}$$

$$\simeq 10^{-12} - 10^{-11} \left(\frac{\mathrm{photons}}{\mathrm{cm}^2\,\mathrm{s}} \right) \left(\frac{W}{10^{50}\,\mathrm{erg}} \right) \left(\frac{n}{1\,\mathrm{cm}^{-3}} \right) \left(\frac{d}{1\,\mathrm{kpc}} \right)^{-2}. \tag{6.4}$$

This is a PeVatron flux well within reach of the current generation of atmospheric gamma ray telescopes; has it been detected?

Looking for them in the highest energy survey of the Galactic plane is evident and points to the Milagro experiment (Abdo, 2007). Their survey in the $\sim 10\,\mathrm{TeV}$ band revealed a subset of sources located within nearby star-forming regions in Cygnus and in the vicinity of Galactic latitude $l = 40$ degrees. Subsequently, directional air Cherenkov telescopes were pointed at three of the sources, revealing them as PeVatron candidates with an E^{-2} energy spectrum that extends to tens of TeV without evidence of a cutoff (Djannati-Atai, 2007; Albert, 2008) and gamma ray fluxes in the range estimated above.

Interestingly, some of the sources cannot be readily associated with known supernova remnants, or with any non-thermal source observed at other wavelengths. These are likely to be molecular clouds illuminated by the cosmic-ray beam accelerated in young remnants located within about 100 pc. Indeed, one expects that multi-PeV cosmic rays are accelerated only over a short time period when the shock velocity is high, i.e., when the remnant transitions from free expansion to the beginning of the Sedov phase. The high-energy particles can produce photons and neutrinos over much longer periods when they diffuse through the interstellar medium to interact with nearby molecular clouds (Gabici, 2007). An association of molecular clouds and supernova remnants is expected, of course, in star-forming regions. In this case, any confusion with synchrotron photons is unlikely.

Particle physics dictates the relation between pionic gamma rays and neutrinos and basically predicts the production of a $\nu_\mu + \bar{\nu}_\mu$ pair for every two gamma rays seen by Milagro. This calculation can be performed using the formalism introduced in the previous section with approximately the same outcome.

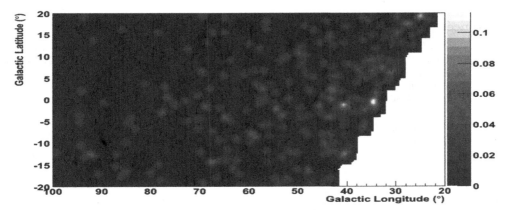

Figure 8. Simulated sky map of IceCube in Galactic coordinates after 5 years of operation of the completed detector. Two Milagro sources are visible with 4 events for MGRO J1852+01 and 3 events for MGRO J1908+06 with energy in excess of 40 TeV. These, as well as the background events, have been randomly distributed according to the resolution of the detector and the size of the sources.

The quantitative statistics can be summarized as follows. For average values of the parameters parametrizing the flux, we find that the completed IceCube detector should confirm sources in the Milagro sky map as sites of cosmic-ray acceleration at the 3σ level in less than one year and at the 5σ level in three years (Gonzalez-Garcia, 2009); see Fig. 8. This assumes that the source extends to 300 TeV, or 10% of the energy of the cosmic rays near the knee in the spectrum. These results agree with previous estimates (Halzen, 2008). There are intrinsic ambiguities of an astrophysical nature in this estimate that may reduce or extend the time required for a 5σ observation (Gonzalez-Garcia, 2009). Also, the extended nature of some of the Milagro sources represents a challenge for IceCube observations that are optimized for point sources. In the absence of an observation of TeV-energy supernova neutrinos by IceCube within a period of 10 years, the supernova origin of cosmic rays in the Galaxy will be challenged.

7. Conclusion: Stay Tuned

In summary, IceCube was designed for a statistically significant detection of cosmic neutrinos accompanying cosmic rays in five years. Here we made the case that, based on multiwavelength information from ground-based gamma ray telescopes and cosmic-ray experiments, we are indeed closing in on supernova remnants, GRBs (if they are the sources of cosmic rays) and GZK neutrinos. The discussion brought to the forefront the critical role of improved spectral gamma-ray data on candidate cosmic-ray accelerators. The synergy between CTA (see CTA), IceCube, and KM3NeT as well as other next-generation neutrino detectors is likely to provide fertile ground for future progress.

8. Acknowledgements

This research was supported in part by the U.S. National Science Foundation under Grants No. OPP-0236449 and PHY-0969061; by the U.S. Department of Energy under Grant No. DE-FG02-95ER40896; by the University of Wisconsin Research Committee with funds granted by the Wisconsin Alumni Research Foundation.

References

Abbasi, R., *et al.* (HiRes Collaboration), 2008, *Phys. Rev. Lett.* 100, 101101, astro-ph/0703099 ; Abraham, J., *et al.* (Auger collaboration), 2008, *Phys. Rev. Lett.* 101, 061101, astro-ph/08064302; and Tokuno, H., *et al.*, *"The Status of the Telescope Array Experiment,"* *J. Phys. Conf. Ser.* 293:012035 (2011)

Abbasi, R., *et al.* (IceCube Collaboration), 2009, *Phys. Rev.* D **79** 102005

Abbasi, R., *et al.* (IceCube Collaboration), 2010, *Astropart. Phys.* **34** 48

Abbasi, R., *et al.* (IceCube Collaboration), 2011, *Phys. Rev.* D **83** 012001

Abbasi, R., *et al.* (IceCube Collaboration), 2011, *Phys. Rev.* D **84** 082001

Abbasi, R., *et al.* (IceCube Collaboration), 2012, *Nature* 484, 351; astro-ph.HE/1204.4219

Abdo, A. A., *et al.* (Fermi-LAT Collaboration), 2011, *Astrophys. J.* 734, 28; astro-ph.HE/1103.5727v1

Abdo, A. A., *et al.* (Milagro Collaboration), 2007, *Astrophys. J.* 658, L33; astro-ph/0611691

Achterberg, A., *et al.* (IceCube Collaboration), 2007 *Phys. Rev.* D 76, 042008, *Erratum* ibid. 77, 089904(E); astro-ph/07051315

See Achterberg, A., 2007 cited above; Ackermann, M., *et al.* (IceCube collaboration), 2008, *Astrophys. J.* 675 , 1014, astro.ph/0711.3022 ; Abbasi, R., *et al.* (IceCube collaboration), 2010, *Phys. Rev.* D 82, 072003, astro.ph.CO/1009.1442v1 ; Aynutdinov, V., *et al.* (Baikal collaboration), 2006, *Astropart. Phys.* 25, 140, astro-ph/0508675 ; Martens, K., *et al.* (HiRes collaboration), 2007, *Proceedings of 23rd Lepton-Photon Conference, Daegu, Korea,* astro-ph/0707.4417 ; Kravchenko, I., *et al.* (RICE collaboration), 2006, *Phys. Rev.* D 73, 082002, astro-ph/0601148 ; Barwick, S. W., *et al.* (ANITA collaboration), 2006, *Phys. Rev. Lett.* 96, 171101, astro-ph/0512265 ; Lehtinen, N. G., Gorham, P. W., Jacobson, A. R., & Roussel-Dupre, R. A., 2004, *Phys. Rev.* D 69, 013008, astro-ph/0309656 ; Gorham, P. W., Hebert, C. L., Liewer, K. M., Naudet, C. J., Saltzberg, D., & Williams, D., 2004, *Phys. Rev. Lett.* 93, 041101, astro-ph/0310232 ; Anchordoqui, L. A., Feng, J. L., Goldberg, H., & Shapere, A. D., 2002, *Phys. Rev.* D 66, 103002, hep-ph/0207139; and Abbasi, R., *et al.* (IceCube collaboration), 2011,*Phys. Rev.* D 83, 012001, astro-ph.HE/1010.3980

Aggouras, G., *et al.* (NESTOR Collaboration), 2005, *Astropart. Phys.* 23, 377

Aguilar, J. A., *et al.* (ANTARES Collaboration), 2006, *Astropart. Phys.* 26, 314

Ahlers, M., Anchordoqui, L. A., Gonzalez-Garcia, M. C., *et al.*, 2010, *Astropart. Phys.* 34, 106; astro-ph.HE/1005.2620

Ahrens, J., *et al.* (IceCube Collaboration), 2004, *Astropart. Phys.* 20, 507, astro-ph/0305196

Albert, J. *et al.*, 2008; astro-ph/08012391

Alvarez-Muniz, J., & Halzen, F., 2002, *Astrophys. J.* 576, L33, astro-ph/0205408; for a more recent discussion of the formalism, see Becker, J. K., Halzen, F., & OMurchadha, A, Olivo, M., 2010, *Astrophys. J.* 721, 1891, astro-ph.HE/0911.2202

Baade, W. & Zwicky, F., 1934, *Phys. Rev. D* 46, 76

Babson, J. Barish, R., Becker-Szendy, R., *et al.*, 1990, *Phys. Rev. D* 42, 3613

Balkanov, V. A., *et al.* (BAIKAL Collaboration), 2003, *Nucl. Phys. B Proc. Suppl.* 118, 363

Barr, G. D., *et al..*, 2004, *Phys. Rev. D* **70** 023006, 2004

Becker, J. K., 2008, *Phys. Rep.* 458, 173; astro-ph/07101557

Butt, Y., 2009, *Nature* 460, 701; astro-ph.he/10093664

Cherenkov Telescope Array (CTA), http://www.cta-observatory.org

Daum, K., *et al.*, 1995, *Zeitschrift für Physik* C **66** 417

Djannati-Atai, A., *et al.* (H. E. S. S. Collaboration), 2007, *30th ICRC, Merida, Mexico* OG2.2 2, 1316; astro-ph/07102418

Enberg, R., Reno, M. H., & Sarcevic, I., 2008, *Phys. Rev. D* **78** 043005

Gabici, S., & Aharonian, F. A., 2008, *Astrophys. J.* 665 2, 431; astro-ph/0705.3011

&Gaisser, T. K., 1997, *OECD Megascience Forum, Taormina, Italy*, astro-ph/9707283 ; the discussion was revisited in Ahlers, M., Anchordoqui, L. A., & Goldberg, H., *et al.*, 2005, *Phys. Rev. D* 72, 023001, astro-ph/0503229

Gaisser, T. K., Halzen, F., & Stanev, T., 1995, *Phys. Rept.*, 258, 173, *Erratum* 271, 355, hep-ph/9410384 ; Learned, J. G., & Mannheim, K., 2000, *Ann. Rev. Nucl. Part. Science*, 50, 679 ; Halzen, F., & Hooper, D., 2002, *Rep. Prog. Phys.*, 65 , 1025, astro-ph/0204527 ; and Katz, U. F., & Spiering, C., 2012, *Prog. Part. Nucl. Phys.*, 67, 651, astro-ph.HE/1111.0507

Gonzalez-Garcia, M. C., Maltoni, M., & Rojo, J., 2006, *J. High Energy Phys.* **10** 75

Gonzalez-Garcia, M. C., Halzen, F., & Mohapatra, S., 2009, *Astropart. Phys.* 31, 437; astro-ph.HE/09021176

Halzen, F., 2006, *Eur. Phys. J. C* 46, 669; astro-ph/0602132

Halzen, F., 2011, *Acta Phys. Polon. B* **42**, 2525; arXiv:1111.1131 [hep-ph]

Halzen, F., Kappes, A., & OMurchadha, 2008, *Phys. Rev. D* 78, 063004, *see also* Nucl. Instr. and Meth. A 602, 117 (2009); astro-ph/0803.0314v2

Halzen, F., & Klein, S. R., 2008, *Physics Today*, 61N5, 29

Halzen, F. & Klein, S. R., 2010, *Rev. Sci. Instrum.* 81, 081101; astro-ph.HE/1007.1247

IceCube Collaboration, 2001, available at http://www.icecube.wisc.edu/science/publications/pdd/pdd.pdf

Ishihara, A., 2012, *Proceedings of Neutrino 2012, Kyoto, Japan*

Katz, U. & Spiering, C., 2011, *Prog. Part. Nucl. Phys.* 67, 651; astro-ph.HE/1111.0507

Markov, M. A., 1960, *Proc. 1960 Intl. Conf. on High Energy Physics* 578

Migneco, E., 2008, *J. Phys. Conf. Ser.* 136, 022048

Reines, F. & Cowan, Jr., C. L., 1956, *Nature*, 17, 446

Rhode, W., *et al.* (Fréjus Collaboration), 1996, *Astropart Phys.* 4, 217

Roberts, A., 1992, *Rev. Mod. Phys.*, 64, 259

Sommers, P., & Westerhoff, S., 2009, *New J. Phys.* 11, 055004, astro-ph/08021267 Hillas, A. M., 2006, astro-ph/0607109v2 ; and Berezinsky,V, 2008, *J. Phys. Conf. Ser.* 120, 012001, astro-ph/08013028

Spiering, C., 2009, *AIP Conf. Proc.* 1085, 18; astro-ph/0811.4747

Sullivan, G., 2012, *Proceedings of Neutrino 2012, Kyoto, Japan*

Waxman, E., 1995, *Phys. Rev. Lett.* 75, 386, astro-ph/9701231 ; Vietri, M, 1998, *Phys. Rev. Lett.* 80, 3690, astro-ph/9802241 ; and Bottcher, M., & Dermer, C. D., 1998, *Astrophys. J. Lett.* 499, L131, astro-ph/9801027v2

Waxman, E. & Bahcall, J., 1998, *Phys. Rev.* D59, 023002

Astrophysics from Antarctica
Proceedings IAU Symposium No. 288, 2012
M. G. Burton, X. Cui & N. F. H. Tothill, eds.

© International Astronomical Union 2013
doi:10.1017/S1743921312016730

The Path from AMANDA to IceCube

Albrecht Karle

Department of Physics and Wisconsin IceCube Particle Astrophysics Center, University of
Wisconsin-Madison, Madison, WI 53706, USA

Abstract. In May 2011, IceCube, a neutrino telescope with one cubic kilometer instrumented
volume started full operation with 5,160 sensors. The plan to build an experiment of this scale
was based in part on the successful realization of a prototype experiment, the Antarctic Muon
and Neutrino Detector Array. Here, we will review some of the major challenges and milestones.

Keywords. Neutrino

1. Introduction

In May 2011, IceCube, a neutrino telescope with one cubic kilometer instrumented
volume started full operation with 5,160 sensors. The plan to build an experiment of
this scale was based on a decade of research and the demonstration that ice was a
suitable medium. The vision for neutrino astronomy was laid out in the early 1960s.
After some pioneering efforts to build neutrino detectors in water, similar efforts were
staged at the South Pole to build and deploy a Cherenkov neutrino detector. First, in the
1990s, the Antarctic Muon and Neutrino Detector Array (AMANDA) was built. Then,
based on AMANDA as a proof of concept, the full kilometer-scale IceCube neutrino
telescope was constructed and completed by 2010 (see Fig. 1). Today, the South Pole
has become the premier site for neutrino astronomy. In a historical perspective, the idea
to use neutrinos for astronomy was born not long after the neutrino was discovered by
Reines and collaborators in 1956. For that, a very large target volume was needed, on the
order of one cubic kilometer. The other requirement for Cherenkov detectors is a good
transparency of the medium. A detailed review of the current state of neutrino astronomy
was provided by Spiering & Katz (2012). Here we will focus on neutrino astronomy at
the South Pole.

2. Overview

In the 1990–91 austral season, the first exploratory effort was made at the South
Pole to deploy photomultipliers in ice at a shallow depth. This would be the first of 13
polar seasons that involved hot water drilling with the goal of deploying photomultipliers
in ice and advancing AMANDA and later IceCube. It was preceded by an important
exploration of the idea to deploy PMTs in natural ice in Greenland in 1990. The result,
the 'Observation of muons using the polar ice cap as a Cerenkov detector' was published
by Lowder *et al.* (Nature, 1991) and marks an important milestone. The authors conclude
"Our results suggest that a full-scale Antarctic ice detector is technically quite feasible,"
and preparations began for an exploration at the South Pole.

A first test array with 80 optical modules on four strings (not shown) was deployed in
1993-94, at depths between 800 and 1,000 m. While the absorption length of blue light was
determined to be exceptionally large, on the order of 300 m, it was found that the effective
scattering length L_{eff} was extremely small, between 40 cm at 830 m depth and 80 cm

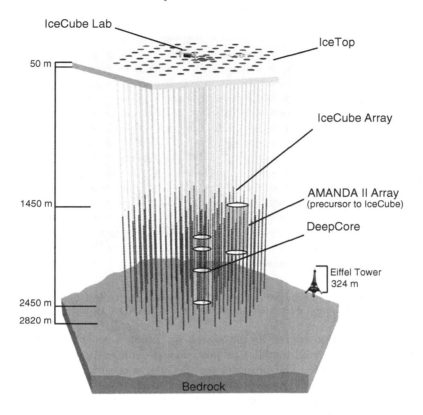

Figure 1. Schematic view of IceCube.

at 970 m (Askebjer *et al.* 1995). At these depths, scattering of light is caused primarily by air bubbles trapped in the ice. The density and size of the air bubbles decreases with increasing pressure and age of the ice at greater depths. This trend, together with evidence from ice cores at other locations, suggested that below about 1,350 m air bubbles disappear and that the air is absorbed in ice clathrate crystals. This was confirmed with a second 4-string array which was deployed in 1995–96. The remaining scattering, averaged over 1,500 to 2,000 m depth, corresponds to $L_{eff} \approx 20$ m and is caused primarily by dust. The absorption length was still very good, with more than 100 m. The ice was suitable for the reconstruction of muon tracks and thus for neutrino astronomy.

Over the course of three more construction seasons, the AMANDA-II array was gradually completed by the 1999–2000 season. Already with data from the 10-string configuration completed in 1997 (AMANDA-B10), the collaboration was able to demonstrate detection of atmospheric neutrinos consistent with expectations, an important milestone. There were 188 upward-going muons from neutrinos found in a livetime of 138 days, a little more than one event per day. Figure 2 shows a graphical depiction of the 10-string configuration together with an observed muon neutrino. The result marked another big milestone that was documented in a publication in *Nature* (Andres *et al.* 2001). In the meantime, by 2000, the full AMANDA-II array was completed and in operation. The collaboration had grown to more than 20 institutions worldwide with an author list exceeding 100.

The completed AMANDA-II array was a fully functioning neutrino telescope consisting of 19 strings and 677 photomultipliers of 20 cm diameter. In seven years of operation

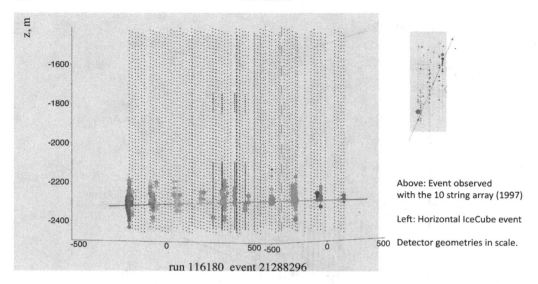

Above: Event observed with the 10 string array (1997)

Left: Horizontal IceCube event

Detector geometries in scale.

run 116180 event 21288296

Figure 2. A horizontal muon event of energy 10 TeV recorded with IceCube. On the right: one of the first upward-going muon events observed with the AMANDA 10-string array set to scale.

it collected 6,595 neutrino events in the search for point sources of neutrinos, about 5 events/day. The drilling and installation process was sufficiently developed that plans could be made for IceCube. The proposal for IceCube was submitted in 1999, before AMANDA-II was completed. The experience gained with respect to improved drilling, detector module design and data analysis with the full AMANDA-II would still be critical in preparation for IceCube. It is important to mention string 18, which was instrumented with a prototype of the Digital Optical Module (DOM). The prototype DOM digitized the waveforms and provided an in-situ demonstration of the concept which was then chosen for IceCube's design. The sensors were hybrid modules with simultaneous analog transmission via multimode optical fibers, the latter being the baseline for AMANDA-II.

The Digital Optical Module 3 is built around a 25 cm diameter hemispherical photomultiplier. Electronics in the sensor are designed to digitize and time stamp the signals. Cherenkov light signals are recorded over a dynamic range from 1 to more than 10,000 photoelectrons. All sensors are synchronized with the master clock to a precision of 1 nsec, a resolution comparable to the spacial extension of the sensors. Each module is equipped with 12 on-board blue LEDs which can be used for verification of spatial detector geometry and timing system as well as for precise measurements of the optical properties of the ice. A glass pressure housing protects the sensor from pressures of up to 500 bar recorded in the deep ice and during the freeze-in process. There are 60 sensors connected to each of the 86 cables that provide power and communication. Maintenance of the full detector was an important consideration for the technology choice. IceCube DOMs and data acquisition system are designed to allow for automatic self-calibration of important parameters at the start of each run. The power consumption of each sensor with its more than 1,000 electronic parts and built in high voltage supply is about 5 W. The system power for IceCube is around 55 kW.

IceCube construction began in earnest in 2004–2005. The commissioning of the more powerful 5-MW IceCube hot water drill proved to be a challenge. The system gathered data from about 300 electronic sensors, and all pumps, heaters and valves were computer controlled and had safety systems. Safety was a big challenge when melting and

Season	Campaign	Sensors cum.	Strings	Depth (m)	Neutrinos per day	Resol. @100TeV
1991–1992	Exploratory	few		Shallow	-	
1992–1993						
1993–1994	AMANDA-A	80	4	800–1000	-	
1994–1995						
1995–1996	AMANDA-B4	86	4	1500–1950	∼ 0.01	
1996–1997	AMANDA-B10	206	6/10	1500–1950	∼ 1	4°
1997–1998						
1998–1999	AMANDA-II	306	3/13	1500–1950		
1999–2000	AMANDA-II	677	6/19	1500–1950	∼ 5	2°
2001–2002						
2002–2003						
2003–2004	IceCube prep.					
2004–2005	IceCube 1	60	1/1	1450–2450		
2005–2006	IceCube 9	540	8/9	1450–2450		
2006–2007	IceCube 22	1320	13/22	1450–2450	18	1.5°
2007–2008	IceCube 40	2400	18/40	1450–2450	40	0.8°
2008–2009	IceCube 59	3540	19/59	1450–2450	120	0.6°
2009–2010	IceCube 79	4740	20/79	1450–2450	180	0.4°
2010–2011	IceCube 86	5160	7/86	1450–2450	>200	0.4°

Table 1. The table summarizes the deployment of optical sensors at the South Pole. The cumulative number of sensors deployed per year is shown (324 IceTop sensors deployed with IceCube are not included). The angular resolution is shown for the reference analysis for point source searches.

circulating more than 800 l/min of hot water at 90 °C and at a pressure of 70 bar. The drill head was 20 m long and initially even small changes to the nozzle diameter could lead to instabilities. Despite difficulties, an all-important first hole was successfully drilled and a functioning first string deployed in that season. Table 1 shows the rapid increase in the number strings installed per season. Drilling became an engineering success of its own, with finely tuned operations performed by a well-trained and motivated drill crew.

Figure 3 shows the overlay of the depth-versus-time profiles of 20 holes drilled in one season. The average time for completing a hole of ∼ 55 cm diameter and 2,500 m depth was about 32 hours. In total, more than 900 tons of cargo and fuel have been delivered to the South Pole for IceCube. More than 300 Hercules LC-130 aircraft flights delivered the last leg of transportation from McMurdo to the South Pole over seven years. IceCube construction was organized in three shifts around the clock, with a total of 50 personnel on-ice from the middle of November to early February. It was clear that drilling would and should determine the schedule. That meant that more than 1,000 DOMs needed to be built, tested and shipped on time to meet the deadlines for vessel shipments.

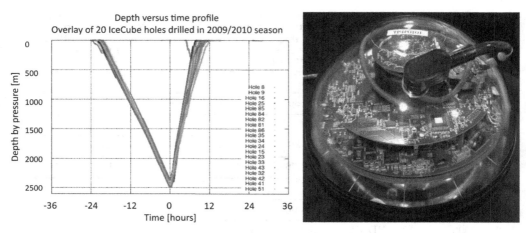

Figure 3. Left: The depth versus time profile is shown for 20 consecutive holes drilled with the IceCube Enhanced Hot Water Drill. The average drill time was 32 hours. Right: The IceCube Digital Optical Module; about 99% of the sensors are working since deployment in the ice.

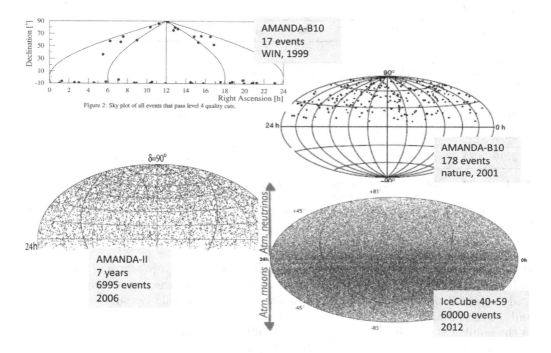

Figure 4. Skymaps at major steps of neutrinos astronomy as indicated in the figure.

The reliability and successful installation of the sensors was a critical requirement. In an initial count, only 84 sensors out of 5,160 did not commission successfully. The reliability after commissioning has been very high. Fewer than 20 sensors failed since commissioning during an accumulated lifetime of 20,000 sensor years. The rates appear to still be dropping. If one were to assume a constant failure rate, it would result in a mean time between sensor failures on the order of 1,000 years.

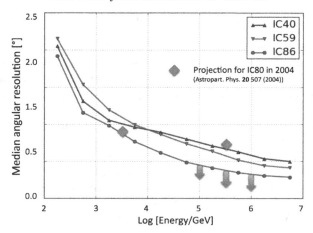

Figure 5. Angular resolution of IceCube as a function of energy. Further improvements are expected especially at higher energies.

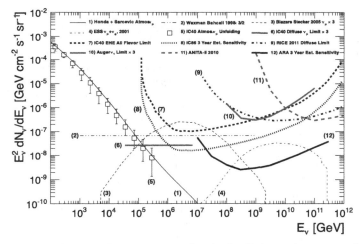

Figure 6. The measured atmospheric neutrino flux by IceCube is shown together with several predictions of neutrino fluxes and upper limits by experiments: (1) Atmospheric neutrino flux by Honda *et al.* (2007) + prompt by Sarcevic *et al.* (2008), (2) Diffuse neutrino flux (Waxman & Bahcall 1995), (3) AGN Blazars (Stecker 2005), (4) Cosmogenic neutrino flux (Engel, Seckel & Stanev 2001), (5) IceCube atmospheric neutrino flux unfolded measurement (Abbasi *et al.* 2011b), (6) IceCube 40, 1-year upper limit to diffuse neutrino flux (Abbasi *et al.* 2011a), (7) IceCube 40, 1-year upper limit to extremely high-energy neutrinos (Abbasi *et al.* 2011c), (8) IceCube 86, rough estimate of 3-year sensitivity (before this conference), (9) RICE upper limit (Kravchenko *et al.* 2012), (10) Auger 2-year limit x 3 (Auger 2012), (11) ANITA upper limit (Gorham *et al.* 2010), and (12) the Askaryan Radio Array (ARA) estimated 3-year sensitivity (Allison *et al.* 2012). Differential limits are corrected for energy binning and flavor differences.

Figure 4 shows the development of neutrino sky maps over 15 years of neutrino astronomy at the South Pole. The number of neutrinos per year increased from 17 to almost 10^5 while the angular resolution improved from about 5° to less then 0.4°. The angular resolution has been tested with real data using the Moon shadow of cosmic rays as a calibration source. Figure 5 shows the angular resolution for detector configurations of 40, 59 and 86 strings was a function of energy. Also shown are the originally projected resolution for IceCube, which has been well exceeded already. New reconstruction algorithms currently in development are expected to improve the resolution, especially at energies above $100\,\mathrm{TeV}$, to levels $\sim 0.1°$.

All performance parameters are meeting or exceeding initial expectations. The high rate of atmospheric muons of $3\,\mathrm{kHz}$ is being processed at the IceCube laboratory in real time. A set of filters reduces the data rate from about $2,000\,\mathrm{GB/day}$ to a more manageable rate of $100\,\mathrm{GB/day}$ of events that are potentially interesting for physics searches. This data set is transmitted by satellite to the Northern Hemisphere on a daily basis. The sky maps offer only a superficial summary of the physics topics that can be covered with IceCube. The current state of measurements of atmospheric neutrinos by AMANDA and IceCube, some models for predictions of astrophysical neutrinos from GRBs and AGNs, as well as IceCube's limits on diffuse fluxes are shown in Figure 6. With 40 strings IceCube has measured the atmospheric neutrino spectrum up to energies of about $300\,\mathrm{TeV}$, an energy where a hard component of astrophysical neutrinos is expected. At much higher energies, above $10^{17}\,\mathrm{eV}$, IceCube has already placed the best limits on the cosmogenic (GZK) neutrino flux.

References

Abbasi, R., et al. (IceCube Collaboration) 2012, Astropart. Phys. 35, 615

Abbasi, R., et al. (IceCube Collaboration) 2011a, ApJ 732, 18

Abbasi, R., et al. (IceCube Collaboration) 2011b, Phys. Rev. D 83, 012001

Abbasi, R., et al. (IceCube Collaboration) 2011c, Phys. Rev. D 83, 092003

Allison, P., et al. (ARA Collaboration) 2012, Astropart. Phys. 35, 457

Akhmedov, E. Kh., Razzaque, S., & Smirnov, A. Yu. 2012, submitted; arXiv: 1205.7071

Andres, E., et al. (AMANDA Collaboration) 2001, Nature 410, 441

Askaryan, G. A. 1962, Soviet Physics, JTEP 14, 441

Askebjer, P., et al. (AMANDA Collaboration) 1995, Science 267, 1147

Auger Collaboration 2011, Phys. Rev. D 84, 122005; Erratum: Phys. Rev. D, 85, 029902(E)

Avrorin, A. V., et al. (Baikal Collaboration) 2009, Astron. Lett., 35, 651

Engel, R., Seckel, D., & Stanev, T., 2001, Phys. Rev. D 64, 093010

Gaisser, T., these proceedings

Gorham, P. W., et al. (ANITA Collaboration) 2010, Phys. Rev. D 82, 022004

Honda, M., et al. 2007, Phys. Rev. D 75, 043006

Kravchenko, I., et al. 2012, Phys. Rev D, in press; arXiv:1106.1164

Lowder, D. M., et al. 1991, Nature 353, 331

Sarcevic, I., et al. 2008, Phys. Rev. D 78, 043005

Spiering, C. & Katz, U., 2012, Prog. Part. Nucl. Phys. 67, 651

Stecker, F. W., 2005, Phys. Rev. D 72, 107301

Waxman, E. & Bahcall, J., 1998, Phys. Rev. D 59, 023002

Discussion

QUESTION: What is the main difference between water and ice?

KARLE: The noise environment may be the biggest difference. The sensors in IceCube run at a noise rate of about $500\,\mathrm{Hz}$, and that is just the noise from the remnant radioactivity in the glass pressure housings of the sensors. This is why IceCube is sensitive to Supernova core collapse. The noise rate in the Mediterranean Sea is at the level of 100 to $1,000\,\mathrm{KHz}$ for similar sensors, mostly from K40-decay and bioluminescence. Otherwise, absorption and scattering lengths are somewhat different, yet it does not make a big difference for detector designs and performance.

Astrophysics from Antarctica
Proceedings IAU Symposium No. 288, 2012
M. G. Burton, X. Cui & N. F. H. Tothill, eds.

© International Astronomical Union 2013
doi:10.1017/S1743921312016742

The IceCube Neutrino Telescope

Thomas K. Gaisser[1] for the IceCube Collaboration[2]

[1] Bartol Research Institute and Dept. of Physics and Astronomy
University of Delaware, Newark, DE 19716 USA
email: `gaisser@bartol.udel.edu`
[2] See http://icecube.wisc.edu

Abstract. Construction of IceCube at the Amundsen-Scott South Pole Station was completed at the end of 2010 after eight construction seasons. The detector consists of 5,160 digital optical modules on 86 cables with 60 modules each, viewing in total a cubic kilometer of ice between 1,450 and 2,450 meters below the surface. IceCube includes a sub-array called DeepCore consisting of 8 special cables, and providing a more densely instrumented region with a lower energy threshold in the deep center of the array. IceCube also includes an air shower array called IceTop directly above the deep detector. Optical modules in all three components of the detector are fully integrated into a single data acquisition system. Data taking and analysis began during construction and continues with the completed detector. This paper describes recent results from IceCube.

Keywords. Neutrinos, cosmic rays

1. Introduction

The idea of detecting neutrinos in a large instrumented volume of water dates back to work of Markov (1960), Greisen (1960) and Reines (1960). The idea is to use the Cherenkov light generated by charged particles produced when neutrinos interact in a large, transparent volume of water (or ice). Development of the original idea followed two paths. One was densely instrumented detectors aimed at the GeV region motivated originally by the search for proton decay. This effort paid off in a big way with the observation of neutrinos from SN1987A by Kamioka (Hirata *et al.*, 1987) and IMB (Bionta *et al.*, 1987) and the discovery of neutrino oscillations by Super-Kamiokande (Fukuda *et al.*, 1998) and SNO (Jelley, MacDonald & Robertson, 2009).

The other path is motivated by the quest to detect high-energy neutrinos of astrophysical origin above the steeply falling spectrum of neutrinos produced locally by interactions of cosmic rays in the atmosphere. For this objective, the instrumented volume should be as large as possible and the detectors placed as far apart as allowed by transparency of the medium. IceCube is in this line, started in the 1970's by the DUMAND Project, an effort to deploy an array of photomultipliers in the ocean near Hawaii. Although DUMAND itself was realized only with the deployment of a single string from a ship for several days in 1987 (Babson *et al.*,1990), the DUMAND effort in the seventies and eighties set the stage for high energy neutrino astronomy. The Lake Baikal detector in Siberia (Balkanov *et al.*, 1999) and the ANTARES detector in the Mediterranean (Ageron, *et al.*, 2011) are the two large neutrino telescopes currently operating in water.

One of the first papers to discuss the possibility of using ice rather than water as the detector medium (Halzen *et al.*, 1989) was presented in 1989 at a conference on prospects for astrophysics in Antarctica (AIP, 1989). The meeting was hosted by Martin Pomerantz and the Bartol Research Institute at Delaware with support from the NSF Office of Polar Programs. Plans for AMANDA (Antarctica Muon and Neutrino Detector

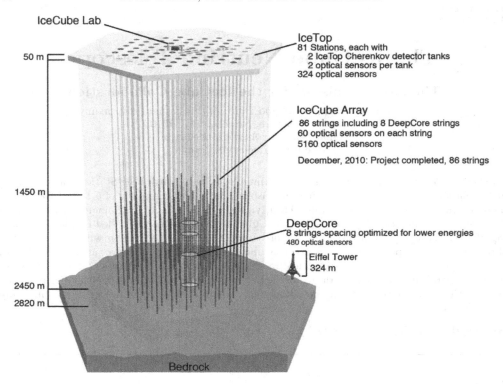

Figure 1. The IceCube Neutrino Telescope.

Array) developed in the decade following this meeting. It is interesting to note that the Center for Astrophysical Research in Antarctica (CARA) for millimeter and submillimeter astronomy at the South Pole traces its origin to the same meeting. Five years later the Martin A. Pomerantz Observatory (MAPO) was inaugurated at the South Pole to house the astronomy experiments and AMANDA. The South Pole Telescope and IceCube are both descendants of that era. Further description of AMANDA (IceCube, 2009a) as the predecessor of IceCube are presented in a previous paper in these Proceedings (Karle, 2012).

Construction of IceCube was achieved in eight Antarctic seasons, starting in 2003/04 with the first shipments of equipment (Karle, 2012). The first deep hole was drilled and the first IceCube cable deployed in January, 2005 along with the first 4 stations of IceTop. The completed IceCube has been in operation since May 2011 with 86 strings between 1.45 and 2.45 km below the surface. Each string carries 60 digital optical modules (DOMs). On the surface, near the top of each string is a pair of tanks filled with clear ice and instrumented with two DOMs each. These 162 tanks at 81 stations constitute a square kilometer air shower array that is fully integrated into the IceCube data acquisition system. IceCube is thus a three-dimensional cosmic-ray detector as well as a neutrino telescope, as illustrated in Fig. 1.

2. Motivation: neutrinos as a probe of cosmic-ray origin

The principal purpose of IceCube is to find high-energy neutrinos of astrophysical origin to probe the origin of cosmic rays. In the Milky Way, this means searching for neutrinos associated with supernova remnants, active star forming regions or other possible sources

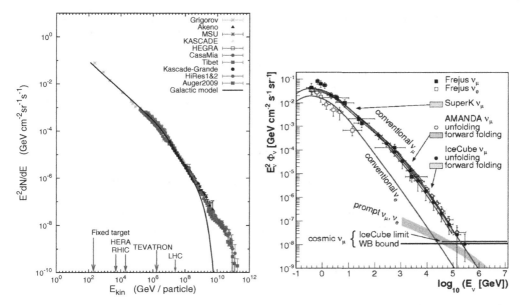

Figure 2. Left: The high-energy cosmic-ray spectrum. Solid line shows a possible model for fraction of observed cosmic rays from sources in the Milky Way. The plot is adapted from Gaisser (2012) where the references for the data are given. Right: Atmospheric ν_μ spectrum and astrophysical limits.

of galactic cosmic rays. In the case of extra-galactic cosmic rays, likely sources are gamma-ray bursts (GRB) and active galaxies. Neutrinos and gamma-rays, being electrically neutral, would point back to the sources of the processes that produce them. Protons and nuclei generate both photons and neutrinos if they interact in the sources or with the surrounding medium to produce pions. Neutral pions decay to photons, while charged pions produce neutrinos. Neutrinos can only be of hadronic origin, so they directly reflect cosmic-ray acceleration. The relation of photons to cosmic rays is more complex. Since electrons are more efficient radiators than protons, they are often the direct progenitors of gamma-rays. In addition, photons are likely to cascade on their way out of the sources and during propagation. To summarize, photons are abundant but complicated to interpret, while neutrinos are clean but rare.

The simplicity of the connection between neutrinos and cosmic-rays makes it possible to estimate from the cosmic-ray spectrum the level at which high-energy astrophysical neutrinos from extra-galactic sources may be expected (Halzen, 2012). Figure 2 (left) shows a summary of measurements of the cosmic ray spectrum. The solid line shows a model of the galactic cosmic rays (Gaisser, 2012). The model is not unique, but it illustrates one class of models by Berezinsky *et al.* (2006) in which the galactic contribution of cosmic rays ends before the ankle in the spectrum around 3×10^{18} eV. Another possibility (Allard *et al.*, 2011) is that the galactic contribution extends somewhat higher in energy so that the ankle marks the transition to the extragalactic contribution.

In either case, by 10^{19} eV, the observed spectrum is expected to be dominated by sources outside the Milky Way. This assumption is also consistent with the nearly isotropic distribution of cosmic rays of this energy and their lack of correlation with the structure of our galaxy. This energy is also below the energy of 5×10^{19} eV above which energy losses by pion photoproduction of protons and photodisintegration of nuclei become important. We can therefore use the observed energy flux in this energy range to estimate the energy content of the extragalactic cosmic rays and hence the power required of their sources. Reading off the plot, the energy flux at 10^{19} eV is

$$E \frac{dN}{d \ln E} \approx 2 \times 10^{-8} \, \text{GeV} \, \text{cm}^{-2} \text{sr}^{-1} \text{s}^{-1}, \tag{2.1}$$

which corresponds to an energy density of $\frac{4\pi}{c} E \frac{dN}{d \ln E} \approx 1.3 \times 10^{-20}$ erg cm^{-3}, and to a differential power required of the sources,

$$\frac{dL}{d \ln E} \approx 10^{36} \, \text{erg} \, \text{Mpc}^{-3} \text{s}^{-1}. \tag{2.2}$$

The simple estimate of Eq. 2.2 is obtained by dividing the observed energy density by the Hubble time, neglecting possible effects of source evolution. The total energy required is $\sim 10^{37}$ erg Mpc^{-3}s^{-1} or greater because the source spectrum must extend over several orders of magnitude. This leads to the estimated power per individual source (Table 1).

	Source density or rate	av. power required per src.
Galaxies	3×10^{-3} Mpc^{-3}	5×10^{39} erg/s
Galaxy clusters	3×10^{-6} Mpc^{-3}	5×10^{42} erg/s
AGN	1×10^{-7} Mpc^{-3}	1×10^{44} erg/s
GRB	300 GRB/(Gpc^3 yr)	10^{51} erg/burst

Table 1. Power requirements for some potential sources of ultra-high energy cosmic rays.

In addition to the energy requirement, potential sources also need to satisfy the Hillas (1984) relation, which requires the product of magnetic field and size of the source to be large enough to accelerate particles to $> 10^{20}$ eV. Active galaxies and GRBs are considered the most likely candidate sources. It is important to note that active galaxies appear on the Hillas plot in two ways. Both active galactic nuclei (AGN) and giant radio galaxies are potential sources. The acceleration could occur inside the jets of AGN or far out in the intergalactic medium at the termination of the jets. The expectation for neutrinos would be quite different in the two cases, with little production if the acceleration occurs in the diffuse intergalactic medium.

If the acceleration occurs inside the jets of AGN or GRB, then one possible scenario is that the protons being accelerated are contained in the acceleration region by the magnetic fields required for the acceleration mechanism to work. Protons would interact with the intense radiation fields by photoproduction, with the main channels being

$$p + \gamma \quad \rightarrow \quad \Delta^+ \rightarrow p + \pi^0 \rightarrow p + \gamma\gamma \quad \text{and} \tag{2.3}$$
$$p + \gamma \quad \rightarrow \quad \Delta^+ \rightarrow n + \pi^+ \rightarrow n + \mu^+ + \nu_\mu$$

followed by muon decay to $e^+ \bar{\nu}_\mu \nu_e$. On average, each neutrino would carry 7% of the energy of the neutron (Ahlers et al. 2005). The neutrons may escape and decay to produce the extragalactic cosmic rays. Attributing the full observed flux of Eq. 2.1 to the escaping neutrons then predicts

$$E_\nu \frac{dN_\nu}{d \ln E_\nu} \approx 1.4 \times 10^{-9} \, \text{GeV} \, \text{cm}^{-2} \text{sr}^{-1} \text{s}^{-1} \tag{2.4}$$

per flavor assuming oscillations equalizes the three different neutrino flavors.

3. Neutrinos in IceCube

IceCube classifies neutrino events in two categories, track-like and cascades. Track-like events are produced by charged current interactions of muon neutrinos. The event rate in this channel is a convolution of the neutrino flux with the neutrino cross section, the detector response and the range of the muon. At high energy the Earth becomes opaque to neutrinos, first for vertically upward neutrinos (~ 30 TeV) and then for more horizontal events ($\sim$PeV). The rate of neutrino induced events is a convolution of the flux with the effective area of the detector. For ν_μ in the charged current channel

$$A_{\text{eff}}(\theta, E_\nu) = \epsilon(\theta)\, A(\theta)\, P_\nu(E_\nu, E_\mu) \exp\{-\sigma_\nu(E_\nu)N_A X(\theta)\}, \qquad (3.1)$$

where $X(\theta)$ is the slant depth (g/cm^2) along a zenith angle $\theta > 90°$, N_A is Avogadro's number, σ_ν is the neutrino cross section and $\epsilon(\theta)$ a reconstruction efficiency.

$$P_\nu(E_\nu, E_\mu) = N_A \int_{E_\mu}^{E_\nu} dE_\mu^* \frac{d\sigma_\nu(E_\nu)}{dE_\mu^*} R(E_\mu^*, E_\mu)$$

is the probability that a muon produced with energy E_μ^* reaches the detector with energy E_μ sufficient to trigger the detector. The typical ranges are ~ 5 and ~ 15 km for muons with 10 TeV and 1 PeV respectively. Because of the large muon range, the efficiency for detecting ν_μ is significantly greater than for the cascade channel. The energy deposited in the detector by a throughgoing muon is only a fraction of the energy of the parent ν_μ, so the visible energy is only statistically related to the neutrino energy. Above a TeV, stochastic energy losses become dominant for muons, so the deposited energy is proportional on average to the energy of the muon at the detector.

Cascade events can be neutral current interactions of any flavor or charged current interactions of ν_e or ν_τ. In this case, the muon range in Eq. 3.1 is replaced by a fraction of the linear dimension of the instrumented volume, sufficiently < 1 km to satisfy a containment criterion. Thus, for equal fluxes, rates of cascade events will be correspondingly lower than track-like events. If the cascade is a charged current interaction of a ν_e or ν_τ, however, the visible energy in the cascade gives a direct measurement of the neutrino energy. The $< \Gamma c\tau >$ for a τ lepton at 1 PeV is 50 m, compared the string spacing IceCube of 125 meters. Therefore it should be possible to discern the characteristic double bang structure (Learned & Pakvasa, 1995) as an elongation of the cascade somewhere above this energy.

Most of the events in IceCube are $\sim$ TeV muons produced by cosmic-ray interactions in the atmosphere above the detector. To find neutrinos, the Earth is used as a filter by selecting upward-moving tracks. The rate of events from atmospheric neutrinos producing muons from below the horizon is at the level of one part per million compared to the downward background. There is therefore a large contamination of mis-reconstructed muon events in the initial online selection. After applying cuts on track quality to reject accidentally coincident downward muons and other badly reconstructed events, a clean sample of neutrinos is found. These are mostly atmospheric, neutrino-induced muons.

3.1. *Atmospheric neutrinos*

Figure 2 (right) shows the fluxes of atmospheric $\nu_\mu + \bar\nu_\mu$ obtained from two analyses of data from IceCube when it was half finished (IC-40). One uses an unfolding method to relate the observed distribution of visible energy to the muon and hence to the neutrino that produced it (IceCube, 2011a). The other ((IceCube, 2011b) uses a forward folding method with the same physics but fitting parameters that describe the neutrino spectrum to the observed distribution of energy deposited in the detector. The parameters are the

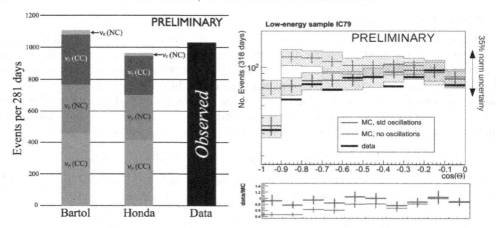

Figure 3. Neutrinos in DeepCore. Left: expectation for cascade events classified by channel compared to observation. Right: Effect of oscillations in the atmosphere on low-energy track-like events.

normalization and slope of the atmospheric neutrino spectrum, the normalization of the prompt neutrinos from decay of charm and the normalization of an astrophysical component assumed to have an E^{-2} spectrum. The solid lines show upper limits on astrophysical neutrinos that will be discussed below. Also shown in the figure are results from IceCube's predecessor, AMANDA (IceCube, 2009b) and (IceCube, 2010b) along with earlier results from Frejus and SuperKamiokande.

In the GeV region there are approximately twice as many muon neutrinos produced in the atmosphere as electron neutrinos as a consequence of the $\pi \to \nu_\mu + \mu \to \bar{\nu}_\mu + \nu_e + e$ decay chain. The observed ratio of GeV neutrinos is closer to one as a consequence of oscillations. Above several GeV, however, the abundance of ν_e decreases rapidly as the higher energy muons reach the ground before decaying. In the TeV range and above, the intensity of atmospheric ν_e is only 3-4% of ν_μ, the primary channel for their production being one of the semileptonic decays of K_L. Because of their low rates and the requirement for containment, it has been difficult to identify cascade like events in the multi-TeV range with the partially completed IceCube. Preliminary results on high-energy cascades in IceCube will be discussed below.

In the last two construction seasons a subarray of more closely spaced sensors with higher quantum efficiency was deployed in the deep central region of IceCube (IceCube, 2012a). With this subarray, called DeepCore, it is now possible to measure the flux of electron neutrinos in the 0.1-1 TeV energy range using the increased sensitivity of the denser subarray to low-energy events. Preliminary results (Ha, 2012) are shown in the left panel of Fig. 3. Quantitative evaluation of the intensity of ν_e will depend on accumulating sufficient statistics to subtract the neutral current contributions and the background of low-energy, charged-current ν_μ events remaining in the sample. The ratios of the different contributions are obtained from simulation starting from two different calculations of the flux of atmospheric neutrinos, by Barr *et al.* (2004) and by Honda *et al.* (2007).

The DeepCore array also provides better sensitivity for short muon tracks, leading to the possibility of reconstructing the directions of charged current ν_μ events with energies between 10 and 100 GeV. The right panel of Fig. 3 compares the angular dependence of these events to what is expected without oscillations (red) and with oscillations assuming the standard oscillation parameters (black). The suppression of ν_μ with long pathlengths through the Earth ($\cos(\theta)$ near -1) is clearly seen (Gross, 2012).

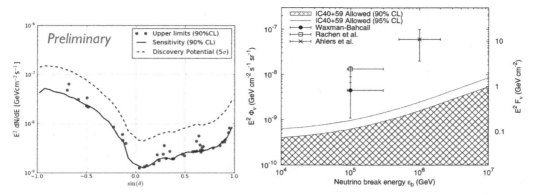

Figure 4. Left: Limits on neutrinos from selected point sources (vs sine of declination). Right: Upper limits on neutrinos from GRBs.

3.2. *Search for point sources of astrophysical neutrinos*

The most direct approach to searching for sources of neutrinos with IceCube is to map the sky in neutrino-induced muons (IceCube, 2011c). A sky map of the significance relative to background is created for the whole sky. For the Northern sky (directions from below the horizon at the South Pole) the background after event selection is produced by atmospheric ν_μ with contamination from mis-reconstructed events at the per cent level. The Southern sky is dominated by the high rate of atmospheric muons from above, so the selection criteria are quite different and include the veto probability from IceTop. In both cases an unbinned maximum likelihood approach is used that accounts for the features of the backgrounds, which are different for the Northern and Southern skies. The likelihood procedure takes into account known features of the background, including the steep energy spectrum and the angular distribution of the atmospheric neutrinos and muons.

Results are also given for a selected list that includes 13 galactic sources such as supernova remnants and microquasars and 30 extragalactic sources, mostly AGNs. The left panel of Fig. 4 shows the limits for the selected sources obtained with two years of the partially completed IceCube running with 40 strings in 2008-09 and with 59 strings in 2009-2010 (Aguilar, 2012). Although no highly significant sources have been identified yet, it is interesting to note that the limits summed over the Northern sky are approaching the level of 10^{-9} GeV cm^{-2}sr^{-1}s^{-1} at which neutrinos are expected if they are produced at the level suggested by Eq. 2.4.

There are also dedicated searches for flaring sources (IceCube, 2012b), both looking for correlations with flaring activity observed in gamma-rays and by looking for sequences of neutrinos correlated in time. A system has been set up for sending alerts for rapid followup by optical telescopes when two or more neutrinos from nearly the same direction occur within a short time interval (IceCube, 2012c).

3.3. *Neutrinos from gamma-ray bursts*

Gamma ray bursts provide sharp time stamps and in most cases a direction in which to look for accompanying neutrinos. IceCube looked for high energy neutrinos in association with 300 bursts that occurred during the same two year period mentioned above (IC-40 and IC59). No neutrinos were seen (IceCube, 2012d). The right panel of Fig. 4 shows the flux upper limits as a function of the break energy in the neutrino spectrum, which depends on the assumed bulk Lorentz factor of the GRB jet and the observed gamma-ray spectrum of the burst. The limit is significantly lower than predictions (Rachen & Meszaros, 1998, Ahlers *et al.*, 2011) normalized by the assumption that GRBs are the source of the extragalactic cosmic rays, including the original calculation of Waxman & Bahcall (1997). The result also rules out a flux at the level predicted by the model of Guetta *et al.* 2004. However, the parameters of this model have since been questioned by Hummer, Baerwald & Winter (2012) who suggest the neutrino flux prediction could be an order of magnitude lower.

The right ordinate of the GRB limit in Fig. 4 shows upper limits on the fluence accumulated during the 300 time windows around each burst. The average duration in the sample was 28 seconds per burst. Assuming 667 GRBs per year in the whole sky, a limit on the diffuse flux from all gamma-ray bursts is calculated and shown on the left ordinate. This more general limit depends on the shape of the model but not its normalization. It is interesting that it is at the expected level of Eq. 2.4.

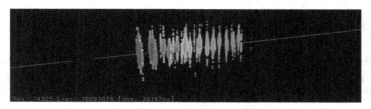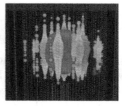

Figure 5. Left: Large track event; Right: Large cascade event.

3.4. *Neutrinos from unresolved sources*

Because neutrinos are not absorbed during propagation, the flux at Earth will receive equal contributions from each spherical shell out to the Hubble radius. It is therefore important to look for neutrinos from all directions (Lipari, 2008). This "diffuse" analysis is the same that produces the atmospheric neutrino spectrum as a byproduct (Fig. 2, right). The normalization and slope of the atmospheric spectrum and the fraction of prompt neutrinos are fitted nuisance parameters of the analysis. The upper horizontal, solid line in Fig. 2 (right) is the upper limit for IceCube in 2009-10 (IC-59) (Schukraft, 2012). The lower line is the Waxman-Bahcall limit on the flux of astrophysical neutrinos (Waxman, 2011), which is essentially at the level of the observed cosmic-ray energy content of Eq. 2.1. The bound is an upper limit derived by Waxman & Bahcall (1998) from the observation that for a transparent (efficient) source there should not be more energy in neutrinos than in the cosmic-rays that produced them. The IceCube upper limit is a factor of two higher than the sensitivity of the analysis. One of the largest events from this analysis is shown in the left panel of Fig. 5.

3.5. *Cosmogenic neutrinos*

If the energy spectrum of extragalactic cosmic rays extends above the threshold for photo-pion production on the microwave background, neutrinos will be produced as the cosmic rays propagate over cosmic distances (Berezinsky & Zatsepin, 1969). The paper of Kotera, Allard & Olinto (2011) is an overview of the possible levels of neutrino production, which depend on factors such as the cosmological evolution of the power of the sources and the spectrum and composition of the accelerated particles. Expectations for event rates in the completed IceCube are at the level of one event per year (IceCube, 2011d).

A search with two years of data (June 2010 - May 2012, IC-79 and IC-86) (Ishihara, 2012) found no events in the energy range expected for cosmogenic neutrinos (10 to 1000 PeV). However, two cascade-like events were found above one PeV, slightly above the threshold in total observed photo-electrons for the analysis. Preliminary energies estimated for the cascades are 1.1 and 1.3 PeV with a total uncertainty (statistical plus systematic) of 35%. The 1.1 PeV event is shown in the right panel of Fig. 5. A study is under way in preparation for unblinding the entire high energy data sample to investigate the spectrum below the threshold for the cosmogenic analysis.

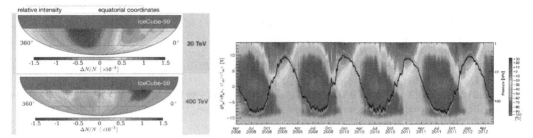

Figure 6. Left: Anisotropy in the Southern hemisphere; Right: Seasonal variations of IceCube rates over four years.

4. Broader aspects of IceCube and Outlook

As a large particle detector, IceCube has a broad scientific scope ranging from studies of cosmic-ray spectrum (IceCube, 2012e) and composition (IceCube, 2012f) to searches for dark matter (Icecube, 2012g) and for new physics such as magnetic monopoles (Ice-Cube, 2012h) and violation of Lorentz invariance (IceCube, 2010a). Counting rates in IceTop tanks provide novel spectral information about particles from solar flares that produce ground level events (IceCube, 2008). Counting rates of DOMs in the stable, quiet, cold environment of the deep ice continuously monitor for ∼ 10 MeV neutrinos from supernovas in our galaxy (IceCube, 2011e)

As the largest deep detector, IceCube detects TeV atmospheric muons at an unprecedented rate. With the completed IceCube, the rate of events that trigger 8 or more DOMs is 2.2 kHz with a seasonal variation of ±9% after a 10% correction for accidental coincidences. Directions and energies are reconstructed online to sufficient accuracy to map cosmic-ray anisotropies that reflect scales of the outer heliosphere and nearby interstellar medium (IceCube, 2012i). The same muons can be used to probe seasonal and short-term variations in the stratosphere with unprecedented precision as the atmosphere expands and contracts in response to changes in temperature (IceCube, 2011f). Muon maps and seasonal variations are illustrated in Fig. 6.

Data taking with the completed IceCube detector began in May 2011, and analysis of data with the full detector is just getting fully underway. Indications of high-energy astrophysical neutrinos are already beginning to appear. The detector is stable with negligible loss of hardware in the ice so far, so it should be able to operate for the extended period of a decade or more needed to accumulate a good sample of neutrinos.

Acknowledgment This work was supported in part by NSF-ANT-0856253.

References

Astrophysics in Antarctica 1989, *AIP Conf. Proc. 198* (ed. Dermott J. Mullan, Martin A. Pomerantz & Todor Stanev).

Ageron, M. *et al.* 2011, arXiv:1104.1607v2.

Aguilar, J.A. 2012 (for the IceCube Collaboration), *Proc. 9th Workshop on Science with the New Generation of High Energy Gamma-ray Experiments* (SciNeGHE 2012).

Ahlers, M., *et al.* 2005, *Phys. Rev.* D72, 023001.

Ahlers, M., M. Gonzalez-Garcia & F. Halzen 2011, *Astropart. Phys.* 35, 87-94.

Allard, D. *et al.* 2011, *JCAP* 10-033.

Babson, J. *et al.* 1990, *Phys. Rev.* D42, 3613-3620.

Balkanov, V. A. *et al.* 1999, *Astropart. Phys.* 12, 75-86.

Barr, G. D. *et al.* 2006, *Phys. Rev.* D74, 094009.

Berezinsky, V.S. & G.T. Zatsepin, *Phys. Lett.* 28 B, 423.

Berezinsky, V., A. Gazizov & S. Grigorieva 2006, *Phys. Rev.* D74, 043005.

Bionta, R. M. *et al.* 1987, *Phys. Rev. Lett.* 58, 1494-1496.

Fukuda, Y. *et al.* 1998, *Phys. Rev. Lett.* 81, 1562-1567.

Gaisser, T. K. 2012, *Astropart. Phys.* 35, 801-806.

Greisen, K. 1960, *Ann. Rev. Nuclear Sci.* 10, 63-108.

Gross, A. 2012 (for the IceCube Collaboration), Neutrino 2012.

Guetta, D. *et al.* 2004, *Astropart. Phys.* 20, 429.

Ha, C. 2012 (for the IceCube Collaboration), arXiv:1209.0698v1.

Halzen, F., J. Learned & T. Stanev 1989, in *AIP Conf. Proc.* 198, 39-51.

Halzen, F. 2012, these Proceedings.

Hirata, K. *et al.*, 1987, *Phys. Rev. Lett.* 58, 1490-1493.

Honda, M. *et al.* 2007, *Phys. Rev.* D75, 043006.

Hummer, S., P. Baerwald & W. Winter 2012, *Phys. Rev. Letters* 108, 231101.

IceCube Collaboration 2008 (R. Abbasi *et al.*), *Ap.J.* 689, L65-68.

IceCube Collaboration 2009a (R. Abbasi *et al.*), *Phys. Rev.* D79, 062001.

IceCube Collaboration 2009b (R. Abbasi *et al.*), *Phys. Rev.* D79, 102005.

IceCube Collaboration 2010a (R. Abbasi *et al.*), *Phys. Rev.* D82, 112003.

IceCube Collaboration 2010b (R. Abbasi *et al.*), *Astropart. Phys.* 34, 48-58.

IceCube Collaboration 2011a (R. Abbasi *et al.*), *Phys. Rev.* D83, 012001.

IceCube Collaboration 2011b (R.Abbasi *et al.*), *Phys. Rev.* D84, 082001.

IceCube Collaboration 2011c (R. Abbasi *et al.*), *Ap.J.* 732, 18.

IceCube Collaboration 2011d (R. Abbasi *et al.*), *Phys. Rev.* D83, 092003.

IceCube Collaboration 2011e (R. Abbasi *et al.*), *Astronomy & Astrophysics* 535, A109.

IceCube Collaboration 2011f (R. Abbasi *et al.*), arXiv:1111.2735 (Proc. 32nd ICRC, Beijing).

IceCube Collaboration 2012a (R. Abbasi *et al.*), *Astropart. Phys.* 35, 615-624.

IceCube Collaboration 2012b (R. Abbasi *et al.*), *Ap.J.* 744, 1.

IceCube Collaboration 2012c (R. Abbasi *et al.*), *Astron. & Astrophys.* 539, A60.

IceCube Collaboration 2012d (R. Abbasi *et al.*), *Nature* 484, 351-354.

IceCube Collaboration 2012e (R. Abbasi *et al.*), arXiv:1202.3039 (submitted to Astropart. Phys.)

IceCube Collaboration 2012f (R. Abbasi *et al.*), arXiv:1207.3455 (submitted to Astropart. Phys.).

IceCube Collaboration 2012g (R. Abbasi *et al.*), *Phys. Rev* D85, 042004.

IceCube Collaboration 2012h (R. Abbasi *et al.*), arXiv:1208.4861 (submitted to Phys. Rev. D).

IceCube Collaboration 2012i (R. Abbasi *et al.*), *Ap.J.* 745, 45.

Ishihara, A. 2012 (for the IceCube Collaboration) to appear in *Proc. Neutrino 2012*.

Jelley, N., A. B. McDonald & R. G. H. Robertson 2009, *Ann. Revs. Nucl. Part. Sci.* 59, 431-465.

Karle, A. 2012, these Proceedings.

Kotera, K., D. Allard & A. V. Olinto 2011, *JCAP* 10-013.

Learned, J. G. & S. Pakvasa 1995, *Astropart. Phys.* 3, 267-274.

Markov, M. A. 1960, *Proc. Annual International Conference on High Energy Physics at Rochester*.

Lipari, P. 2008, *Phys. Rev.* D 78, 083011.

Rachen, J.P., & P. Meszaros 1998 in *Fourth Huntsville Gamma-Ray Burst Symposium* (A.I.P. Conf. Proceedings 428, eds. C.A. Meegan, R.D. Preece & T.M. Koshut) 776-780.

Reines, F. 1960, *Ann. Rev. Nuclear Sci.* 10, 1-26.

Schukraft, A. 2012 (for the IceCube Collaboration), presented at NOW2012.

Waxman, E. & J. Bahcall 1997, *Phys. Rev. Letters* 78, 2292.

Waxman, E. & J. Bahcall 1998, *Phys. Rev.* D59 012002.

Waxman, E. 2011 arXiv:1101.1155v1.

Astrophysics from Antarctica
Proceedings IAU Symposium No. 288, 2012
M. G. Burton, X. Cui & N. F. H. Tothill, eds.

© International Astronomical Union 2013
doi:10.1017/S1743921312016754

The Askaryan Radio Array

Kara D. Hoffman

Physics Department, University of Maryland,
College, Park, MD 20742 U.S.A.
email: kara@umd.edu

Abstract. Ultra high energy cosmogenic neutrinos could be most efficiently detected in dense, radio frequency (RF) transparent media via the Askaryan effect. Building on the expertise gained by RICE, ANITA and IceCube's radio extension in the use of the Askaryan effect in cold Antarctic ice, we are currently developing an antenna array known as ARA (The Askaryan Radio Array) to be installed in boreholes extending 200 m below the surface of the ice near the geographic South Pole. The unprecedented scale of ARA, which will cover a fiducial area of ≈ 100 square kilometers, was chosen to ensure the detection of the flux of neutrinos suggested by the observation of a drop in high energy cosmic ray flux consistent with the GZK cutoff by HiRes and the Pierre Auger Observatory. Funding to develop the instrumentation and install the first prototypes has been granted, and the first components of ARA were installed during the austral summer of 2010–2011. Within 3 years of commencing operation, the full ARA will exceed the sensitivity of any other instrument in the 0.1-10 EeV energy range by an order of magnitude. The primary goal of the ARA array is to establish the absolute cosmogenic neutrino flux through a modest number of events. This information would frame the performance requirements needed to expand the array in the future to measure a larger number of neutrinos with greater angular precision in order to study their spectrum and origins.

Keywords. neutrinos, diffuse radiation

---

1. Introduction

One of the most tantalizing questions in astronomy and astrophysics, namely the origin and the evolution of the cosmic accelerators that produce the highest energy (UHE) cosmic rays, may be best addressed through the observation of UHE cosmogenic neutrinos. Astrophysical proton acceleration sites must also be neutrino emitters, since protons must undergo pion photo production in order to escape the magnetic field of their source (Ahlers *et al.*, 2011). The decay chain of the daughter pions will contain gamma rays and neutrinos. Unlike the parent proton, neutrinos travel from their source undeflected by magnetic fields and largely unimpeded by interactions with intervening material. At high energies (above 10^{16} eV), neutrinos could be most efficiently detected in dense, radio frequency (RF) transparent media via the Askaryan (Askaryan, 1962) effect. The abundant cold ice covering the geographic South Pole, with its exceptional RF clarity, has been host to several pioneering efforts to develop this approach, including RICE (Kravchenko *et al.*, 2003) and ANITA (Gorham, 2006).

Building on the expertise gained in these efforts, and the infrastructure developed in the construction of the IceCube optical Cherenkov observatory, we are currently installing prototype instrumentation for an array, known as ARA (The Askaryan Radio Array), in the deep ice near the geographic South Pole. Above $\approx 10^{15}$ eV the power emitted in RF excedes that in optical, and the attenuation length of RF in cold ice is long – on the order of a kilometer – allowing coverage with a larger, sparse array at low cost. South Polar ice is perhaps the most extensively studied on Earth. The combination of ice thickness and favorable radiofrequency dielectric characteristics (see Figure 3), as well as the excellent scientific infrastructure, makes the site unparalleled for the location of this array.

With an instrumented area of an unprecedented ≈ 100 km^2, ARA's size was chosen to ensure the detection of the flux of neutrinos implied by the observation of a steep drop in the cosmic ray spectrum at $\approx 10^{19.5}$ eV (Greisen, 1966, Zatsepin & Kuzmin, 1966) by HiRes (Abbasi et al., 2008) and the Pierre Auger Observatory (Abraham et al., 2008). Greisen, Zatsepin and Kuzmin (GZK) predicted the existence of this cutoff, which they attribute to the scattering of cosmic ray protons on the cosmic microwave background (CMB) photons via:

$$p + \gamma_{2.7K} \rightarrow \Delta^+ \rightarrow n + \quad \pi^+$$
$$\hookrightarrow \quad \mu\nu_\mu$$
$$\hookrightarrow e\nu_e\nu_\mu.$$

The presence of neutrinos in the decay of the resulting Δ resonance and the fact that ultra-high energy cosmic rays (UHECR) have been observed, and almost certainly include a significant proton fraction, implies the existence of neutrinos at $10^{17} - 10^{19}$ eV, unless, coincidentally, this energy corresponds to a limit in the energy that can be attained by cosmogenic accelerators. The observation of ultra high energy cosmogenic neutrinos is needed to confirm that the observed dip is the result of the GZK process. The spectrum of astrophysical neutrinos is expected to follow the spectrum of cosmic rays, whose rates fall exponentially as a function of energy. However, if the drop in flux is attributable to the GZK process, the precise magnitude of the resultant neutrino flux will depend on the proton/iron composition of the cosmic rays, as well as the distribution of their sources in space. Thus, a measurement of ultra high energy neutrinos would also give an indirect measurement of the cosmic ray species.

Within 3 years of commencing operation, the full ARA will exceed the sensitivity of any other neutrino observatory in the 0.1-10 EeV energy range by an order of magnitude. Because the antennas will be deployed in boreholes extending below the firn layer to 200 m depth, it will have the ability to distinguish surface noise from sources originating in the ice cap, otherwise not possible in the ballon borne approach employed by ANITA (ANITAinstr). Even under the extreme assumption that UHE cosmic rays are pure iron, ARA will have sufficient sensitivity to establish the presence or absence of the secondary UHE neutrinos produced in the GZK interaction. Such an observatory would also provide an unique probe of long baseline high energy neutrino interactions unattainable with any man-made neutrino beam.

2. Array Baseline Design

The baseline design of ARA consists of 37 antenna clusters or "stations" arranged 2 km apart on a triangular grid, as shown in Figure 1. The primary goal of the ARA array is to establish the absolute cosmogenic neutrino flux through a modest number of gold plated events. We have therefore adopted a geometry in which a single localized cluster may act as a standalone array, which trades precise angular resolution for increased coverage. Because each local cluster has the ability self-trigger, this also avoids the complications of long trigger windows and deep buffers that would otherwise arise in triggering a sparse, large scale array.

The configuration of the ARA clusters is shown in Figure 2. Each cluster will consist of four strings of receiver antennas and associated amplifiers installed in boreholes of

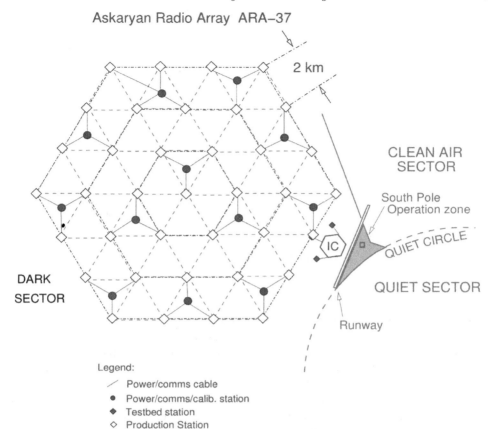

Askaryan Radio Array ARA–37

2 km

CLEAN AIR
SECTOR

South Pole
Operation zone

QUIET CIRCLE

DARK
SECTOR

QUIET SECTOR

Runway

Legend:
⟋ Power/comms cable
● Power/comms/calib. station
◆ Testbed station
◇ Production Station

Figure 1. Planned layout of the 37 ARA stations with respect to the South Pole Station and associated sectors. The footprint of IceCube, which covers a square kilometer, is shown by the blue hexagon. The geographic South Pole is marked by the magenta square in the South Pole operation zone.

approximately 6 inches (15 cm) in diameter and extending 200 m deep into the icecap. The depth was chosen to allow the instrumentation to be placed below the first 100 m of the icecap, known as the "firn" layer, which has a varying index of refraction due to the gradual compactification of snow into ice. The firn layer would reflect away much of the power and complicate the ray tracing used in event reconstruction. The drilling technology and fuel constraints limit the depth and diameter of the borehole.

Each string will contain 2 pairs of antennas, each pair consisting of a horizontally polarized antenna and a vertically polarized antenna. The antenna pairs will be separated by a distance of 10–50 m for depth discrimination. By combining information from the two antenna polarizations, the entire Cherenkov cone may be reconstructed in the data analysis, allowing for more precise vertex reconstruction and additional background rejection. Each antenna will be coupled to a low noise amplifier with low and high pass filters that select the sensitive frequency range, which extends from 100 MHz to 900 MHz. The digitization electronics is a descendent of those developed for ANITA (Gorham *et al.*, 2009, Varner *et al.*, 2007). The cluster of four strings, along with the surface trigger logic, and two additional antennas to transmit calibration pulses, form a station.

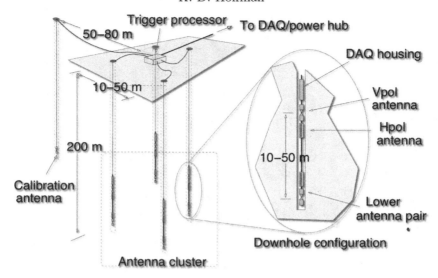

Figure 2. ARA Station layout and antenna cluster geometry.

3. Prototype Construction and Performance

One of the goals of the first season of ARA construction was to test prototypes of the hardware. This prototype hardware can, in turn, be used to probe the noise environment at a distance of one kilometer from the IceCube array. Toward this goal, a "test bed" of shallow antennas was installed during the austral summer of 2010–2011. It consists of pairs of antennas, a "batwing" for horizontal polarization and a discone for vertical polarization, installed in trenches 3 m below the surface of the snow, as well as two low frequency antennas installed near the surface. An additional four pairs of antennas were installed in 6 inch boreholes drilled 40 m into the ice. The design of the antenna pairs for the narrow boreholes, which served as prototypes for the ARA stations, were constrained by the size of the borehole, and the requirement that the power and communications cable has to pass through the center of the antenna. The deeper 40 m boreholes also served to test different drilling technologies. Because we intend to build a sparse, large scale array, any drill used must be portable, and must be capable of drilling to the desired depth in a matter of hours.

Another goal of the first season of ARA was to measure the RF properties of ice to a sufficient degree to understand the performance of the array. The RF attenuation length decreases as a function of temperature, and the temperature of the ice at the South Pole has been precisely surveyed and has been found to vary from $-50°$ C at the surface to $-10°$ C at 2,800 m depth (Price et al., 2002). In addition, the refractive index varies as a function of density, resulting in a lensing effect near near the top of the polar ice cap, where the snow is gradually compacting into ice. The RF propagation could also conceivably be effected by layering in the ice that could act as a waveguide. This "birefringence", if it is present, could also complicate the analysis of data. To investigate the RF properties of the ice, the ARA collaboration took advantage of the drilling of deep holes for the installation of the final IceCube strings to parasitically install three 5 kV broadband pulser and transmitter antenna assemblies at depths of 1,400 and 2,500 meters. The antenna was designed to wrap around the IceCube cable to eliminate uncertainty in the transmission pattern from cable shadowing effects. The siting of these transmitters will allow the RF properties of the deep ice to be probed without the

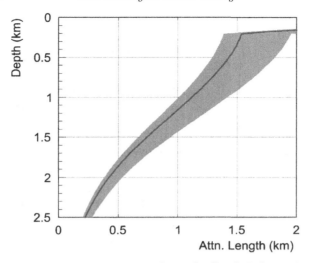

Figure 3. Ice Attenuation measurements made at the South Pole in 2011 measured from direct transmission from a source 2,500 m in depth along a 3.2 km slant depth.

uncertainties associated with the bottom reflection, and the high voltage of these pulsers will allow them to be detected from many attenuation lengths away – well beyond the extent of IceCube, at the testbed, and within range of a large fraction of the planned ARA stations. The deepest transmitter is located at a slant depth of 3.2 km from the testbed, allowing a direct measurement of ice attenuation length over an unprecedented distance. The resulting measurement is shown in Figure 3. This will serve as a valuable calibration tool for ARA.

The testbed and high voltage transmitters have been operating for nearly two years, and the full results have been published (Allison *et al.*, 2012). The noise environment has been found to be very favorable, with the spectral noise density plotted for the low frequency surface antennas shown in Figure 4. The measurement agrees with the ambient thermal noise above 150 MHz. Below 150 MHz, it is dominated by the galactic thermal noise. A modulation of the galactic noise is seen as the horizontally oriented antenna aligns and anti-aligns with the plane of the galaxy, which is inclined at $\approx 63°$ with respect to the South Pole, over a period of a sidereal day. Sources of manmade interference include weather balloon launches that employ a 400 MHz transponder for data telemetry, and a 129.3 MHz communication channel used for incoming and departing aircraft. The amount of interference decreases significantly over the winter. Overall, we expect to loose 3% lifetime due to interference.

Local calibration pulsers installed within the testbed have been used to determine the geometry of the array *in situ*. By cross correlating the transmitted reference signal in the Vpol antennas, the timing of the transmissions for an ensemble of pulses can be measured to 4.5 ps, with a standard deviation of 136 ps. This corresponds to a 1 mm error in the position of the antennas.

In the austral summer following the installation of the testbed, one full prototype ARA station was installed. It closely resembles the ARA stations as proposed, except that boreholes of only 100 m were achieved due to the ongoing development of the drill. An additional two prototypes are planned for installation during the austral summer of 2012–2013 which include upgrades to the data-acquisition system, which will be near the final design.

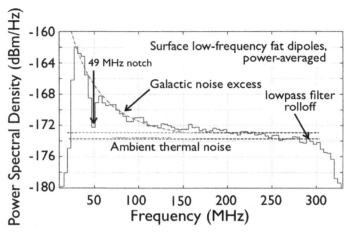

Figure 4. Spectrum of the noise density for surface dipoles measured at the ARA testbed.

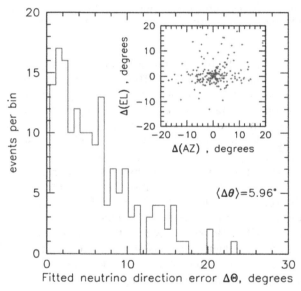

Figure 5. Distribution of fractional range errors in single-antenna-cluster reconstruction of neutrino interaction vertex. Distribution of polar angle errors for full reconstruction of the incoming neutrino direction, using vertex reconstruction, amplitude and polarization information. Inset: the 2-D distribution of reconstructed directions relative to true neutrino direction.

4. Performance of the Proposed Array

4.1. *Angular resolution*

Because the goal of the ARA array is to make the first detections of neutrinos via the Askaryan effect, and establish the neutrino flux, angular resolution was not emphasized in the design. However, good up-down discrimination is essential in vetoing surface anthropogenic noise. The simulated angular resolution of the ARA baseline array is shown in Figure 5, with a median angular resolution of 5.96°. This resolution assumes a single station detection, and is dominated by the azimuthal error. The error in the zenith angle reconstruction is about one degree.

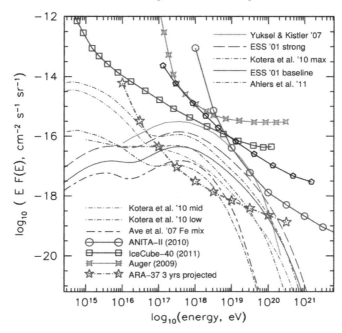

Figure 6. Compilation of sensitivity estimates from existing instruments, published limits, and a range of GZK neutrino models, along with the expected 3 year ARA sensitivity.

4.2. *Sensitivity to the neutrino flux*

The limit on the cosmogenic neutrino flux from ARA after 3 years of operation of the full 37 station array is shown by the blue stars in Figure 6. Also shown are the expected flux upper limits from several competing experiments including RICE (Kravchenko *et al.*, 2006, pentagons), ANITA (Gorham, 2010, circles), Auger (Abraham *et al.*, 2009, stars) and IceCube (Abbasi *et al.*, 2011, squares). The reason that the sensitive energy range of ARA was chosen to be lower than ANITA's was to optimize the instrument so that its peak sensitivity lies in the region where the cosmogenic neutrino flux is greatest, thereby maximizing the probability of detection. Also shown are several theoretical flux calculations based on different assumptions about the cosmic ray composition and the source evolution, including strong source evolution models (Engel *et al.*, 2001, Yuksel & Kistler, 2007), mixed iron composition models (Kotera *et al.*, 2010, Ave *et al.*, 2005), and models incorporating constraints from the Fermi diffuse gamma ray flux (Ahlers *et al.*, 2010). ARA has the greatest sensitivity of any neutrino telescope in the energy interval from $10^{17} - 10^{19.5}$ eV. ARA can either detect or exclude a cosmogenic neutrino flux even in the more pessimistic scenario of an all iron cosmic ray spectrum (Ave *et al.*, 2005). ARA is intended as a discovery instrument.

The prototyping phase for ARA is nearly complete. After funding to build the balance is secured, the array could be completed in five years. Once the flux of cosmogenic neutrinos is measured, this could inform the design of an observatory class array with precise angular resolution to study the origin and evolution of the highest energy cosmic ray accelerators.

5. Acknowledgements

We thank the National Science Foundation for their generous support through Grant NSF ANT-1002483. We are very grateful to Raytheon Polar Services Corporation and the Lockheed Martin Antarctic Support Division for their excellent field support at Amundsen-Scott Station, and the IceCube drillers and specialists who helped with our installation.

References

Abbasi, R. *et al.* (2008). *Phys. Rev. Lett.*, 100, 101101.

Abbasi, R. *et al.* (2011). *Phys. Rev.*, D83, 093003.

Abraham, J. *et al.* (2008). *Phys. Rev. Lett.*, 101, 061101.

Abraham, J. *et al.* (2009). *Phys. Rev.*, D79, 102001.

Ahlers, M., Anchordoqui, L., Gonzalez-Garcia, M., Halzen, F. & Sarkar, S. (2010). *Astropart. Phys.*, 34, 106–115.

Ahlers, M., Gonzalez-Garcia, M. & Halzen, F. (2011). *Astropart. Phys.*, 35, 87–94.

Allison, P., Auffenberg, J., Bard, R., Beatty, J., Besson, D. *et al.* (2012). *Astropart. Phys.*, 35, 457–477.

Askaryan, G. A. (1962). *JETP Lett.*, 14, 441.

Ave, M., Busca, N., Olinto, A. V., Watson, A. A. & Yamamoto, T. (2005). *Astropart. Phys.*, 23, 19–29.

Engel, R., Seckel, D. & Stanev, T. (2001). *Phys. Rev.*, D64, 093010.

Gorham, P. W. *et al.* (2009). *Astropart. Phys.*, 32, 10–41.

Gorham, P. W. (2006). *Int. J. Mod. Phys.*, A21S1, 158–162.

Gorham, P. W. (2010). *Phys. Rev.*, D82, 022004.

Greisen, K. (1966). *Phys. Rev. Lett.*, 16, 748–750.

Kotera, K., Allard, D. & Olinto, A. (2010). *JCAP*, 1010, 013.

Kravchenko, I. *et al.* (2003). *Astropart. Phys.*, 19, 15–36.

Kravchenko, I. *et al.* (2006). *Phys. Rev.*, D73, 082002.

Price, P. B., *et al.* (2002). *Proc. Nat. Acad. Sciences USA*, 99, 7844–7847.

Varner, G. S., Ruckman, L. L., Gorham, P. W., Nam, J. W., Nichol, R. J., Cao, J. & Wilcox, M. (2007). *Nucl. Instrum. Meth.*, A583, 447–460.

Yuksel, H. & Kistler, M. D. (2007). *Phys. Rev.*, D75, 083004.

Zatsepin, G. T. & Kuzmin, V. A. (1966). *JETP Lett.*, 4, 78–80.

Astrophysics from Antarctica
Proceedings IAU Symposium No. 288, 2012
M. G. Burton, X. Cui & N. F. H. Tothill, eds.

© International Astronomical Union 2013
doi:10.1017/S1743921312016766

Cometary dust in Antarctic micrometeorites

Naoya Imae

Antarctic Meteorite Research Center, National Institute of Polar Research,
10-3, Midori-cho, Tachikawa, Tokyo 190-8518, Japan
email: `imae@nipr.ac.jp`

Abstract. Cometary nuclei consist of aggregates of interstellar dust particles less than $\sim 1\,\mu$m in diameter and can produce rocky dust particles as a result of the sublimation of ice as comets enter the inner solar system. Samples of fine-grained particles known as chondritic porous interplanetary dust particles (CP-IDPs), possibly from comets, have been collected from the Earth's stratosphere. Owing to their fine-grained texture, these particles were previously thought to be condensates formed directly from interstellar gas. However, coarse-grained chondrule-like objects have recently been observed in samples from comet 81P/Wild 2. The chondrule-like objects are chemically distinct from chondrules in meteoritic chondrites, possessing higher MnO contents (0.5 wt%) in olivine and low-Ca pyroxene. In this study, we analyzed AMM samples by secondary electron microscopy and backscattered electron images for textural observations and compositional analysis. We identified thirteen AMMs with characteristics similar to those of the 81P/Wild 2 samples, and believe that recognition of these similarities necessitates reassessment of the existing models of chondrule formation.

Keywords. chondrule, chondrule-like object, micrometeorite, comet, forsterite, enstatite

1. Introduction: A Review of Cometary Dust

The internal structure of cometary nuclei has long been considered, theoretically at least, to be the result of aggregation of interstellar dust particles with diameter $< 1\,\mu$m (Greenberg & Hage 1990). Previous studies have modeled the internal structure of these particles, which are now known to have a silicate core mantled with organics and covered by ice. Their weight ratio of silicate, organics, and ice according to the solar elemental abundance is nearly 1 : 1 : 1 (Greenberg 1998). The interstellar dust particles are thought to have formed by condensation from interstellar gas, and the condensed phases have been predicted from chemical equilibrium thermodynamics using the solar elemental abundance (Anders & Grevesse 1989). These dust particles contain forsterite and enstatite as the major silicates (Davis & Richter 2003) and both water and carbon monoxide ice (Yamamoto 1983).

Dust particles from comets can reach Earth because many rocky dust particles are produced by the sublimation of ice when a comet approaches Earth. Chondritic porous interplanetary dust particles (CP-IDPs) consist of fluffy and fine-grained anhydrous Mg-rich silicates (forsterite and enstatite; Klöck *et al.* 1989; Bradley 2003) and have been assumed to originate from comets owing to their unusually high Mn contents.

Chondrules are small silicate droplets that cooled rapidly from silicate melt in the solar nebula and consist mainly of coarse-grained crystals of olivine, pyroxene and Fe-Ni metal; they are the main constituents of unequilibrated chondrites. Chondrule diameters vary between chemical groups of chondrites, with mean values of 0.15, 0.3, 1.0, 0.7, 0.2, 0.3, 0.5 and 0.6 mm for CO3, CM2, CV3, CR2, EH3, H3, L3, and LL3, respectively (Scott & Krot 2005). Fig. 1 illustrates representative examples of the different chemical groups.

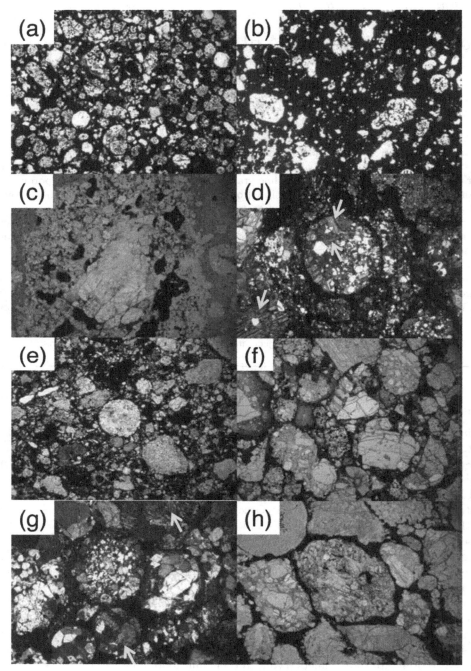

Figure 1. Photographs of unequilibrated chondrites under an optical microscope. The side of each photo is 2.3 mm. (a)–(c): Irregularly shaped chondrules are abundant. (d)–(h): Rounded chondrules are abundant. (a) Yamato-81020 CO3 chondrite. Open nicol. (b) Yamato 980051 CM2 chondrites. Open nicol. (c) Yamato-86751 CV3 chondrite. Open nicol. (d) Yamato-793495 CR2 chondrite. Polysynthetic twinning of pyroxene is observed. Cross nicol. (e) Yamato-691 EH3 chondrite. Open nicol. (f) Yamato-82038 H3 chondrite. Open nicol. (g) Yamato-74191 L3 chondrite. Polysynthetic twinning of pyroxene is observed. Cross nicol. (h) Allan Hills-764 LL3 chondrite. Open nicol.

Recently, cometary dust particles from comet 81P/Wild 2 were successfully recovered by the Stardust mission (Brownlee *et al.* 2006), and chondrule-like objects consisting of massive coarse grains were found to comprise many of the recovered particles (Nakamura *et al.* 2008). Additionally, many Antarctic micrometeorites (hereafter, AMMs) have been shown to be texturally and compositionally similar to these chondrule-like objects. Such massive coarse-grained particles have been collected only from polar regions: interplanetary dust particles collected from the stratosphere are generally $\leqslant 20 \,\mu\mathrm{m}$ in diameter and are predominantly fine grained.

In this study, we analyzed the texture and composition of AMM samples, and used a comparison between these samples and those from comet 81P/Wild 2 to verify the appropriateness of existing models of chondrule formation.

2. Methods

AMMs sampled in this study were recovered from the Tottuki icefield (28 AMMs bearing relicts among identified 99 AMMs; Iwata & Imae 2002) and the water well at Amundsen-Scott Station (68 MMs bearing relicts among identified 373 AMMs, Taylor *et al.* 1998). The diameters of collected AMMs were primarily 50–100 $\mu\mathrm{m}$, which is within the usual size range for micrometeorites reaching Earth from space, based on the size and mass distributions of extraterrestrial particles obtained from a recovered artificial satellite (Love & Brownlee 1993). This implies that AMMs represent the major cosmic components on Earth.

Candidate AMM samples were lined individually with adhesive carbon tape. Then, the samples were observed and analyzed by secondary electron microscopy with an attached energy dispersive system (SEM-EDS). The EDS detected the coexistence of magnesium, silicon and iron, suggesting a cosmic origin; this is supported by backscattered electron images (BSEs) that display a welded surface morphology due to heating during atmospheric entry. Polished sections of the AMM samples were made to allow observation of internal texture and analysis of constituent minerals (using an electron probe microanalyzer: JEOL, JXA-8200). Based on textural observations and compositional analyses, AMMs and their relict phases were determined.

3. Results and Discussion

3.1. *Size and Composition of Chondrule-like Objects in Cometary Dust*

The mean size of grains constituting chondrule-like objects in AMMs was found to be slightly lower ($\sim$10–20 $\mu\mathrm{m}$) than that of chondrules in meteorites ($\geqslant 30 \,\mu\mathrm{m}$; Nakamura *et al.* 2008), i.e., the chondrule-like objects in our study exhibit smaller diameters than chondrules in chondritic meteorites.

Compositional differences (particularly in terms of olivines and pyroxenes) have been recognized between chondrule-like objects from comets and chondrules in chondritic meteorites: MnO contents in forsterites and enstatites in the chondrule-like objects of comets are distinctly higher than those in chondritic meteorites (Fig. 2). Although the chemical formula of pure forsterite (Mg end-member of olivine) is $\mathrm{Mg_2SiO_4}$ and that of pure enstatite (Mg end-member of pyroxene) is $\mathrm{MgSiO_3}$, small amounts of $\mathrm{Fe^{2+}}$, $\mathrm{Ca^{2+}}$, $\mathrm{Mn^{2+}}$, $\mathrm{Al^{3+}}$ and $\mathrm{Cr^{3+}}$ can be substituted for $\mathrm{Mg^{2+}}$ in forsterite and enstatite to form a solid solution series. The compositions of forsterites and enstatites in 81P/Wild 2 are clearly different from those in chondrites (Figs. 2a–2d): forsterites and enstatites in cometary dust are slightly enriched in Mn ($\mathrm{MnO} \geqslant 0.5 \,\mathrm{wt\%}$) compared to chondritic meteorites

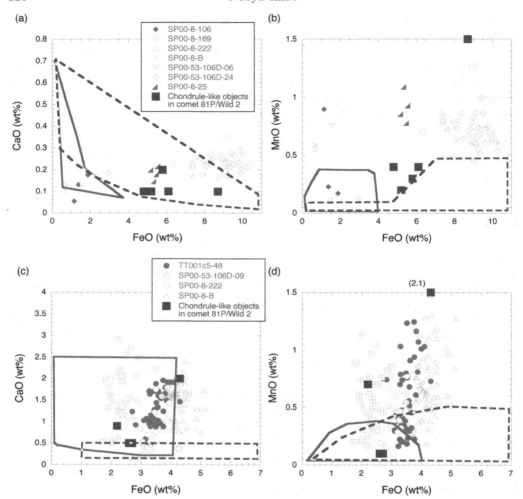

Figure 2. Compositions of Antarctic micrometeorites from comet and comet 81P/Wild 2 with the compositional range of carbonaceous chondrites (indicated as the area enclosed by the solid line) and unequilibrated ordinary chondrites (indicated as the area enclosed by a dotted line). (a) FeO-CaO plot for forsterite. (b) FeO-MnO plot for forsterite. (c) FeO-CaO plot for enstatite. (d) FeO-MnO plot for enstatite.

(MnO $\leqslant$ 0.5 wt%). The slightly ferroan forsterites and enstatites (FeO = 2–8 wt%) in chondrule-like objects of comet 81P/Wild 2 are also distinguishable from those in carbonaceous chondrites (FeO = 1–2 wt%).

Of the collected AMMs, we identified 13 consistent with chondrule-like objects in comet 81P/Wild 2; four of these are shown in Fig. 3. The chondrule-like objects are characterized by coarse-grained and poikilitic textures (Figs. 3a–3d); similar textures have been observed in the chondrule-like objects of comet 81P/Wild 2 (Nakamura *et al.* 2008). Furthermore, we have noted close similarities between high-Mn AMMs and chondrule-like objects in terms of ferroan olivine and ferroan low-Ca pyroxenes, and the detail will be described with the present Mg-rich phases elsewhere.

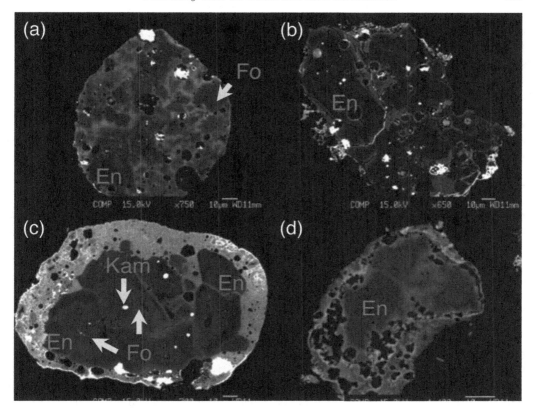

Figure 3. Back-scattered electron images of Antarctic micrometeorites from comets. (a) SP00-8-222. (b) SP00-53-106D-09. (c) SP00-8-B. (d) TT001c5-48. Fo = forsterite. En = enstatite. Kam = kamacite (Fe-Ni alloy).

3.2. *A New Constraint on Chondrule Formation*

Mn, which is slightly more volatile than magnesium and iron, is enriched in cometary dust; thus, chondrule formation in the asteroid belt would have occurred under refractory conditions, while chondrule formation in the Kuiper belt would have occurred under Mn-rich conditions.

The different compositions of precursors for chondrules could be ascribed to different regions of chondrule formation for comets and chondritic meteorites. If this was the case, chondrules would have formed in both the Kuiper belt region and the asteroidal region, which contradicts models in which chondrule formation accompanies the X-wind (Shu *et al.* 1996; Liffman 1996), although the small grain size of chondrule-like objects is as expected based on the X-wind model. Rather, the results of the present study correspond more closely to those expected based on the shock heating model (e.g., Nakamoto *et al.* 2005).

3.3. *On Temperatures of Chondrule Formation*

Clinoenstatite (or clinopyroxene) and high-temperature orthoenstatite (or orthopyroxene), which are polymorphs of $MgSiO_3$, can be distinguished based on calcium contents in enstatites: the CaO content of clinoenstatite is less than $1\,wt\%$, while that of high-temperature orthoenstatite is 1–$3\,wt\%$ (Carlson 1988; Ohi *et al.* 2008). Clinoenstatite was inverted from protoenstatite, which is stable at high temperatures ($\geqslant 1000°C$), and shows polysynthetic twinning under cross-polarized light (Figs. 1d and 1g). Clinoenstatite is common in CR2 (Fig. 1d), EH3 (Fig. 1e), H3 (Fig. 1f), L3 (Fig. 1g) and LL3 (Fig. 1h). High-temperature orthoenstatite occurs in carbonaceous chondrites and comets, but less so in unequilibrated chondrites and enstatite chondrites. Clinoenstatite formed by transition from protoenstatite at temperatures lower than $\sim$1000°C. Protoenstatite crystallizes at higher temperatures ($\leqslant$ 1553°C) than high-temperature orthoenstatite ($\sim$1370–1450°C) (Carlson, 1988). This suggests that the maximum temperature for chondrule formation for both carbonaceous chondrites and comets was lower than that for unequilibrated ordinary chondrites and enstatite chondrites. Chondrules are likely to have formed repeatedly in the solar nebula (e.g., Hewins 1996). The results of the present study suggest that the maximum temperature of repeated chondrule formation events in the Kuiper belt was lower than that for events in the asteroidal region.

The shape of chondrules can be interpreted in the context of pyroxene polymorphs. Clinopyroxenes showing polysynthetic twinning commonly occur in chondrules in CR2 (Fig. 1c), EH3 (Fig. 1e), H3 (Fig. 1f), L3 (Fig. 1g) and LL3 (Fig. 1h). The rounded shapes of chondrules in these chemical groups are clearer than those in CO3 (Fig. 1a), CM2 (Fig. 1b) and CV3 (Fig. 1d), in which orthoenstatite is common and clinopyroxene inverted from protopyroxene is unusual. This may be related to the temperature of chondrule formation: the degree of melting is greater for chondrules crystallizing protopyroxene than for those crystallizing orthopyroxene.

3.4. *Orthoenstatite, the Mn Carrier Phase*

Calcium is positively correlated with manganese in enstatite (Figs. 2c and 2d), and the CaO content of enstatite with MnO $\geqslant$ 0.5 wt% is 1–3 wt% (Fig. 2). Clinoenstatite inverted from protoenstatite is stable in the range of up to 1 wt% CaO content, but high-temperature orthoenstatite is stable in the range 1–3 wt% CaO content (Carlson 1988; Ohi *et al.* 2008). Therefore, high-temperature orthoenstatite is the carrier of the Mn component. It is likely that the coexistence of high-Mn forsterite and high-Mn orthoenstatite may be due to elemental diffusion into forsterite from melt during chondrule formation and occurs owing to the speed at which the process occurs.

4. Conclusions

In this study, we compared textural observations and compositional analyses of chondrules in AMMs with the characteristics of chondrule-like structures in comet 81P/Wild 2. Thirteen of the sampled AMMs displayed chondrule-like structures that were texturally similar to those in the 81P/Wild 2 samples, and the compositional analyses indicated striking similarities in Mn content for the two sources. Our results provide support for models proposing that chondrules could have formed in both the Kuiper belt and the asteroidal belt, and suggest that maximum temperatures of chondrule formation events were lower in the Kuiper belt. We believe that our results necessitate a thorough reconsideration of existing models of chondrule formation.

Acknowledgments

I am grateful to Susan Taylor for providing the SP00 micrometeorites and to Naoyoshi Iwata for collaboration in sampling at the Tottuki bare icefield. I am also grateful to Akira Miyake for discussions regarding enstatite polymorphs. The preparation of this paper was supported by an NIPR publication grant.

References

Bradley, J. P. 2003, in: A. M. Davis (ed), *Treatise on Geochemistry*, vol. 1 (Elsevier), pp. 689–711

Brownlee, D., *et al.* 2006, *Science* 314, 1711

Carlson, W. D. 1988, *Am. Min.* 73, 232

Greenberg, J. M. & Hage, J. 1990, *Ap. J.* 361, 260

Greenberg, J. M. 1998, *A & A*, 330, 375

Hewins, R. H. 1996, in: R. H. Hewins, R. H. Jones & E. R. D. Scott (eds), *Chondrules and the Protoplanetary Disk* (Cambridge), pp. 3–9

Iwata, N. & Imae, N. 2002, *Ant. Met. Res.* 15, 25

Klöck, W., Thomas, K. L., McKay, D. S., & Palme, H. 1989, *Nature* 339, 126

Liffman, K. & Brown, M. J. I. 1996, in: R. H. Hewins, R. H. Jones & E. R. D. Scott (eds), *Chondrules and the Protoplanetary Disk* (Cambridge), pp. 285–302

Love, S. G. & Brownlee, D. E. 1993, *Science*, 262, 550

Nakamura, T., Noguchi, T., Tsuchiyama, A., Ushikubo, T., Kita, N. T., Valley, J. W., Zolensky, M. E., Kakazu, Y., Sakamoto, K., Mashio, E., Uesugi, K., & Nakano, T. 2008, *Science* 321, 1664

Nakamoto, T., Hayashi, M. R., Kita, N. T., & Tachibana S. 2005, in: A. N. Krot, E. R. D. Scott & B. Reipurth (eds), *Chondrites and the Protoplanetary Disk*, ASP Conference Series, vol. 341, pp. 883–892

Ohi, S., Miyake, A., Shimobayashi, N., Yashimma, M., & Kitamura, M. 2008, *Am. Min.* 93, 1682

Scott, E. R. D. & Krot, A. N. 2005, in: A. N. Krot, E. R. D. Scott & B. Reipurth (eds), *Chondrites and the Protoplanetary Disk*, ASP Conference Series, vol. 341, pp. 15–53

Shu, F., Shang, H., & Lee, T. 1996, *Science* 271, 1545

Taylor, S., Lever, J. H., & Harvey, R. P. 1998, *Nature* 392, 899

Yamamoto, T. 1985, *A & A*, 142, 31

Astrophysics from Antarctica
Proceedings IAU Symposium No. 288, 2012
M. G. Burton, X. Cui & N. F. H. Tothill, eds.

© International Astronomical Union 2013
doi:10.1017/S1743921312016778

Antarctic meteorites and the origin of planetesimals and protoplanets

Akira Yamaguchi

National Institute of Polar Research 10-3 Midori-cho, Tachikawa, Tokyo 190-8518, Japan
email: `yamaguch@nipr.ac.jp`

Abstract. Almost all meteorites (about 99% by number) are samples from a few hundreds of asteroids that are leftover from planetesimals and protoplanets formed within several million years after the birth of the Solar System. These meteorites record detailed evolutionary history of dust to planets, including condensation, accretion, aqueous alteration, thermal metamorphism, partial and total melting.

Keywords. Meteorites, planetesimals, protoplanets, asteroids, metamorphism, differentiation

1. Introduction

It is widely accepted that most meteorites were derived from asteroids on the basis of evidence such as the infrared spectra of meteorites being similar to those of asteroids, observations of some meteor falls tracking back to asteroid regions, and cosmogenic ages consistent with asteroidal origin. A small fraction of meteorites have originated from the Earth's Moon and from Mars (about 0.5% by number). We have collected rocks from several hundreds of asteroids (and comets?), the Moon and Mars as meteorites, providing us with samples directly available for a variety of research. As of August 2012, the number of approved meteorites is 44,263 (Meteoritical Society, 2012). About 70% of these meteorites (30,712) were recovered from Antarctica.

Meteorites fall with equal frequency on the Earth. However, the number of recovered meteorites from Antarctica is much greater than those from other areas. On December 10th, 1969, Japanese Antarctic Research Expedition (JARE-10) discovered 9 meteorites with at least 5 distinct groups (E, H, L chondrites, a carbonaceous chondrite, and a diogenite) on blue ice fields from the Yamato Mountains (e.g., Yoshida, 2010). Their discovery implies that some kind of meteorite concentration mechanism operated on the Antarctic ice sheet. The importance of the meteorite discovery was immediately recognized, and many expeditions teams have subsequently been sent to Antarctica.

One might ask why so many meteorites are found in Antarctica? Meteorites fall on the ice, are covered by snow and frozen into the ice. Glaciers that carry meteorites flow toward mountains where the movements of glaciers are stopped, and where ice layers are then removed by ablation (abrasion and sublimination) leaving the meteorites behind. Not all blue ices bear meteorites; special conditions may be needed (Harvey, 2002). It should be noted that since early 1990's, many meteorites have been recovered from other areas, especially hot deserts (e.g., Sahara, Arabian Peninsula), but Antarctica is still the most fertile area for meteorite recovery. Detailed studies of many Antarctic meteorites expand the understanding of the origin of planetesimals and protoplanets.

2. Classification of asteroidal meteorites

Asteroidal meteorites are mainly classified into two groups, chondrites (primitive or undifferentiated meteorites) and achondrites (differentiated meteorites) (e.g., Weisberg *et al.*, 2006) (Figure 1). Chondrites are the most primitive rocks in the Solar System. They have chemical compositions broadly similar to the composition of the Sun, except for hydrogen, helium and other highly volatile elements. Chondrites are composed of four major components: chondrules, refractory inclusions (Ca-Al rich inclusions – CAI), amoeboid olivine aggregates and fine-grained matrix. These components are products of different parts in the solar nebular. Currently, there are 15 chondrites groups, including carbonaceous (CI, CM, CO, CV, CK, CR, CH, CB), ordinary (H, L, LL), enstatite (EH, EL), and R and K chondrites. Ordinary chondrites are the most common types. It is plausible that the members of each group come from a single parent body. Most chondritic meteorites have been affected to some degree by later events such as impacts and aqueous alteration.

The achondrites lack chondritic textures, and are igneous rocks or breccias of igneous rocks, or iron and stony-iron meteorites that formed by melting events in the parent bodies. The degree of melting varies from partial melting to total melting that produced a magma ocean. The latter meteorites may have been derived from protoplanets. Some meteorites have achondritic (igneous or metamorphic) textures but have similar chemical compositions to those of chondrites. These meteorites are called primitive achondrites. Achondrites consist of differentiated achondrites (angrites, aubrites, howardites-eucrites-diogenites (HEDs), mesosiderites, 3 groups of pallasites, groups of iron meteorites and primitive achondrites (ureilites, brachinites, acapulcoites, lodranites), plus many ungrouped achondrites.

3. Metamorphism in planetesimals

At the first stage, primordial components aggregated into planetesimals. Planetesimals experienced various types of secondary processing caused by internal heating. One of the most plausible heat sources is the heat caused by rapid decay of ^{26}Al (half live = 0.7 million years) that existed in the early Solar System. ^{26}Al was incorporated into planetesimals during accretion, and the planetesimals were heated by varying degrees. The degree of heating depends on several factors, including the initial abundance of ^{26}Al,

Figure 1. (a) Ordinary chondrite (LL3), Yamato-790448. The cut surface shows a typical chondritic texture. (b) Achondrite (gabbroic eucrite), Yamato-980433. This meteorite is partly covered by fa usion crust (black). Broken surfaces show an igneous texture (upper middle). The size of the black cube on the lower right is 1 cm.

the sizes of the bodies, and the timing of accretion. Impact events played an important role for the geologic history of asteroids as seen in shocked and brecciated meteorites and heavily cratered surfaces of asteroids. Impacts have been suggested for the source of early heating events. However, impact events are not effective for global heating for the sizes of planetesimals (Keil et al., 1997). Development of regolith layers by repetitive impacts on the surface, which are poor conductors of the heat, will insulate the interior from it. Large impact events would have destroyed planetesimals, causing rapid cooling of the hot interiors.

Chondrites experienced secondary processing to varying degrees. Thermal metamorphism of chondritic precursors caused recrystallization that blurred original chondritic textures and chemical homogenization of mineral such as olivine ($(Mg,Fe)_2SiO_4$) and pyroxene ($(Mg,Fe,Ca)SiO_3$). Many chondrites are primarily classified into six petrologic types (type 1–6). Types 1 and 2 are only represented by some groups of carbonaceous chondrites, with a different kind of processing – aqueous alteration. The subsequent reactions between solid and fluid on parent bodies led to the formation of secondary minerals such as carbonates and magnetite, the alteration of primary silicate minerals (olivine, pyroxene) to phyllosilicates, and the oxidation of metal and sulfide grains.

Ordinary chondrites are subdivided into four petrologic types (types 4–6) on the basis of the degree of thermal metamorphism. Peak temperatures of 750–950°C are suggested for type 6 chondrites (the most metamorphosed) and 400–600°C for those of type 3 chondrites (e.g., McSween and Patchen, 1989). Cooling rates are estimated to be a few tens of degrees per million years. Heating by ^{26}Al decay of planetesimals caused concentric zones of different metamorphic grades, i.e., high metamorphic grade in centre, and low metamorphic grades near the surface. This structure is referred to as the onion shell model. Most ordinary chondrites are breccias composed of various metamorphic grades. This implies that the parent bodies are mixtures of various metamorphic grades from various depths of the onion shell bodies, which is referred to as the rubble-pile model. The presence of type 6 (most metamorphosed) fragments derived from the centre indicates that the parent bodies are totally disrupted and reassembled. Such structures are inferred from direct observations of asteroids. Itokawa is composed of LL chondrite like material, but the density is very low ($1.9 g/cm^3$), indicating a rubble pile structure with significant pore spaces inside (Fujiwara et al., 2006).

Primitive achondrites experienced more intense heating than did chondrites. These meteorites have roughly similar bulk chemical compositions to those of chondritic precursors, but experienced some degrees of partial melting of silicate and FeNi-FeS. They were heated at around 1000–1100°C and experienced partial melting. Primitive achondrites with the lowest metamorphic grades in some cases are remnants of chondrules, whereas the highest ones are completely recrystallized.

4. HED meteorites and other asteroidal igneous rocks

Howardites, eucrites, diogenites meteorites (HEDs) are a suite of achondrites that are genetically related, and are among the largest group of achondrites, and probably derived from a large asteroid Vesta (530 km in diameter). Eucrites are basalts or gabbros and diogenites are orthopyroxenites and harzburgites (pyroxene-olivine rocks). Howardites are mechanical mixtures mainly composed of eucrites and diogenites.

Several lines of mineralogical, geochemical and isotopic evidence suggest that the differentiation of Vesta was triggered by a magma ocean in its early history (e.g., Hewins and Newsom, 1988; Greenwood et al., 2005; Takeda, 1997). In the magma ocean model, eucrites were residual liquids after extensive fractional crystallization, and diogenites were cumulate rocks accumulated beneath the magma ocean (e.g., Righter and Drake, 1997; Takeda, 1997). A metallic FeNi core may have formed before the silicate fractionation (Hewins and Newsom, 1988).

However, the increasing diversity of diogenites revealed by recent finds has challenged this view. First, the minor and trace element abundances indicate that diogenites crystallized from multiple magmatic bodies (e.g., Barrat *et al.*, 2008, 2010 and references therein). Some diogenites have geochemical evidence for remelting of some cumulate lithologies crystallized in a magma ocean and interactions of the outer eucritic crust. Also, several diogenites probably crystallized near the surface arguing against a deep crustal origin (Yamaguchi *et al.*, 2010). Thus, it can be inferred that parental melts of some diogenites intruded the eucritic crust (post-magma ocean volcanism). This view is broadly consistent with results from the recent DAWN mission (Russell, *et al.*, 2012).

To date, five anomalous basaltic asteroidal basalts from distinct asteroids have been identified, HED meteorites (Vestan crustal rocks, see above), NWA 011 (and paired meteorites), Dho 700, Ibitira (e.g., Yamaguchi *et al.* 2002; Scott *et al.* 2010; Greenwood *et al.*, 2012). These rocks are petrologically very similar to eucrites, but from a distinct origin mainly on the basis of oxygen isotopic compositions. Presumably, parent bodies of these meteorites experienced similar igneous, metamorphic, and impact histories as did asteroid 4 Vesta. The presence of andesitic meteorites (GRV06128 and 06129) implies a different style of early melting and fractionation (e.g., Day *et al.*, 2009)

Iron meteorites provide more evidence for the presence of differentiated asteroids. Magmatic iron meteorites are believed to have been cores of differentiated planetesimals or protoplanets. Wasson *et al.* (2012) argued that at least 26 extensively differentiated asteroids were disrupted in the inner asteroid belt on the basis of iron meteorite data. Yang *et al.* (2007) suggested that some iron meteorites originated from cores of protoplanets that formed several million years after the birth of Solar System. Thus, numerous differentiated bodies similar to Vesta might have existed in the early Solar System.

5. Summary

Almost all meteorites were derived from asteroids (and comets?) that are remnants from the building blocks of the Solar System. These asteroidal meteorites are very old, formed less than a few tens million years after the formation of Solar System, allowing us to understand the early geologic processes that took place in planetesimals and protoplanets. In comparison to other extraterrestrial samples available for direct studies, meteorites are robots with respect to their sizes and masses. Among the recovery sites, Antarctica is the most fertile area for meteorite recovery. Continued efforts in meteorite recovery from Antarctica, as well as from other areas such as hot deserts, and space missions will help us to understand the origin of the Solar System.

References

Barrat, J. A., Yamaguchi, A., Greenwood, R. C., Beniot, M., Cotten, J., Bohn, M., & Franchi, I. A. 1995, *Meteor. Planet. Sci.*, 30, 490

Barrat, J. A., Yamaguchi, A., Zanda, B., Bollinger, C., & Bohn, M. 1993, *Geochim. Cosmochim. Acta*, 74, 6218

Day, J. M., Ash, R. D., Liu, Y., Bellucci, J. J., Rumble III, D., McDonough, W., Walker, R., & Taylor, L. A. 2009, *Nature*, 457, 179

Fujiwara A. *et al.* 2006, *Science*, 312, 1330

Greenwood, R. C., Franchi, I. A., Jambon, A., & Buchanan, P. C. 2006, *Nature*, 435, 916

Greenwood, R. C., Barrat, J. A., Scott, E. R. D., Janots, E., Franchi, I. A., Hoffman, B., Yamaguchi, A., & Gibson, J. M. 2012, *Lunar Planet. Sci.*, 43, 2771

Harvey, R. 2003, *Chem. Erde*, 63, 93

Hewins, R. H. & Newsom, H. E. 1998, *Meteorite and Early Solar System, ed. by Kerrige J.F. & Matthews, M.S.*, 631, 976

Keil K., Stoffler, D., Love, S. G., & Scott, E. R. D. 1997, *Meteor. Planet. Sci.*, 32, 349

McSween, H.Y., Jr., & Patchen, A.D., 1989 *Meteoritics*, 24, 219

Righter, K. & Drake, M. J. 1997, *Meteor. Planet. Sci.*, 32, 929

Russell, C. T., *et al.* 2012, *Science*, 336, 684

Scott E. R. D., Greenwood, R. C., Franchi, I. A., & Sanders, I. S. 2009, *Geochim. Cosmochim. Acta*, 73, 5835

Takeda H. 1997, *Meteor. Planet. Sci.*, 32, 841

Wasson, J. T. 2012, *Lunar Planet. Sci.*, 43, 2931

Weisberg, M. K., McCoy, T. J., & Krot, A. N. 2012, *Meteorite and Early Solar System II. ed. Lauretta, D.S., & McSween Jr., H.Y.,*, 43, 2931

Yamaguchi A., *et al.* 2002, *Science*, 296, 334

Yamaguchi, A., Barrat, J. A., Ito, M., & Bohn, M. 2011 . 2011, *J. Geophys. Res.*, 116, 334, E08009, doi:10.1029/2010JE003753

Yang, J., Goldstein, J. I., & Scott, E. R. D. 2007, *Nature*, 446, 888

Yoshida, M. 2010, *Polar Science*, 3, 272

Astrophysics from Antarctica
Proceedings IAU Symposium No. 288, 2012
M. G. Burton, X. Cui & N. F. H. Tothill, eds.

© International Astronomical Union 2013
doi:10.1017/S174392131201678X

THz Observations of the Cool Neutral Medium

John M. Dickey

University of Tasmania, School of Maths and Physics
Private Bag 37, Hobart, TAS 7001 Australia
email: john.dickey@utas.edu.au

Abstract. The astrophysical drivers for far-infrared spectroscopy of the Galactic interstellar medium using a 15m class telescope on Dome A are compelling. For the diffuse, atomic phase, the most important lines in the far-IR spectrum are OI at 63μm and CII at 158μm. These are the dominant cooling lines of the cool, neutral medium, and they show rich spectral structure in Herschel observations at low latitudes. But theory predicts that they should both be highly sub-thermal in excitation, so that the level populations are not in equilibrium with the kinetic temperature of the gas. A large single dish telescope or an interferometer may be able to study the absorption and emission to determine the optical depth and column density of atoms and the physical conditions in the emission regions. Comparison of Herschel CII spectra with 21-cm absorption spectra indicates that a significant fraction of the 158μm flux may be coming from the atomic rather than the molecular phase.

Keywords. line: profiles, ISM: clouds, ISM: atoms, ISM: structure, Galaxy: disk, infrared: ISM

1. Introduction

Unlike oxygen and nitrogen, the ionization potential of carbon is slightly less than that of hydrogen, 11.26 vs. 13.59 eV. These 2 eV make a huge difference to the energy balance of the interstellar medium (ISM), because photons that can ionize carbon fly freely through most of the volume of the Galactic disk, and are only blocked by the high column densities of dense, cool clouds. So C^+ is the dominant ionization state of carbon in the diffuse phases of the ISM, including most cool, neutral medium clouds (see Draine 2011 §31.7 and fig. 31.2). Observing the boundary between CII and CI is an effective way to trace the interface between the atomic and molecular phases of the medium. In a classical photo-dissociation region (PDR, Hollenback & Tielens, 1997), the neutral, atomic carbon is in a thin layer somewhat deeper inside the molecular cloud than the photodissociation front where hydrogen goes from atomic to molecular. But in the general ISM, different conditions may cause these two layers to reverse their order. This is important to find out, because CII is such an efficient coolant for the atomic medium.

2. The Fine Structure Lines

Studying the aggregate spectrum of the interstellar radiation field in the Milky Way and other spiral galaxies shows that a significant fraction of the total luminosity (0.1 to 1% of the far-IR continuum flux) comes out in the 158μm line (Stacey *et al.* 2010). Studies of cooling processes in the diffuse molecular and atomic phases of the ISM show that this line carries away most of the kinetic energy of the medium, since it is easily excited at kinetic temperatures of a few tens to hundreds of K (Wolfire *et al.* 1995, Draine 2011 §30.3 and Figure 30.1). The 63μm line of OI is the second most important coolant,

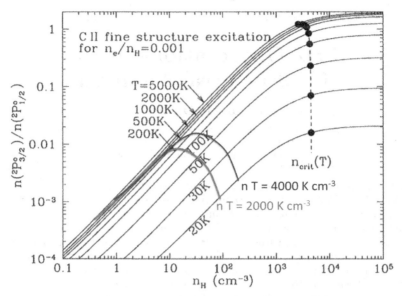

Figure 1. Level populations for the 158µm transition from Draine (2011, fig. 17.4). A typical range of kinetic pressures in the CNM and WNM are plotted. The implication is that the line will have very sub-thermal excitation, so that there may be a large reservoir of C^+ in the ground state, untraced by emission in the line.

and between them these two lines account for the most of the energy flow through the atomic ISM, at temperatures less than 10^4 K. So mapping the emission in these lines from the atomic medium is an important astrophysical diagnostic of conditions in the warm, neutral atomic medium (WNM) and the cool, neutral atomic medium (CNM), even though the total luminosity of the Galaxy in the lines may be dominated by HII regions and their nearby PDRs. In particular, the ratio of the strengths of the two lines is a sensitive diagnostic of the transition between the WNM and CNM, which is hard to trace by other means.

An important consideration for the far-IR fine-structure lines is their excitation temperature, i.e. the ratio of the populations of the upper and lower quantum levels expressed as a temperature through the Boltzmann equation. Figure 1 (taken from Draine 2011 fig. 17.4) shows the ratio of the level populations for the 158µm line as a function of density, with typical kinetic pressures for the CNM and WNM superposed. The density is two to three orders of magnitude below the critical density, n_{crit}, so the vast majority of the atoms are in the lower level. This means that observing the line in emission misses a huge reservoir of C^+ atoms that stay in their ground state. The 63µm OI line is similarly sub-thermally excited. Thus collisional deexcitation is very rare, and almost every collisional excitation of these two lines results in emission of a photon that carries energy away. But the long column densities of atoms in the ground state suggest that the optical depth of the line could be significant. Developing a way to measure the absorption in the line would be helpful. This technical challenge is a driver for future THz telescope design and construction. The Herschel and SOFIA observatories can begin the search for absorption in the 158µm line, but ultimately a large single dish telescope or an interferometer at the Antarctic Dome A site may be needed.

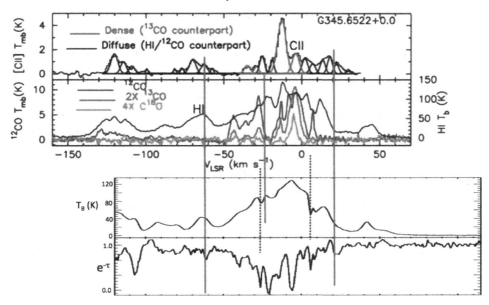

Figure 2. Spectra from Langer *et al.* (2010) showing the 158μm line (top panel) with comparison emission spectra in the 21-cm line (HI, and three CO isotopic species with different optical depths (^{12}CO, ^{13}CO, and C^{18}O) in the second panel. For comparison, the lower two panels show the 21-cm line emission and absorption toward a nearby continuum source, from Strasser (2006). Many of the CII line components show 21-cm absorption but no CO emission at all.

3. Molecular and Atomic Clouds

One of the most detailed studies of the 158μm line in the Galactic plane is that of Langer *et al.* (2010) made using the Herschel telescope. A typical line of sight, at (ℓ, b)=(345.65,0.0), is shown in Figure 2 (based on Langer *et al.*, 2010 fig. 3). Super-posed on Figure 2, with the same velocity scales, are HI spectra from Strasser (2006). These show the 21-cm line of atomic hydrogen in emission (third panel) and absorp-tion in the nearby direction of a bright continuum background source at (ℓ, b)=(345.4,+0.2), taken from the Southern Galactic Plane Survey (McClure-Griffiths *et al.* 2005). As indicated by the vertical solid lines, there is good correspondence between the velocities of the molecule-free CII lines and the 21-cm absorption lines. Absorption in the 21-cm line occurs only in the CNM; the WNM contributes to the emission but not significantly to the optical depth (Dickey *et al.* 2003). Overall, the correspondence of the feature strengths in the CII spectrum with the 21-cm absorption is at least as good as with the ^{12}CO emission, and better than with the ^{13}CO and C^{18}O features.

Two interesting features of the 21-cm absorption spectrum are the deep but very narrow lines at $v_{LSR} = -27$ and $+8$ km s^{-1}. These components show corresponding dips in the 21-cm emission spectrum, indicating that the clouds are cold, so that they absorb the emission from gas at the same velocity in the background. Such features are common at low latitudes, and are often called HI self-absorption (HISA). HISA clouds may contain a significant fraction of the HI mass of the ISM (Gibson *et al.* 2005). They can be nearly invisible in 21-cm emission line surveys, but they appear clearly in absorption toward background continuum sources. Studying such clouds is one of the main goals of the GASKAP survey, planned for the Australian Square Kilometre Array Prototype telescope (Dickey *et al.* 2013).

4. Astronomy from Kunlun Station

The valiant work done by Chinese Antarctic explorers to develop and test the astronomical potential of the Dome A site is the foundation for several future telescopes that will operate at the Kunlun Observatory. Among the planned telescopes is the 5m diameter DATE5 THz/far-IR telescope. At $\lambda 158\mu$m the beam size will be about $7''$. This small beam will allow on-minus-off observations toward bright continuum sources that can resolve away the confusion due to spatial variations in the emission, leading to very precise absorption spectra. If the 158μm line is optically thick in one or more phases of the ISM, the Kunlun Observatory may be the first to map the optical depth of the transition, throughout the inner Galaxy. This would be a breakthrough in ISM astrophysics. In addition to the possibility of detecting absorption at $\lambda\lambda 158$ and 63μm, the ratio of the emission brightness in these two lines will be a powerful diagnostic for distinguishing the different phases in the atomic and diffuse molecular ISM. Although the atmosphere will be a severe limitation on observing at these wavelengths, for some fraction of the time the lines may be accessible from Dome A, unlike any other site on the surface of the Earth. It is therefore well worth the attempt to study them from Kunlun Observatory.

References

Dickey, J. M., McClure-Griffiths, N., Gaensler, B., & Green, A. 2003, *ApJ.*, 585, 801.

Dickey, J. M., McClure-Griffiths, N., Gibson, S. J., Gomez, J. F., Imai, H., *et al.* 2012, *Pub. Astr. Soc. Aust.* in press, arXiv 1207.0891

Draine, B. T. 2011, Physics of the Interstellar and Intergalactic Medium, (Princeton: Princeton University Press).

Gibson, S. J., Taylor, A. R., Higgs, L. A., Brunt, C. M., & Dewdney, P. E. 2005, *ApJ.*, 626, 195.

Hollenbach, D. J. & Tielens, A. G. G. M. 1997, *Ann. Rev. Astron. Astrophys.*, 35, 179.

Langer, W. D., Velusamy, T., Pineda, J. L., Goldsmith, P. F., Li, D., & Yorke, H. W. 2010, *Astron. Astrophys.*, 521, L17.

McClure-Griffiths, N. M., Dickey, J. M., Gaensler, B. M., Green, A. J., Haverkorn, M., & Strasser, S. 2005, *ApJ. Supp.*, 158, 178.

Stacey, G. J., Hailey-Dunsheath, S., Ferkinhoff, C., Nikola, T., Parshley, S. C., Benford, D. J., Staguhn, J. G., & Fiolet, N. 2010, *ApJ.*, 724, 957.

Strasser, S. T. 2006, Ph. D. Thesis, University of Minnesota.

Wolfire, M. G., Hollenbach, D., McKee, C. F., Tielens, A. G. G. M., & Bakes, E. L. O. 1995, *ApJ.* 443, 152.

Astrophysics from Antarctica
Proceedings IAU Symposium No. 288, 2012
M. G. Burton, X. Cui & N. F. H. Tothill, eds.

© International Astronomical Union 2013
doi:10.1017/S1743921312016791

The Exploration of the ISM from Antarctica

Mark G. Wolfire

Astronomy Department
University of Maryland
College Park, MD 20742
email: mwolfire@astro.umd.edu

Abstract. Antarctica presents a unique environment for the exploration of the interstellar medium. The low column of water vapor opens windows for sub-mm and THz astronomy from ground and sub-orbital observatories while the stable atmosphere holds promise for THz interferometry. Various current and potentially future facilities occupy a niche not available to current space or stratospheric instruments. These allow line and continuum observations addressing key questions in e.g., star formation, galactic evolution, and the life-cycle of interstellar clouds. This review presents scientific questions that can be addressed by the suite of current and future Antarctic observatories.

Keywords. ISM: general, ISM: molecules, infrared: ISM

1. Introduction

One of the great advantages of observing from Antarctica is that it is so dry. This is extremely important for sub-mm and THz observations. At Dome A, good conditions can reach $\sim 25\%$ transmission for the [C II] fine-structure transition at 158 μm and $\sim 80\%$ at the [C I] fine-structure transition at 310 μm (Lawrence 2004). High altitude balloon flights over Antarctica from McMurdo Station can see essentially the entire THz submillimeter range with little intervening opacity. Many diagnostic emission lines from a range of interstellar medium environments are accessible in the observing windows opened up by the dry conditions. For example, there are [O I], [C II], and [C I] fine-structure lines. These are collisionally excited in warm neutral gas and come from photodissociation regions or "PDRs". There are also high-J rotational transitions (e.g., CO 7-6, CO 13-12, CO 29-28) that also arise in PDRs and in shocks. There are ionized gas lines, such as [N II] that comes from compact and diffuse H II regions. Light hydrides (e.g., H_3O^+, SH) might be seen in emission or absorption and come from PDRs, and there is also the dust continuum emission.

What science questions can be addressed with observations from Antarctica? For example:

- What is the energy budget of the Galactic interstellar medium (ISM) gas as a function of position and scale height? How does radiative heating compare to turbulent heating?

- What is the distribution of "phases" and thermal pressure in the ISM. How do these compare to global models of star formation and the ISM.

- What is the distribution and physical properties of ionized gas in the Galaxy?

- What is the carbon budget? Where is carbon mainly C^+, C and CO?

- What is the life-cycle of clouds from formation to photoevaporation?

- What are the physical conditions (density and temperature) and far-ultraviolet radiation fields across giant molecular clouds (GMCs) ?

- What is the distribution and mass of "dark molecular gas" in the Galaxy and LMC/SMC?

- What are the important channels for production of molecules in diffuse gas?

A common requirement running through these sample science projects is good atmospheric transmission in the THz and sub-mm bands. Several require large scale survey mapping which can best be done from Antarctica.

2. Photodissociation Regions

Many of the THz and sub-mm emission lines arise in PDRs and we next consider in more detail the structure of PDRs. A working definition is that a photodissociation region is a gas phase in which far-ultraviolet (FUV; $6\,\mathrm{eV} \leqslant h\nu \leqslant 13.6\,\mathrm{eV}$) radiation plays a role in the heating or chemistry. This range of energies is responsible for dissociating molecules and also dominates the heating process. PDRs are hydrogen neutral regions and lie outside of H II regions. Historically, the radiation field has been measured in units of the integrated FUV interstellar radiation field. G_0 has become a standard notation for a Habing interstellar field and χ generally indicates a Draine field and they differ by a factor 1.7. Fields that are typically encountered range from $G_0 = 1$ for the ISRF to $G_0 = 10^5$ for the Orion Trapezium PDR.

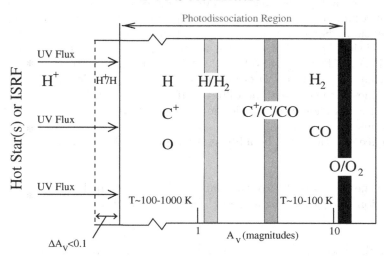

Figure 1. Schematic of PDR as a function of A_V into a cloud. Note that the peak O_2 abundance is now known to occur closer to the cloud surface (Hollenbach *et al.* 2009).

PDRs are found where OB stars shine on molecular clouds. The FUV that escapes the H II region produces a PDR on the cloud surface that can extend to an $A_V \sim 8$ though the gas. The mean column density of Galactic molecular clouds is also $A_V \sim 8$ and thus much of the mass in molecular gas is in PDRs. Another place PDRs are found is in the diffuse medium. The same PDR physics that works in the molecular cloud surfaces is also at work in diffuse clouds but with generally lower FUV field and lower column densities.

In addition, PDRs may contribute to emission in protostellar outflows. van Kempen *et al.* (2010) suggest that PDRs can contribute to the high-J CO emission in the protostellar walls of HH46. Gorti & Hollenbach (2009) find that in the outer parts of protostellar disks, PDRs can contribute to the heating, chemistry, and structure.

A schematic of a PDR as a function A_V is shown in Figure 1. In the outer portions of the cloud the gas is mainly neutral atomic hydrogen, helium and oxygen, and single ionization states of metals (e.g., C^+, Si^+, Mg^+, Fe^+). The dominant heating process is photoelectric heating from grains. Most of the FUV photon energy is absorbed by grains and radiated as IR continuum. As much as $\sim 5\%$ of the photon energy can be converted to gas heating via the grain photoelectric effect. The heating is balanced by [C II] 158 μm, and [O I] (63 μm, 145 μm) fine-structure line emission which can be observed from Antarctica. The balance between heating and cooling typically gives temperatures between 100 and 1000 K, but both higher and lower temperatures are possible.

Deeper into the cloud H_2 starts to form on grain surfaces and C^+ recombines to neutral carbon. The depth of the transition from the cloud surface depends on the ratio of G_0/n (Wolfire *et al.* 2010). Photoelectric heating dominates balanced by [C II], [O I], and [C I] (610 μm, 370 μm) line cooling. Additional diagnostics include the rotational and vibrational excitation of molecular hydrogen. We have also have light hydrides forming, many observable from Antarctica including H_3O^+, and SH (Gerin *et al.* 2010, Neufeld *et al.* 2012).

After C^+ recombines to C, CO starts to form and high-J CO line emission is produced followed by low-J CO line emission. Deep into the cloud the heating could be photoelectric heating, but can also be dominated by cosmic rays or coupling with warm dust. The cooling is from molecules with typical temperatures between 10 and 100 K. We next consider a few of the scientific questions that were raised in the introduction.

3. Galactic ISM Energy Budget

COBE FIRAS found a total [C II] all sky luminosity of $10^{7.7}$ $L_\odot$ (Wright *et al.* 1991), a luminosity that is ~ 4 times brighter than the combined [N II] (122 μm, 205 μm) lines and ~ 100 times brighter than the [C I] (370 μm, 610 μm) lines. We still do not know entirely which phase produces the [C II] emission which is important to know how to apportion the luminosity among the various gas phases. There have been a number of suggestions over the years and it must surely depend on where you look in the Galaxy. Heiles (1994) suggested that it might be the diffuse ionized gas (WIM) that also produces the [N II] lines. Bennett *et al.* (1994), suggested that it might be the cold neutral atomic clouds (CNM). Or it could be mainly associated wth PDRs on surfaces of GMCs (Stacey *et al.* 1985, Shibai *et al.* 1991, Cubick *et al.* 2008). In regions where the [C II] line dominates the gas cooling (CNM and $n < 3 \times 10^3$ cm^{-3} GMC surfaces) the total cooling rate measured by [C II] provides the total energy input into the gas.

Another reason to sort out the [C II] emission components is for extragalactic observations. For example in a [C II] map of NGC 3521 using Herschel PACS the 10″ beam subtends a size of about 500 pc (Kennicutt *et al.* 2011). Each beam contains a mix of components and it is difficult to untangle the [C II] emission. One approach is to use the Galaxy as a template to sort out the contribution from various emitting components.

Another problem for the energy budget comes from the extragalactic [O I] 63 μm maps (e.g., NGC 3521; Kennicutt *et al.* 2011). Where there are bright H II regions the [O I] is probably radiatively heated and comes from PDRs on GMC surfaces. Emission is seen,

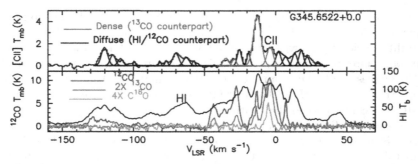

Figure 2. [C II] spectra obtained with Herschel/HIFI for the GOTC+ program at $l = 345.65°$ and $b = 0°$, along with the CO data from the Mopra telescope and H I surveys (Langer *et al.* 2010).

however, between the arms in 250 pc beams. The emission could be unresolved PDRs, but there are suggestions that the [O I] comes from diffuse gas that is mechanically heated in shocks or turbulence (Beirão *et al.* 2012). Is the wide spread [O I], radiatively or mechanically heated? One way to check is to look for [O I] emission in the Galactic diffuse ISM from Antarctica.

How would one carry out these sorts of studies? First consider a plot of thermal pressure versus density in thermal equilibrium. Along a line of constant pressure there is a stable phase of temperature ~ 8000 K (the WNM) and a stable phase of temperature ~ 100 K (the CNM) and in between there is a thermally unstable phase (Wolfire *et al.* 2003). The phase diagram along with heating and cooling rates can be used to estimate the cooling rate in the cold gas. The [C II] line cooling in the cold gas (CNM) ($\sim 4 \times 10^{-26}$ erg cm^{-2} s^{-1} H^{-1}) is more than a factor of 10 higher than in the warm gas (WNM). Even gas that is thermally unstable does not produce much [C II]. This means the [C II] isolates the cold clouds in emission which were previously seen only in absorption from H I absorption or FUV absorption spectroscopy.

Next one needs a large scale, velocity resolved (< 1 km s^{-1}) map in [C II] of the Galactic plane. There is a proposed Antarctic balloon project GUSSTO to do just that. GUSSTO can look along a line of sight and separate the various [C II] emitting components. The GOTC+ team has started to do this using Herschel HIFI. GOTC+ is an Open Time Herschel Key Project to observe [CII] along individual lines of sight. Figure 2 shows a GOTC+ spectrum with [C II] in the top panel, and H I and CO in the lower panel. Individual [C II] emitting components are seen. Components which show H I but no [C II] are probably the WNM. Velocities with H I and [C II] but no CO are probably CNM components. Where there is [C II], H I and CO, there is probably a molecular cloud. But note these are single lines of sight and without mapping it is hard to know if the line of sight passes a diffuse cloud, or the edge of a molecular cloud, which could be C$^+$ and H I, or C$^+$ and H$_2$, or if CO is observed the line-of-sight might be skimming along the edge of the molecular portion. With [C II] and resolved maps in H I and CO, it is possible to make a 3D decomposition of the [C II] emitting components and determine where the [C II] luminosity is coming from.

4. Ionised Gas

The COBE FIRAS maps show that the [N II] lines are quite bright being second to that of [C II]. If both the 205 μm and 122 μm lines can be obtained, then they can be used as indicators of the electron density. Figure 3 shows that the line ratio is sensitive to densities between ~ 10 and 10^4 cm^{-3} which are high diffuse cloud densities to classic

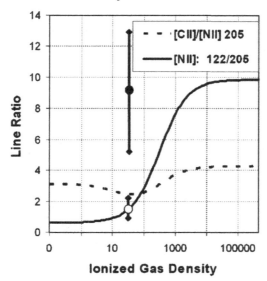

Figure 3. Ratio of [C II]/[N II] 205 μm and [N II] 122 μm/[N II] 205 μm line intensities versus density (Oberst *et al.* 2006).

H II regions. The [N II] lines can then map out the ionized gas distribution and thermal pressure in the Galaxy. Figure 3 also shows the [C II]/[N II] ratio provides an estimate of the fraction of [C II] that arises in the ionized phase. Observations can also zoom in on individual H II regions and map out the electron density and ionizing photon distribution as was done for the Carina Nebula by Oberst *et al.* (2011) using AST/RO.

5. The Dark Molecular Gas

Both observations (Smith & Madden 1997) and models (van Dishoeck & Black 1988) of cloud surfaces usually find a layer in which H turns to H_2 before C turns to CO. Using the CO alone to estimate the molecular mass misses the mass contained in the H_2 layer. This has been called the "dark molecular gas" although it is not really dark at all and emits in [C II], [C I] and dust continuum – all of which are observable from Antarctica.

Wolfire *et al.* (2010) presented PDR models of the dark molecular gas in which they 1) added a turbulent density distribution, 2) applied global GMC properties, and 3) calculated the cloud masses. These are 1D models but used the median density expected from turbulence. The median of the log-normal density distribution is given by $\langle n \rangle_{\mathrm{med}} = \bar{n}\exp(\mu)$ with $\mu = 0.5\ln(1+0.25\mathcal{M}^2)$ (Padoan *et al.* 1997). The volume averaged density distribution ($\bar{n} \propto 1/r$) is the first global GMC property. The sound speed that enters in the Mach number, $\mathcal{M}$ is calculated from the PDR model output while the turbulent velocity is given by the GMC size-linewidth relation $v \propto R^{0.5}$.

Figure 4 shows the model abundances as a function of A_V into the cloud. The dashed line marked $A_V(R_{H2})$ is where gas is half molecular, and $A_V(CO)$ is where the optical depth in CO (1-0) is equal to one. The layer between the dashed lines is the dark molecular gas (which emits in [C II], [C I] and IR continuum).

If $M(R_{H_2})$ is the molecular mass within the radius where the molecular fraction is 0.5, and $M(CO)$ is the mass within the radius of the $\tau_{CO} = 1$ surface, then the dark gas fraction can be defined as $f_{DG} = [M(R_{H_2}) - M(R_{CO})]/M(R_{H_2})]$. Figure 5 shows the calculated f_{DG} for several GMC mass and FUV radiation fields. We find a rather constant fraction of about $\sim 30\%$. This value agrees reasonably well with the EGRET

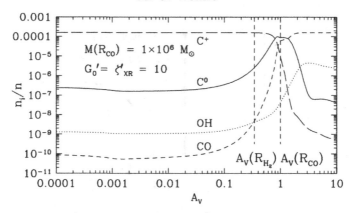

Figure 4. Abundances of C^+ (long-dashed curve), C^0 (solid curve), OH (dotted curve) and CO (short-dashed curve) as functions of optical depth into the cloud for $M = 1 \times 10^6 \, M_\odot$, Galactic dust and metals, and incident radiation field $G'_0 = \chi = 10$. Note that we do not include freeze out of H_2O on grain surfaces, which would affect OH abundances at $A_V > 3$ (Hollenbach et al. 2009).

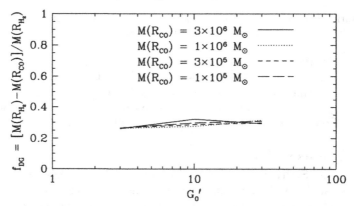

Figure 5. Dark-gas fraction f_{DG} versus incident radiation field normalized to the local interstellar field, $G'_0 = \chi = G_0/1.7$.

gamma ray observations (Grenier et al. 2005), but slightly lower than the Fermi gamma ray observations (Abdo et al. 2010) and those based on Planck IR continuum observations (Planck Collaboration et al. 2011). A simple expression for the dark gas fraction is given by

$$f_{DG} = 1 - \exp\left(\frac{-4.0\Delta A_V}{\bar{A}_V}\right), \qquad (5.1)$$

where ΔA_V is the optical depth in the dark gas layer and $\bar{A}_V = 5.26 Z \bar{N}_{22}$ is the mean optical depth through the CO portion. Here, Z is dust abundance relative to the local Galactic conditions, and $\bar{N}_{22}$ is the column density in units of 10^{22} cm^{-2} through the CO portion. The higher f_{DG} values determined by Fermi and Planck may be due to lower cloud columns sampled by these surveys.

6. Terahertz Interferometry

An intriguing prospect is to conduct Terahertz Interferometry from Antarctica made possible in part due to the longer coherent times that can be achieved. One project might be to revisit the Herschel PACS [C II] and [O I] extragalactic maps with greater spatial resolution to separate the emitting components. The PACS observations are $\sim 10''$ and with ~ 20 times greater resolution at 11 Mpc, the GMCs will start to be resolved in [C II] emission. Another project might be to observe the protostellar outflows as modeled in van Kempen *et al.* (2010) and Visser *et al.* (2012). The rotational line emission of CO is observed to very high-J (~ 33) levels covering the mm through THz bands. The models suggest that the low-J mm lines arise in a passively heated dust envelope, the mid-J sub-mm lines arise in a PDR in the outflow walls, while the high-J THz lines arise in C-shocks. Thus the CO emitting volume and kinematics vary with wavelength and it might be possible to test the theoretical predictions.

7. Conclusions

Three short conclusions are 1) good atmospheric transmission at Antarctica allows for studies in many different sub-mm and THz lines and continua probing a wide range in ISM studies. 2) Antarctica allows for large area survey or mapping modes that are not available elsewhere. 3) THz interferometry, if possible, opens unique opportunities for science exploration from Antarctica.

Acknowledgements

M.G.W. acknowledges partial support from NASA award number NNX10AM78G "The Stratospheric Terahertz Observatory".

References

Abdo, A. A., *et al.* 2010, *ApJ*, 710, 133
Beirão, P., Armus, L., Helou, G., *et al.* 2012, *ApJ*, 751, 144
Bennett, C. L., Fixsen, D. J., Hinshaw, G., *et al.* 1994, *ApJ*, 434, 587
Cubick, M., Stutzki, J., Ossenkopf, V., Kramer, C., & Röllig, M. 2008, *A&A*, 488, 623
Gerin, M. *et al.* 2010, *A&A*, 518, L110
Gorti, U. & Hollenbach, D. 2009, *ApJ*, 690, 1539
Grenier, I. A., Casandjian, J.-M., & Terrier, R. 2005, *Science*, 307, 1292
Heiles, C. 1994, *ApJ*, 436, 720
Hollenbach, D., Kaufman, M. J., Bergin, E. A., & Melnick, G. J. 2009, *ApJ*, 690, 1497
Kennicutt, R. C., Calzetti, D., Aniano, G., *et al.* 2011, *PASP*, 123, 1347
Langer, W. D., Velusamy, T., Pineda, J. L., *et al.* 2010, *A&A*, 521, L17
Lawrence, J. S. 2004, *PASP*, 116, 482
Neufeld, D. A., Falgarone, E., Gerin, M., *et al.* 2012, *A&A*, 542, L6
Oberst, T. E., Parshley, S. C., Stacey, G. J., *et al.* 2006, *ApJL*, 652, L125
Oberst, T. E., Parshley, S. C., Nikola, T., *et al.* 2011, *Apj*, 739, 100
Padoan, P., Jones, B. J. T., & Nordlund, A. P. 1997, *ApJ*, 474, 730
Planck Collaboration, Ade, P. A. R., Aghanim, N., *et al.* 2011, *A&A*, 536, A19
Shibai, H., Okuda, H., Nakagawa, T., *et al.* 1991, *ApJ*, 374, 522
Smith, B. J. & Madden, S. C. 1997, *AJ*, 114, 138
Stacey, G. J., Viscuso, P. J., Fuller, C. E., & Kurtz, N. T. 1985, *ApJ*, 289, 803
van Dishoeck, E. F. & Black, J. H. 1988, *ApJ*, 334, 771
van Kempen, T. A., Kristensen, L. E., Herczeg, G. J., *et al.* 2010, *A&A*, 518, L121
Visser, R., Kristensen, L. E., Bruderer, S., *et al.* 2012, *A&A*, 537, A55
Wright, E. L., Mather, J. C., Bennett, C. L., *et al.* 1991, *ApJ*, 381, 200
Wolfire, M. G., Hollenbach, D., & McKee, C. F. 2010, *ApJ*, 716, 1191
Wolfire, M. G., McKee, C. F., Hollenbach, D., & Tielens, A. G. G. M. 2003, *ApJ*, 587, 278

Astrophysics from Antarctica
Proceedings IAU Symposium No. 288, 2012
M. G. Burton, X. Cui & N. F. H. Tothill, eds.

© International Astronomical Union 2013
doi:10.1017/S1743921312016808

Submillimeter Astronomy
from the South Pole (AST/RO)

Antony A. Stark

Smithsonian Astrophysical Observatory,
60 Garden Street, Cambridge, MA, USA
email: aas@cfa.harvard.edu

Abstract. The Antarctic Submillimeter Telescope and Remote Observatory (AST/RO), a 1.7 m diameter offset Gregorian telescope for astronomy and aeronomy studies at wavelengths between 200 and 2000 μm, saw first light in 1995 and operated until 2005. It was the first radio telescope to operate continuously throughout the winter on the Antarctic Plateau. It served as a site testing instrument and prototype for later instruments, as well as executing a wide variety of scientific programs that resulted in six doctoral theses and more than one hundred scientific publications. The South Pole environment is unique among observatory sites for unusually low wind speeds, low absolute humidity, and the consistent clarity of the submillimeter sky. Especially significant are the exceptionally low values of sky noise found at this site, a result of the small water vapor content of the atmosphere. Multiple submillimeter-wave and Terahertz detector systems were in operation on AST/RO, including heterodyne and bolometric arrays. AST/RO's legacy includes comprehensive submillimeter-wave site testing of the South Pole, spectroscopic studies of 492 GHz and 809 GHz neutral atomic carbon and 460 GHz and 806 GHz carbon monoxide in the Milky Way and Magellanic Clouds, and the first detection of the 1.46 THz [N II] line from a ground-based observatory.

Keywords. surveys, site testing, submillimeter, Galaxy: kinematics and dynamics, ISM: kinematics and dynamics, ISM: molecules, ISM: evolution

1. Introduction

Establishment of the Antarctic Submillimeter Telescope and Remote Observatory (AST/RO, Stark *et al.* 2001) as a year-round, permanently-manned facility was an important step in the development of the South Pole as an observatory. It had long been thought that the Antarctic Plateau would be an exceptional site for high-frequency radio (for example Townes & Melnick 1990; see Indermuehle, Burton, & Maddison 2005 for a review), but the demonstration of this ground truth and the persuasion of skeptics required extensive development and site testing during the latter years of the 20th century. The simultaneous development of the Atacama Large Millimeter Array site in Chile was seen within the astronomical community as a rival for scarce resources; some predicted that winter observations from the Pole would be impossible or unrealistically expensive, and would never surpass what could be better done in Chile. Early attempts at Polar astronomical expeditions in 1984-85 (Pajot *et al.* 1989), 1986-87 (Dragovan *et al.* 1990) and 1988-89 (Meinhold & Lubin 1991, Tucker *et al.* 1993) showed promise, but they also showed the difficulty of setting up summer-only facilities for a few weeks of observing during the warmest and wettest part of the year, only to tear them down a few weeks later in anticipation of the oncoming winter. AST/RO was proposed as stand-alone multi-year observatory to the U.S. National Science Foundation Office of Polar Programs in 1988. It was funded under DPP88-18384 and also supported in part by Bell Laboratories and Boston University. While still under design in 1991, AST/RO was incorporated into

Table 1. AST/RO Personnel and their Institutional Affiliations during AST/RO Operations.

Winter-over Scientists[1]	Senior Personnel	Collaborators	Students[2]
Richard A. Chamberlin	Antony A. Stark[4,3] (PI)	Jürgen Stutzki[5]	Alberto Bolatto[6]
Simon P. Balm	Adair P. Lane[3] (PM)	R. Schieder[5]	Christopher Groppi[7]
Xiaolei Zhang	Christopher K. Walker[7]	Jacob W. Kooi[8]	Maohai Huang[6]
Roopesh Ojha	Robert W. Wilson[4,3]	Peter Zimmermann[9]	A. Hungerford[7]
Rodney D. Marks[10]	Jingquan Cheng[3]	Rüdiger Zimmermann[9]	Henry H. Hsieh[11]
Christopher L. Martin	K.-Y. Lo[12]	Gordon J. Stacey[13]	James Ingalls[6]
Wilfred M. Walsh	Thomas M. Bania[6]	S. Yngvesson[14]	Craig Kulesa[7]
Kecheng Xiao	James M. Jackson[6]	E. Gerecht[14]	Johannes Staguhn[5]
Karina Leppik	Gregory A. Wright[4]	Youngung Lee[3,15]	Gregory Engargiola[12]
Julienne Harnett	Sungeun Kim[3]	Dennis Mumma[4]	
Nicholas F. H. Tothill	John Bally[4]	Karl Jacobs[5]	
Andrea Löhr		Jonas Zmuidzinas[12,8]	

Notes:
1 AST/RO Winter-over Scientists were affiliated with the Smithsonian Astrophysical Observatory during the period of their winterover.
2 Many who were students during the AST/RO project are now distinguished senior scientists.
3 Smithsonian Astrophysical Observatory; Harvard-Smithsonian Center for Astrophysics; 60 Garden St.; Cambridge, MA 02138; USA.
4 AT&T Bell Laboratories; Crawford Hill; Holmdel, NJ 07733; USA.
5 I. Physikalisches Institut, Universität zu Köln; Zülpicher Straße 77; D-50937 Köln; Germany.
6 Boston University; 725 Commonwealth Ave.; Boston, MA 02215; USA.
7 Steward Observatory; 933 N. Cherry Ave.; University of Arizona; Tucson, AZ 85721; USA.
8 Caltech Submillimeter Observatory; Caltech 320-47; Pasadena, CA 91125; USA.
9 Radiometer Physics GmbH; Birkenmaarstraße 10; 53340 Meckenheim, Germany.
10 Rodney Marks died while wintering over in 2000, under circumstances that are still not fully known.
11 Harvard University; Department of Astronomy; 60 Garden St.; Cambridge, MA 02138; USA.
12 University of Illinois; 1002 W. Green St.; Urbana, IL 61801; USA.
13 Department of Astronomy; Cornell University; 610 Space Sciences Building; Ithaca, NY 14853; USA.
14 Department of Electrical and Computer Engineering; University of Massachusetts; Amherst, MA 01103; USA.
15 Taeduk Radio Astronomy Observatory, Korea Astronomy Observatory, Whaam-dong 61-1, Yuseong-gu; Taejon 305-348, Korea.

the Center for Astrophysical Research in Antarctica (CARA, cf. Novak & Landsberg 1998), then a newly-formed NSF Science and Technology Center with an 11-year life-span. AST/RO winter-over operations began in 1995 and concluded in 2005. This article will summarize the operations, site testing and science that demonstrated the feasibility of year-round operations and the quality of the Pole as an observatory site.

2. Operations

The design of AST/RO is described in Stark *et al.* (1997a). AST/RO was located on the roof of a dedicated support building across the aircraft skiway in the *Dark Sector* of the United States National Science Foundation Amundsen-Scott South Pole Station, the first of a group of observatory buildings in an area designated to have low radio emissions and light pollution. The AST/RO building was a single story, 4m × 20m, elevated 3m above the surface on steel columns to reduce snow drifts. The interior was partitioned into six rooms, including laboratory and computer space, storage areas, a telescope control room, and a Coudé room containing the receivers on a large optical table suspended from the telescope. The station provided logistical support for the observatory: room and board for on-site scientific staff, electrical power, network and telephone connections, heavy equipment support, and cargo and personnel transport. The station powerplant provided an average 25 kW of power to the AST/RO building.

Aircraft flights to Pole are scheduled from late October to early February, so that the station is inaccessible for nine months of the year. This long *winter-over* period is central to all logistical planning for Polar operations. Plans and schedules were made in March and April for each year's deployment to South Pole: personnel on-site, tasks to be completed, and the tools and equipment needed. All equipment had to be ready for shipment by the end of September. For quick repairs and upgrades during the Austral summer season, it is possible to send equipment between South Pole and anywhere serviced by commercial express delivery in about five days.

AST/RO group members deployed to South Pole in groups of two to six people throughout the Austral summer season, carrying out their planned tasks and returning after stays ranging from 2 weeks to 3 months. Each year there were one or two AST/RO *winter-over scientists* who remained with the telescope for a full year (see Table 1). The winter-over scientist position was usually a three year post-doctoral appointment: one year of preparation and training, one year at South Pole with the telescope, and one year after the winter-over year to reduce data and prepare scientific results. If there were no instrumental difficulties, the winter-over scientist controlled telescope observations through an automated program, carried out routine pointing and calibration tasks, tuned the receivers, and filled the liquid helium dewars. If instrumental difficulties developed, the winter-over scientist carried out repairs with radio or email consultation back to the home institution.

There were five heterodyne receivers mounted on an optical table suspended from the telescope structure in a spacious (5m × 5m × 3m), warm Coudé room:

- 230 GHz SIS receiver, 55–75 K double-sideband (DSB) noise temperature;
- 450–495 GHz SIS waveguide receiver, 200–400 K DSB (Walker *et al.* 1992);
- 450–495 GHz SIS quasi-optical receiver, 165–250 K DSB (Engargiola, Zmuidzinas, & Lo 1994, Zmuidzinas & LeDuc 1992);
- 800-820 GHz fixed-tuned SIS waveguide mixer receiver, 950–1500 K DSB (Honingh *et al.* 1997);
- an array of four 800-820 GHz fixed-tuned SIS waveguide mixer receivers, 850–1500 K DSB, the PoleSTAR array (Groppi *et al.* 2000).

Seven intermediate-frequency bandpasses were processed by acousto-optical spectrometers (AOS; Schieder, Tolls, & Winnewisser 1989):

- two low resolution spectrometers with a bandwidth of 1 GHz (bandpass 1.6–2.6 GHz);
- an array AOS with four low resolution spectrometer channels with a bandwidth of 1 GHz (bandpass 1.6–2.6 GHz) for the PoleSTAR array; and
- one high-resolution AOS with 60 MHz bandwidth (bandpass 60–120 MHz).

The SIS receivers used on AST/RO each required about 2 liters of liquid helium per day continuously throughout the year. The National Science Foundation and its support contractors supplied liquid helium in one or more large (4000 to 12000 liter) storage dewars. Some years this supply lasted the entire winter, but in 1996, 1998, 2000, 2002, and 2004 an insufficiency of supply affected operations—the supply of liquid helium was the main failure mode for AST/RO operations. Newer facilities such as South Pole Telescope and the Keck array make use of closed-cycle cryogenic systems.

3. Site Testing

Submillimeter astronomy requires dry, frigid observatory sites, where the atmosphere contains less than 1 mm of precipitable water vapor (PWV). The South Pole station meteorology office has used balloon-borne radiosondes to measure profiles of temperature, pressure, and water vapor at least once a day for several decades (Schwerdtfeger 1984). These have typically shown atmospheric water vapor values about 90% of saturation for air coexisting with the ice phase at the observed temperature and pressure. The PWV

Figure 1. The AST/RO telescope and building. An insulated steel tower supports the telescope and extends through the building, which houses receivers and electronics in a shirt-sleeve environment. Photo credit A. Lane.

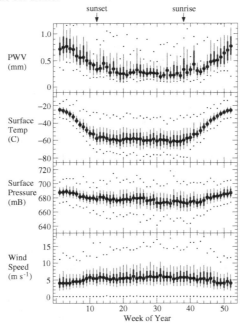

Figure 2. South Pole weather data binned by week of year. The large dot indicates the median value, the thick bar indicates the 25 to 75 percentile, the thin bar indicates 10 to 90 percentile, and the small dots are maximum and minimum measured values. The top panel shows precipitable water vapor as measured by AIR Model 4a weather balloon flights from 1991 to 1996. The bottom 3 panels summarize hourly NOAA weather data from 1977 to 2001. Note the "coreless winter" with typical PWV values of 0.3 mm. The highest recorded wind speed is remarkably small. Analysis and figure by R. Chamberlin.

values consistent with saturation are, however, extremely low because the air is desiccated by the frigid temperatures: at the South Pole's average annual temperature of −49 C, the partial pressure of saturated water vapor is only 1.2% of what it is at 0 C (Goff &

Gratch 1946). Figure 2 shows the PWV averaged by week of the year. *PWV values at South Pole are small, stable and well-understood.*

Water vapor is usually the dominant source of opacity, but thousands of other molecular lines (Waters 1976, Bally 1989) make a *dry air* contribution to the opacity. Chamberlin & Bally (1994, 1995), Chamberlin, Lane, & Stark (1997), and Chamberlin (2001) showed that the dry air opacity is relatively more important at the South Pole than at other sites such as Chajnantor in Chile, since the total atmospheric pressure is higher. Dry air opacity is, however, less variable than the opacity caused by water vapor, and therefore causes less *sky noise* (Lay & Halverson 2000), a factor of 30 to 50 less at Pole than other sites. This is the critical factor for successful observation because constant white noise and constant opacity can simply be overcome by longer observation, whereas sky noise introduces a "$1/f$" component that requires faster switching or perhaps cannot be overcome at all. Antarctic plateau sites are truly exceptional in this respect.

4. Science Results

Many of AST/RO's scientific results were studies of emission lines from dense interstellar gas in the Milky Way and Magellanic Clouds, in particular the $J = 7 \rightarrow 6$ and $J = 4 \rightarrow 3$ transitions of CO, and the $^3P_1 \rightarrow^3 P_0$ and $^3P_2 \rightarrow^3 P_1$ ground-state transitions of C I. When these transitions are mapped in areas of the sky and combined with millimeter-wave observations of $J = 1 \rightarrow 0$ or $J = 2 \rightarrow 1$ transitions of CO and ^{13}CO, these data can be used to model the density and excitation temperature as a function of radial velocity along many lines of sight and determine the mass and kinematics of dense interstellar clouds.

Studies of star-forming clouds. Hsieh (2000), Huang *et al.* (1999), Huang (2001), Kulesa (2002), Groppi (2003), Groppi *et al.* (2004), Kulesa *et al.* (2005), Walsh & Xiao (2005), Kim & Narayanan (2006), Löhr *et al.* (2009), and Tothill *et al.* (2009) used AST/RO's multi-frequency mapping ability to study the properties of interstellar clouds in the process of star formation and their interaction with supernovae and H II regions.

The physical state of high-latitude translucent clouds. Ingalls (1999), and Ingalls *et al.* (2000) studied CO toward molecular clouds near the Sun but out of the plane of the Milky Way, and found conditions indicating that much of the emission originates in tiny (~ 2000AU) cold (8K) fragments within the ~ 100 times larger CO-emitting extent of a typical high-latitude cloud.

Cloud formation and turbulence in the Carina region. Zhang *et al.* (2001) made large-area ($3\,\mathrm{deg}^2$) , fully sampled maps of CO and C I in the Carina molecular cloud complex. They present evidence that the spiral density wave shock associated with the Carina spiral arm is playing an important role in the formation and dissociation of the cloud complex, and also maintaining the internal energy balance of the clouds in this region by feeding interstellar turbulence.

The ^{12}C/^{13}C ratio measured in C I. Tieftrunk *et al.* (2001) made measurements of the ^{12}C/^{13}C abundance ratio from observations of the $^3P_2 \rightarrow^3 P_1$ transitions near 809 GHz. They determined intrinsic ^{12}C/^{13}C ratios of of 23 ± 1 for G 333.0-0.4, 56 ± 14 for NGC 6334A and 69 ± 12 for G 351.6-1.3. The enhancement of ^{13}C towards G 333.0-0.4 may be due to strong isotope-selective photodissociation of the chemical precursor ^{13}CO, outweighing the effects of chemical isotopic fractionation; towards NGC 6334 A and G 351.6-1.3 these effects appear to be balanced.

Physical state and dynamics of galactic center gas. Ojha *et al.* (2001), Staguhn (1996), Staguhn *et al.* (1997), Kim *et al.* (2002), and Martin *et al.* (2004) used AST/RO observations of CO and C I to determine the thermodynamic state of dense gas within a few

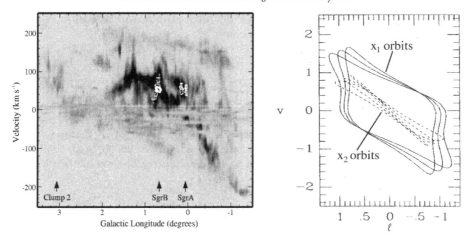

Figure 3. Carbon Monoxide in the Galactic Center Region. Left: the gray scale is $J = 4 \to 3$ CO, and the white contours are $J = 7 \to 6$ CO, shown in a longitude-velocity diagram. The nearly horizontal absorption features are foreground spiral arms. Right: Theoretical calculation of x_1 and x_2 closed orbits from Binney *et al.* (1991). Note the faint CO parallelogram tracing the x_1 orbits and the brighter, denser material on x_2 orbits.

kiloparsecs of the Milky Way center. Much of this gas is located on closed x_1 and x_2 orbits (Binney *et al.* 1991; Figure 3). The gas that has accumulated on x_2 orbits is nearly at a density that will cause tipping into instability, where the ring of gas will coagulate into one or two giant ($\sim 10^7$ M$_\odot$) clouds that undergo starburst (Stark *et al.* 2004).

Dense molecular gas in the Magellanic Clouds. Stark *et al.* (1997b), Bolatto (2001), Bolatto *et al.* (2000a,b), Kim, Walsh, & Xiao (2004), Kim *et al.* (2005), Bolatto, Israel, & Martin (2005), and Kim (2006) mapped star-forming regions in the Magellanic Clouds to determine their physical properties and kinematics. Temperatures as high as 300 K occur within these clouds, which have extended photo-dissociation envelopes compared to galactic clouds.

Terahertz observations. Two Terahertz detector systems were installed on AST/RO as guest instruments: the Terahertz Receiver with NbN HEB Device (TREND) was a heterodyne receiver with a LASER local oscillator (Gerecht *et al.* 1999, 2004, Yngvesson *et al.* 2002); and the South Pole Imaging Fabry-Perot Interferometer (SPIFI, Swain *et al.* 1998, Bradford *et al.* 2002). The principal difficulty in Terahertz operation was that when engineering teams could be brought in during the Austral summer, the sky was not transparent enough even for testing, so an untried system would be left for the winter-over scientists to attempt first light during brief winter periods of marginal Terahertz weather. Nevertheless, 205 μm N II was detected (Oberst *et al.* 2006) and mapped in the Eta Carina region (Oberst *et al.* 2011).

5. Conclusion

AST/RO site testing contributed to the characterization of the South Pole as an observatory site, showing the high transparency and stability of the millimeter-wave sky and the possibility of ground-based Terahertz observations. Routine observations over a decade enhanced our knowledge of submillimeter-wave spectroscopic line emission from interstellar gas, the nature of star formation, and the dynamics of the Galaxy. The AST/RO project demonstrated the feasibility of operating a state-of-the-art high-frequency radiotelescope at the South Pole during the Antarctic winter, and thereby laid the foundation for the instruments to come.

References

Bally, J. 1989, in *Astrophysics in Antarctica*, ed. D. J. Mullan, M. A. Pomerantz, & T. Stanev (New York: American Institute of Physics), 100

Binney, J., Gerhard, O. E., Stark, A. A., Bally, J., & Uchida, K. 1991, *MNRAS*, 252, 210

Bolatto, A. D., Jackson, J. M., Kraemer, K. E., & Zhang, X. 2000a, *ApJ* (Letters), 541, L17

Bolatto, A. D., Jackson, J. M., Israel, F. P., Zhang, X., & Kim, S. 2000b, *ApJ*, 545, 234

Bolatto, A. D, 2001, *The Interstellar Medium in Low Metallicity Environments*, Ph.D. Thesis, Boston University

Bolatto, A. D., Israel, F. P., & Martin, C. L. 2005, *ApJ*, 633, 210

Bradford, C. M., Stacey, G. J., Swain, M. R., Nikola, T., Bolatto, A. D., Jackson, J. M., Savage, M. L., Davidson, J. A., & Ade, P. A. R. 2002, *Appl. Opt.*, 41, 2561

Chamberlin, R. A. & Bally, J. 1994, *Appl. Opt.* 33, 1095

— 1995, *Int. J. Infrared and Millimeter Waves,* 16, 907

Chamberlin, R. A., Lane, A. P., & Stark, A. A. 1997, *ApJ* 476, 428

Chamberlin, R. A. 2001, *J. Geophys. Res.: Atmospheres*, 106 (D17), 20101

Dragovan, M., Stark, A. A., Pernick, R., & Pomerantz, M. 1990, *Appl. Optics*, 29, 463

Engargiola, G., Zmuidzinas, J., & Lo, K.-Y. 1994, *Rev. Sci. Instr.*, 65, 1833

Gerecht, E., Musante, C. F., Zhuang, Y., Yngvesson, K. S., Goyette, T., Dickinson, J., Waldman, J., Yagoubov, P. A., Gol'tsman, G. N., Voronov, B. M., & Gershenzon, E. M. 1999, *IEEE Trans.*, MTT-47, 2519

Gerecht, E., Yngvesson, S., Nicholson, J., Zhuang, Y., Rodriguez-Morales, F., Zhao, X., Gu, D., Zannoni, R., Coulombe, M., Dickinson, J., Goyette, T., Gorveatt, W., Waldman, J., Khosropanah, P., Groppi, C., Hedden, A., Golish, D., Walker, C., Kooi, J., Chamberlin, R., Stark, A., Martin, C., Stupak, R., Tothill, N., & Lane, A. 2004, *Deployment of TREND — A Low Noise Receiver User Instrument at 1.25 THz to 1.5 THz for AST/RO at the South Pole*, Proceedings of the 14th International Symposium on Space Terahertz Technology, eds. C. Walker and J. Payne, pp. 179-188

Goff, J. A. & Gratch, S. 1946, *Trans. Amer. Soc. Heat. and Vent. Eng.,* 52, 95

Groppi, C. E. 2003, *Submillimeter Heterodyne Spectroscopy of Star Forming Regions*, Ph.D. thesis, University of Arizona

Groppi, C., Walker, C., Hungerford, A., Kulesa, C., Jacobs, K., & Kooi, J. 2000, in *Imaging at Radio Through Submillimeter Wavelengths,* ed. J. G. Mangum & S. J. E. Radford, Vol. 217 (San Francisco: ASP Conference Series), 48

Groppi, C. E., Kulesa, C., Walker, C. K., & Martin, C. L. 2004 *ApJ*, 612, 946

Honingh, C. E., Hass, S., Hottgenroth, K., Jacobs, J., & Stutzki, J. 1997, *IEEE Trans. Appl. Superconductivity*, 7, 2582

Hsieh, H. H. 2000, ^{12}CO $J = 4 \rightarrow 3$ and CI $^3P_1 \rightarrow^3 P_0$ *Observations of a Selection of Interstellar Clouds Located Throughout the Galaxy*, Senior Honors Thesis, Harvard University.

Huang, M., Bania, T. M., Bolatto, A. Chamberlin, R. A., Ingalls, J. G., Jackson, J. M., Lane, A. P., Stark, A. A., Wilson, R. W., & Wright, G. A. 1999 *ApJ*, 517, 282

Huang, M. 2001, *Interstellar Carbon Under the Influence of H II Regions* , Ph.D. Thesis, Boston University

Indermuehle, B. T., Burton, M., & Maddison, S. T. 2005, *PASA*, 22, 73

Ingalls, J. G. 1999, *Carbon Gas in High Galactic Latitude Molecular Clouds*, Ph.D. Thesis, Boston University.

Ingalls, J. G., Bania, T. M., Lane, A. P., Rumitz, M., & Stark, A. A. 2000, *ApJ*, 535, 211

Kim, S., Martin, C. L., Stark, A. A., & Lane, A. P. 2002, *ApJ*, 580, 896

Kim, S., Walsh, W. M., & Xiao, K. 2004 *ApJ*, 616, 865

Kim, S., Walsh, W. M., Xiao, K., & Lane, A. P. 2005 *AJ*, 130, 1635

Kim, S. 2006, *PASP*, 118, 94

Kim, S. & Narayanan, D. 2006 *PASJ*, 58, 753

Kulesa, C. A. 2002, *Molecular Hydrogen and Its Ions in Dark Interstellar Clouds and Star-Forming Regions*, Ph.D. Thesis, University of Arizona

Kulesa, C., Hungerford, A., Walker, C. K., Zhang, X., & Lane, A. P. 2005 *ApJ*, 625, 194

Lay, O. P. & Halverson, N. W. 2000 *ApJ*, 543, 787

Löhr, A., Bourke, T. L., Lane, A. P., Myers, P. C., Parshley, S. C., Stark, A. A., & Tothill, N. F. H. 2007 *ApJ* (Suppl. Series), 171, 478

Martin, C. L., Walsh, W. M., Xiao, K., Lane, A. P. Walker, C. K., & Stark, A. A. 2004, *ApJ* (Suppl. Series), 150, 239

Meinhold, P. & Lubin, P. 1991, *ApJ* (Letters), 370, L11

Novak, G. & Landsberg, R. H. 1998 *Astrophysics from Antarctica* ASP Conference Series Vol. 141

Oberst, T. E., Parshley, S. C., Stacey, G. J., Nikola, T., Löhr, A., Harnett, J. I., Tothill, N. F. H., Lane, A. P., Stark, A. A., & Tucker, C. E. 2006 *ApJ* (Letters), 652, L125

Oberst, T. E., Parshley, S. C., Nikola, T., Stacey, G. J., Löhr, A., Lane, A. P., Stark, A. A., & Kamenetzky, J. 2011 *ApJ*, 739, 100

Ojha, R., Stark, A. A., Hsieh, H. H., Lane, A. P., Chamberlin, R. A., Bania, T. M., Bolatto, A. D. Jackson, J. M., & Wright, G. A. 2001 *ApJ*, 548, 253

Pajot, F., Gispert, R., Lamarre, J. M., Peyturaux, R., Pomerantz, M. A., Puget, J.-L., Serra, G., Maurel, C., Pfeiffer, R., & Renault, J. C. 1989, *A&A*, 223, 107

Schwerdtfeger, W. 1984, *Weather and Climate of the Antarctic* (Amsterdam: Elsevier)

Schieder, R., Tolls, V., & Winnewisser, G. 1989, *Exp. Astron.*, 1, 101

Staguhn, J. 1996, *Observations Towards the Sgr C Region Near the Center of Our Galaxy*, Ph.D. Thesis, University of Cologne

Staguhn, J., Stutzki, J., Chamberlin, R. A., Balm, S. P., Stark, A. A., Lane, A. P., Schieder, R., & Winnewisser, G. 1997, *ApJ*, 491, 191

Stark, A. A., Chamberlin, R. A., Ingalls, J., Cheng, J., & Wright, G. 1997a *Rev. Sci. Instr.*, 68, 2200

Stark, A. A., Bolatto, A. D., Chamberlin, R. A., Lane, A. P., Bania, T. M., Jackson, J. M., and Lo, K.-Y. 1997b *ApJ* (Letters), 480, L59

Stark, A. A., Bally, J., Balm, S. P., Bania, T. M., Bolatto, A. D., Chamberlin, R. A., Engargiola, G., Huang, M., Ingalls, J. G., Jacobs, K., Jackson, J. M., Kooi, J. W., Lane, A. P., Lo, K.-Y., Marks, R. D., Martin, C. L., Mumma, D., Ojha, R., Staguhn, J., Stutzki, J., Walker, C. K., Wilson, R. W., Wright, G. A., Zhang, X., Zimmermann, P., & Zimmermann, R. 2001, *PASP*, 113, 567

Stark, A. A., Martin, C. L., Walsh, W. M., Xiao, K., & Lane, A. P., Walker C. K. 2004, *ApJ* (Letters), 614, L41

Swain, M. R., Bradford, C. M., Stacey, G. J., *et al.* 1998, *Proc. SPIE* 3354, 480.

Tieftrunk, A. R., Jacobs, K., Martin, C. L., Seibertz, O., Stark, A. A., Stutzki, J., Walker, C. K., & Wright, G. A. 2001 *A & A*, 375, L23.

Tothill, N. F. H., Löhr, A., Parshley, S. C., Stark, A. A., Lane, A. P., Harnett, J. I., wright, G. A., Walker, C. K., Bourke, T. L., & Myers, P. C. 2009 *ApJ* (Supplement), 185, 98

Townes, C. H. & Melnick, G. 1990 *PASP*, 102 357

Tucker, G. S., Griffin, G. S., Nguyen, H. T., & Peterson, J. B. 1993, *ApJ* (Letters), 419, L45

Walker, C. K., Kooi, J. W., Chan, W., LeDuc, H. G., Schaffer, P. L., Carlstrom, J. E., & Phillips, T. G. 1992, *Int. J. Infrared and Millimeter Waves,* 13, 785

Walsh, W. & Xiao, K. 2005, *Highlights of Astronomy*, 13, 965

Waters, J. W. 1976, in *Methods of Experimental Physics: Astrophysics: Part B: Radio Telescopes,* ed. M. L. Meeks, Vol. 12 (New York: Academic Press), 142

Yngvesson, K. S., Nicholson, J., Zhuang, Y., Zhao, X., Gu, D., Zannoni, R., Gerecht, E., Coulombe, M., Dickinson, J., Goyette, T., Waldman, J., Walker, C. K., Stark, A. A., & Lane, A. P. 2002, *Progress on TREND - A Low Noise Receiver User Instrument at 1.25 THz to 1.5 THz for AST/RO at the South Pole*, in Proceedings of the 13th Intern. Symp. on Space THz Technology, eds. R. Blundell and E. Tong, E., (Harvard Univ., Cambridge, MA), 26-28 Mar. 2002, p. 461

Zhang, X., Lee, Y., Bolatto, A. D., & Stark, A. A. 2001, *ApJ*, 553, 274

Zmuidzinas, J. & LeDuc, H. G. 1992, *IEEE Trans. Microwave Theory Tech.,* 40, 1797

Astrophysics from Antarctica
Proceedings IAU Symposium No. 288, 2012
M. G. Burton, X. Cui & N. F. H. Tothill, eds.

© International Astronomical Union 2013
doi:10.1017/S174392131201681X

The Balloon-borne Large Aperture Submillimetre Telescope (BLAST) and BLASTPol

Enzo Pascale
for the BLAST and BLASTPol collaboration

School of Physics & Astronomy, Cardiff University, 5 The Parade, Cardiff, CF24 3AA, UK

Abstract. Balloon observations from Antarctica have proven an effective and efficient way to address open Cosmological questions as well as problems in Galactic astronomy. The Balloon-borne Large Aperture Submillimetre Telescope (BLAST) is a sub-orbital mapping experiment which uses 270 bolometric detectors to image the sky in three wavebands centred at 250, 350 and 500 μm with a 1.8 m telescope. In the years before *Herschel* launched, BLAST provided data of unprecedented angular and spectral coverage in frequency bands close to the peak of dust emission in star forming regions in our Galaxy, and in galaxies at cosmological distances. More recently, BLASTPol was obtained by reconfiguring the BLAST focal plane as a submillimetric polarimeter to study the role that Galactic magnetic fields have in regulating the processes of star-formation. The first and successful BLASTPol flight from Antarctica in 2010 is followed by a second flight, currently scheduled for the end of 2012.

1. Introduction

The Balloon-borne Large Aperture Submillimeter Telescope (BLAST, Pascale *et al.* 2008) is a sub-orbital surveying experiment designed to study the evolutionary history and processes of star formation in local galaxies (including the Milky Way) and galaxies at cosmological distances. The BLAST continuum camera, which consists of 270 detectors distributed between three arrays, observes simultaneously in broad-band (30%) spectral-windows at 250, 350 and 500 μm. The 1.8 m Ritchey-Chrétien telescope provides resolutions of $36''$, $42''$ and $60''$ at 250, 350 and 500 μm, respectively. BLAST is a forerunner of the Spectral and Photometric Imaging Receiver (Griffin *et al.* 2010) on the 3.5 m *Herschel* space observatory, and with SPIRE it shared similar focal plane technology and scientific goals.

In the years preceding the launch of *Herschel*, BLAST made three flights. A 24 hr test flight from Fort Sumner, and two long-duration science flights – from Kiruna, Sweden in 2005, and from Antarctica in 2006. The BLAST cosmological surveys, centered on the South Ecliptic Pole (Valiante *et al.* 2010) and on the Great Observatories Origins Deep Survey (GOODS-South, see Devlin *et al.* 2009), were conducted during the 2006 flight, at an altitude of 38 km. The GOODS-South region combines a wide-area map of 8.7 square degrees with a deeper, confusion-limited map of 0.8 square degrees. We refer to these fields as BLAST GOODS-S Wide (BGS-Wide), and BLAST GOODS-S Deep (BGS-Deep) respectively. The BGS-Deep survey is dominated by the point-source confusion arising from the $36''$–$60''$ BLAST beams, rather than instrumental noise. The BLAST data, combined with existing photometric measurements taken from the optical to the FIR, and with the available redshift information, have enabled Marsden *et al.* (2009) and Pascale *et al.* (2009) to use stacking analysis to estimate the contribution to the Cosmic Infrared Background (CIB) at wavelengths close to its peak. Patanchon

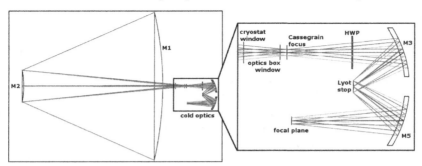

Figure 1. Schematic of the optical layout for the BLAST telescope and receiver shown on the left, with the 1.5 K optics, located within the cryostat, shown in an expanded view on the right. The image of the sky formed at the input aperture is re-imaged onto the bolometer detector arrays at the focal plane. The M4 mirror serves as a Lyot stop, which defines the illumination of the primary mirror for each element of the bolometer detector arrays. The three wavelength bands are separated by a pair of dichroic beam-splitters (not shown here). The sapphire half-wave plate used in BLASTPol is also shown, mounted 19 cm behind the focus of the telescope.

et al. (2009) studied the pixel histogram of the Deep and Wide maps to constrain a parametric model of differential source counts. Viero *et al.* (2009) detected the clustering signal of the submillimetric sources at 250, 350 and 500 μm. A 50 square degree survey on the Vela Molecular Ridge (Netterfield *et al.* 2009) was also conducted in 2006, revealing large numbers of starless and pre-stellar cores, which are the earliest stages of star-formation.

In May 2009, *Herschel* was successfully launched and all BLAST targets have been observed from space at the same wavelengths, but with the improved sensitivity and angular resolution of SPIRE. The BLAST multiband photometer, no longer necessary, was re-configured as a polarimeter, resulting in BLASTPol, a enhanced version of BLAST. With the addition of a polarizing grid in front of each of the 270 feed horns and a stepped achromatic Half Wave Plate (AHWP), BLASTPol is now an instrument designed to measure polarized dust emission from star-forming regions. By mapping polarization from dust grains aligned with respect to their local magnetic field, the field orientation (projected on the sky) can be traced. The magnetic-field strength can also be estimated indirectly using the polarization angular dispersion (Chandrasekhar & Fermi 1953).

During the first flight of BLASTPol in December 2010 from McMurdo, Antarctica, we made sensitive degree-scale maps of several nearby molecular clouds. While the angular resolution ($< 1.5'$) was poorer than planned due to a blocking filter that was melted by the Sun on ascent, our preliminary polarization maps indicate coherent polarization across our target clouds at the few percent level.

This BLASTPol dataset is being used to investigate the role that magnetic fields play in the star-formation process, an important outstanding question in our understanding of how stars form. BLASTPol maps of magnetic fields cover entire Giant Molecular Clouds (GMCs), yet have sufficient resolution to probe fields in dense filamentary sub-structures and molecular cores. The experiment provides a crucial bridge between the large area but coarse resolution polarimetry provided by experiments such as *Planck* and the high resolution but small area of ALMA data.

A second flight is scheduled to occur in December 2012 from Antarctica to improve on the area, depth and angular resolution obtained during the 2010 flight.

2. INSTRUMENT

A detailed description of the BLAST instrument is given in Pascale *et al.* (2008) and Marsden *et al.* (2008). The main features of the optical system are summarized in Fig. 1. The Ritchey-Chrétien telescope has an aluminium primary mirror with a diameter of 1.8 m. The radiation collected is re-imaged by a series of cold (~ 1.5 K) reflecting optical elements arranged into an ideal Offner relay inside a long-duration cryostat. This cryostat uses liquid helium and nitrogen and has a hold time of more than 10 days. The telescope's secondary mirror is actuated, so the system can be refocused in-flight. The light is split into three 30%-wide submillimetric bands respectively, centered at 250, 350 and 500 μm. The BLAST focal plane consists of arrays of 149, 88 and 43 detectors at 250, 350 and 500 μm respectively. The arrays are cooled to 270 mK. Each array element is a silicon nitride micromesh "spiderweb" bolometer (Glenn *et al.* 1998), coupled to the front optics by a smooth-walled conical feed-horn (Chattopadhyay *et al.* 2003).

The BLASTPol instrument (Fissel *et al.* 2010) is a modification of the BLAST telescope that adds linear polarization capabilities. This is achieved by adding a polarizing grid at the mouth of each feed-horn, and a stepped achromatic half-wave plate (AHWP) to modulate the polarization.

The grids are patterned to alternate the polarization angle sampled by 90° from horn-to-horn and thus bolometer-to-bolometer along the scan direction. This arrangement has proved effective in rejecting $1/f$ noise correlated among detectors in an array (array common modes). BLASTPol scans so that a source on the sky passes along a row of detectors, and thus the time required to measure one Stokes parameter (Q or U) is just equal to the separation between bolometers divided by the scan speed. For the 250 μm detector array where the bolometers are separated by 45″, and assuming a typical scan speed of 0.05 °/s , this time is 0.25 s. This timescale is short compared to the characteristic low frequency ($1/f$) noise knee for the detectors at 0.035 mHz (Pascale *et al.* 2008).

The use of a cryogenic, stepped AHWP (Moncelsi *et al.* 2012) allows modulation of the Stokes parameters Q and U such that each detector measures I, Q, and U multiple times in each sky direction. A total of four AHWP position angles are used (at 0, 22.5, 45 and 67.5°), stepped at the end of the telescope's raster-scan on a given target. This mitigates the effect of unbalanced gains between adjacent detectors which would result in a large bias on the estimated Q and U if only detector differences are used to estimate the Stokes parameters.

Each astronomical target was observed with BLAST (or BLASTPol) in a slow raster-scan mode. Slow scanning is preferable to a mechanical chopper for mapping large regions of the sky. The telescope is scanned in azimuth at a constant velocity of $\sim 0.05°/s$. At the end of each azimuthal scan, the elevation is stepped by 1/3 of the array's 7′ field of view (FOV) in elevation (the array FOV is 14′ × 7′).

3. BLAST Observations

The BLAST point source catalog (Devlin *et al.* 2009) lists hundreds of sources detected in the BGS-Wide and BGS-Deep regions at more than 5σ confidence, and thousands at 3σ. However, more than 75% of the astrophysical information is contained in the emission from unresolved sources. For this reason, some of the most important BLAST results emerge from a statistical analysis of the intensity fluctuations in the maps, rather than from a study of individually detected sources.

Information about the distribution of sources versus flux density (the differential source counts) is in the map pixel histogram. Patanchon *et al.* (2009) has successfully used this

"P(d)" analysis to show that parametric models of the differential source counts can be constrained by the map pixel histogram at the three BLAST wavelengths, independently, and at flux densities fainter than could be probed with detected sources only. The retrieved counts fall more rapidly in flux than would be expected for a population of sources showing no evolution in luminosity or density. Recent studies with SPIRE are in good agreement with these early BLAST results, and have extended the source counts both at the faint and bright end of the distribution (Clements *et al.* 2010; Glenn *et al.* 2010).

Stacking analysis using positional information of sources selected at different wavelengths is a powerful tool to estimate the contribution that a given class of object has to the submillimetre background. Even in the case where the stacking catalog has a high surface density, with several sources for each BLAST beam, stacking provides an unbiased estimate of the associated flux, under the assumption that the catalog is not correlated on scales comparable with the beam.

By stacking the BLAST fluxes selected at the positions of the $24\,\mu$m sources detected in deep MIPS surveys, Marsden *et al.* (2009) have shown that the retrieved CIB from 250 to $500\,\mu$m is consistent with the FIRAS measurements. A similar result was obtained by Dole *et al.* (2006) at wavelengths shorter than the peak of the CIB, confirming that these measurements are consistent with $24\,\mu$m selected sources generating the full CIB. Pascale *et al.* (2009) have studied the redshift distribution of the CIB, and have related it to the history of obscured star formation in the Universe, which shows a steep increase between $0 < z < 1$, confirming early results obtained from optical measurements (Lilly *et al.* 1996; Steidel *et al.* 1999).

In the 50 square degree survey of the Vela Molecular Ridge, Netterfield *et al.* (2009) detect more than 1000 compact sources in a range of evolutionary stages. The mass-dependent life-time is found to be longer than has been found in previous surveys of either low or high-mass cores, and significantly longer than free-fall or likely turbulent decay times. This implies some form of non-thermal support for cold cores during the early stage of star formation, which could be provided by the magnetic field.

4. Probing the roles of magnetic fields in star formation with BLASTPol

One of the key goals of modern astrophysics is to understand the details of star formation: how their masses are determined, and what the dominant physical processes that regulate the overall rate of star formation are. Significant progress has been made on these questions in recent years. For example, observations of dust emission and extinction (Nutter & Ward-Thompson 2007) show that the overall distribution of core masses mimics the distribution of stellar masses. Recent *Herschel* observations have shown that molecular clouds present a ubiquitous filamentary structure, in which long thin filaments form first, and then fragment into pre-stellar cores (André *et al.* 2010; Hill *et al.* 2011). However, fundamental questions regarding molecular cloud structure and star formation are still being debated (McKee & Ostriker 2007). For example, some investigators argue that molecular clouds, as well as cores, clumps and filaments inside the clouds, are dynamic structures, whose lifetimes are approximately equal to their turbulent crossing times (Vázquez-Semadeni *et al.* 2006). Others favor longer lifetimes, of order several crossing times (Netterfield *et al.* 2009; Blitz *et al.* 2007; Goldsmith *et al.* 2008). If clouds and their sub-structures do indeed live longer than a crossing time, they require support against gravity. This support could be provided by magnetic fields, which in many

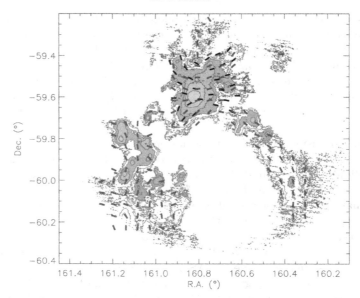

Figure 2. Preliminary results from the BLASTPol 2010 fight. The lines indicate the inferred magnetic field direction. Here we show a map made from 3 hours of BLASTPol data on the Carina Nebula. Total intensity contour levels measured at 500 μm are shown in the background. The red pseudo-vectors are BLASTPol measurements. The black pseudo-vectors are measurements from the SPARO polarimeter at 450 μm (Li *et al.* 2006) showing a good agreement between the two detections.

numerical simulations dramatically affect both the star formation efficiency and the lifetime of molecular clouds (Li *et al.* 2010; Hennebelle *et al.* 2011). However, observationally, the strength and morphology of magnetic fields in molecular clouds have been poorly constrained. Zeeman splitting detections are limited to the brightest Galactic sources (Crutcher *et al.* 2010; Falgarone *et al.* 2008) and optical polarimetry is not possible in these regions of high extinction. The best method for probing these fields is far-IR and submillimeter polarimetry (Hildebrand *et al.* 2000; Ward-Thompson *et al.* 2000, 2009), where the radiation from asymmetric dust grains, aligned by the local magnetic field, is detected in polarization. BLASTPol is the first submillimeter polarimeter with both sufficient mapping speed to trace fields across entire clouds and sub-arcminute spatial resolution to trace the field at the scale of dense cores. This provides a critical link between the *Planck* all-sky polarization maps (with 5′ resolution) and the planned ALMA polarization measurements at ultra-high resolution of small individual sources (though with only a 20″ field of view). BLASTPol data allow the first comprehensive detailed comparisons between observed magnetic fields in molecular clouds, and models derived from numerical simulations (Ostriker *et al.* 2001). Recent observations show that the extended sub-millimeter emission from molecular clouds is indeed polarized (Ward-Thompson *et al.* 2000, 2009; Li *et al.* 2006) and BLASTPol data is in good agreement with some of these early results (see next Section).

BLASTPol observations target the following three key questions in star formation and are discussed by Fissel *et al.* (2010): i) *Is core morphology and evolution determined by large-scale magnetic fields?* ii) *Does filamentary structure have a magnetic origin?* iii) *What is the field strength, and how does it vary from cloud to cloud?*

Table 1. BLASTPol 2010 Observed Targets

Target	Distance (pc)	Obs. Time (hours)	No. of B-vectors, 2010
Vela C	~ 700	64	~ 250
Lupus I	~ 200	62	~ 40
Puppis Cloud Complex	$\sim 1,000$	22	TBD
IRDC filaments	$\sim 2,000 - 4,000$	8	TBD
cool GMCs	$\sim 3,000 - 5,000$	18	TBD
Carina Nebula	$\sim 3,000$	3	~ 100

5. BLASTPol Observations

During our initial fight of BLASTPol in 2010, most components worked flawlessly during the entire 9 day Antarctic flight and we were able to obtain degree-scale polarization maps of several nearby molecular clouds. The quality of the data were unfortunately limited by the failure of one of our spectral blocking filters. The filter was damaged just before launch or during ascent, and the result was a loss of angular resolution as well as a high instrumental polarization ($\sim 5\%$) that varies across the focal plane. In spite of this problem, analysis of four sources has been successfully completed: Vela C , Lupus I, and our bright calibrators Carina Nebula (see Figure 2) and G331.5-0.1. For the latter two targets our results are consistent with those obtained using SPARO at South Pole station (Li *et al.* 2006). In Vela C (Fissel *et al.*, in preparation) and Lupus I (Matthews *et al.*, in preparation), we detect coherent polarization at the level of a few percent for most sightlines, providing valuable new information on the large-scale polarization properties of these nearby star-forming regions. A summary of the full target list for the 2010 flight is given in Table 1 (See also http://blastexperiment.info). A plan has been implemented to better protect the blocking filter that failed in 2010, and we are now preparing for a second BLASTPol flight, planned for December 2012 - January 2013. During this second flight we expect to map more sky area, with much higher angular resolution and better sensitivity, which will lead to an increase in the number of vectors per target cloud by a factor of at least 10 compared to the 2010 observations.

Acknowledgements

The BLAST collaboration acknowledges the support of NASA through grant numbers NAG5-12785, NAG5-13301 and NNGO-6GI11G, the Canadian Space Agency (CSA), the Leverhulme Trust through the Research Project Grant F/00 407/BN, Canada's Natural Sciences and Engineering Research Council (NSERC), the Canada Foundation for Innovation, the Ontario Innovation Trust, the Puerto Rico Space Grant Consortium, the Fondo Istitucional para la Investigacion of the University of Puerto Rico, and the National Science Foundation Office of Polar Programs; C. B. Netterfield also acknowledges support from the Canadian Institute for Advanced Research. We would also like to thank the Columbia Scientific Balloon Facility (CSBF) staff for their outstanding work.

References

André, P., Men'shchikov, A., Bontemps, S., *et al.* 2010, *A&A*, 518, L102
Blitz, L., Fukui, Y., Kawamura, A., *et al.* 2007, in *Protostars and Planets V*, ed. B. Reipurth, D. Jewitt, & K. Keil, 81–96
Chandrasekhar, S. & Fermi, E. 1953, *ApJ*, 118, 113
Chattopadhyay, G., Glenn, J., Bock, J. J., *et al.* 2003, IEEE Trans. Micro. T. Tech

Clements, D. L., Rigby, E., Maddox, S., *et al.* 2010, *A&A*, 518, L8

Crutcher, R. M., Wandelt, B., Heiles, C., Falgarone, E., & Troland, T. H. 2010, *ApJ*, 725, 466

Devlin, M. J., Ade, P. A. R., Aretxaga, I., *et al.* 2009, *Nature*, 458, 737

Dole, H., Lagache, G., Puget, J.-L., *et al.* 2006, *A&A*, 451, 417

Falgarone, E., Troland, T. H., Crutcher, R. M., & Paubert, G. 2008, *A&A*, 487, 247

Fissel, L. M., Ade, P. A. R., Angilè, F. E., *et al.* 2010, in Society of Photo-Optical Instrumentation Engineers (SPIE) Conference Series, Vol. 7741

Glenn, J., Bock, J. J., Chattopadhyay, G., *et al.* 1998, in Proc. SPIE, Advanced Technology MMW, Radio, and Terahertz Telescopes, Thomas G. Phillips; Ed., Vol. 3357, 326–334

Glenn, J., Conley, A., Béthermin, M., *et al.* 2010, *MNRAS*, 409, 109

Goldsmith, P. F., Heyer, M., Narayanan, G., *et al.* 2008, ArXiv e-prints, 802

Griffin, M. J., Abergel, A., Abreu, A., *et al.* 2010, *A&A*, 518, L3

Hennebelle, P., Commerçon, B., Joos, M., *et al.* 2011, *A&A*, 528, A72

Hildebrand, R. H., Davidson, J. A., Dotson, J. L., *et al.* 2000, *PASP*, 112, 1215

Hill, T., Motte, F., Didelon, P., *et al.* 2011, *A&A*, 533, A94

Li, H., Griffin, G. S., Krejny, M., *et al.* 2006, *ApJ*, 648, 340

Li, Z.-Y., Wang, P., Abel, T., & Nakamura, F. 2010, *ApJ*, 720, L26

Lilly, S. J., Le Fevre, O., Hammer, F., & Crampton, D. 1996, *ApJ*, 460, L1+

Marsden, G., Ade, P. A. R., Benton, S., *et al.* 2008, in Society of Photo-Optical Instrumentation Engineers (SPIE) Conference Series, Vol. 7020

Marsden, G., Ade, P. A. R., Bock, J. J., *et al.* 2009, *ApJ*, 707, 1729

McKee, C. F. & Ostriker, E. C. 2007, *ARA&A*, 45, 565

Moncelsi, L., Ade, P., Elio Angile, F., *et al.* 2012, ArXiv astro-ph: 1208.4866

Netterfield, C. B., Ade, P. A. R., Bock, J. J., *et al.* 2009, *ApJ*, 707, 1824

Nutter, D. & Ward-Thompson, D. 2007, *MNRAS*, 374, 1413

Ostriker, C., Stone, J. M., & Gammie, C. F. 2001, *ApJ*, 546, 980

Pascale, E., Ade, P. A. R., Bock, J. J., *et al.* 2008, *ApJ*, 681, 400

Pascale, E., Ade, P. A. R., Bock, J. J., *et al.* 2009, *ApJ*, 707, 1740

Patanchon, G., Ade, P. A. R., Bock, J. J., *et al.* 2009, *ApJ*, 707, 1750

Steidel, C. C., Adelberger, K. L., Giavalisco, M., Dickinson, M., & Pettini, M. 1999, *ApJ*, 519, 1

Valiante, E., Ade, P. A. R., Bock, J. J., *et al.* 2010, *ApJS*, 191, 222

Vázquez-Semadeni, E., Ryu, D., Passot, T., González, R. F., & Gazol, A. 2006, *ApJ*, 643, 245

Viero, M. P., Ade, P. A. R., Bock, J. J., *et al.* 2009, *ApJ*, 707, 1766

Ward-Thompson, D., Kirk, J. M., Crutcher, R. M., *et al.* 2000, *ApJ*, 537, L135

Ward-Thompson, D., Sen, A. K., Kirk, J. M., & Nutter, D. 2009, *MNRAS*, 398, 394

Astrophysics from Antarctica
Proceedings IAU Symposium No. 288, 2012
M. G. Burton, X. Cui & N. F. H. Tothill, eds.

© International Astronomical Union 2013
doi:10.1017/S1743921312016821

Dome Fuji Station in East Antarctica and the Japanese Antarctic Research Expedition

Kazuyuki Shiraishi

National Institute of Polar Research
10-3, Midoricho, Tachikawa, Tokyo 190-8518, Japan
email: kshiraishi@nipr.ac.jp

Abstract. Japanese Antarctic Research Expedition (JARE) commenced on the occasion of International Geophysical Year in 1957–1958. Syowa Station, the primary station for JARE operations, is located along the northeastern coastal region of Lützow-Holm Bay, East Antarctica (69°00′S, 39°35′E), and was opened on 29 January 1957. Since then, JARE have been carrying out research in various fields of earth and planetary sciences and life science. Astronomical science, however, has not been popular in Antarctica. In 1995, JARE established a new inland station, Dome Fuji Station (77°19′S, 39°42′E), which, at 3,810 m a.s.l., is located on one of major domes of the Antarctic ice sheet, some 1,000 km south of Syowa. The climatic conditions at Dome Fuji are harsh, with an annual average air temperature of −54°C, and a recorded minimum of −79°C. In 2007, JARE completed scientific drilling to obtain ice core samples of the Antarctic ice sheet reaching 3,050 m in depth. These ice cores record environmental conditions of the earth extending back some 720,000 B.P. In recent years, it is widely known that the high-altitude environment of inland Antarctica is suitable for astronomical observations and the Japanese astronomy community identified Dome Fuji Station as a potential candidate for a future astronomical observatory. In this article, the history of Japanese Antarctic activities are described in terms of access to the inland plateau of the Antarctic continent. The general scheme and future plans of science objectives and logistics of JARE will also be introduced.

Keywords. Antarctica, Dome Fuji, ice sheet, astronomy

---

1. Introduction

Antarctica is mostly covered by a thick ice sheet which contains more than 70% of the fresh water on Earth. The ice sheet has an average thickness of ∼2,450 m, making Antarctica the highest continent on the earth with mean surface elevation of ∼2,300 m a.s.l. East Antarctica, which occupies mostly the eastern hemisphere of the Antarctic continent, has a higher elevation than west Antarctica. A zone between ∼75–85°S in latitude in East Antarctica makes a high ridge along which three gentle peaks are recognised: Dome A (4,090 m a.s.l.), Dome C (3,260 m a.s.l.) and Dome F (3,810 m a.s.l.). Dome F was named Dome Fuji by the Japanese Antarctic Research Expedition (JARE) after the nearby "Fuji Toge (Pass)", which was traversed in 1968 by the Japanese traverse team to the South Pole. In 1995, JARE established Dome Fuji Station (77°19′S, 39°42′E) for an ice drilling project in order to obtain continuous ice samples down to the base of the ice sheet.

This short paper outlines a brief history of Japanese activities in the inland region of Antarctica, the scientific achievements at Dome Fuji, as well as future plans for JARE operations in inland regions of East Antarctica.

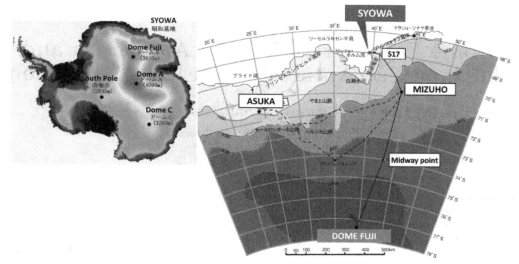

Figure 1. A map showing the stations of the Japanese Antarctic program.

2. A brief history of Japanese Antarctic Activity

Japanese Antarctic Expedition 1910–12. The first attempt to access inland Antarctica was performed by the first Japanese explorer, Nobu Shirase and his private expedition (Dagnell & Shibata, 2011). Shirase with *Kainan-maru*, a tiny ex-fishery boat of 204 tonnes, left Tokyo in November 1910 and arrived at Ross Sea via Wellington, New Zealand in January 1911. Because it was too late to continue the voyage, they returned to Sydney and wintered over there, then returned to the Ross Sea in December 1911. There they encountered the *"Fram"* of the Norwegian Expedition led by R. Amundsen at the Bay of Whales.

On 28 January 1912, the five members of Shirase's Main Landing Party using the dog sledges reached at 80°05′S in latitude and 156°37′W in longitude on the Ross Ice Shelf, where Shirase named the "Yamato Yukihara (Snow plain)". While the Main Landing Party marched to the south, *Kainan-maru* surveyed to the east along the coast which was later named as the Shirase Coast. They collected zoological, botanical and geological specimens and safely returned home.

IGY: International Geophysical Year. After 45 years, the third International Polar Year, that is, International Geophysical Year (IGY) was planned by the International Council for Science (ICSU), with the objective of detailed exploration and scientific investigation of the Antarctic continent. Japan decided to participate in this campaign which received enthusiastic support by the Japanese people. After discussions at the Antarctic sub-committee of the IGY special committee (CSAGI) in Brussels in September 1955, Japan agreed to carry out scientific investigations around Prince Harald Coast, Dronning Maud Land, East Antarctica. The first Japanese Antarctic Expedition (JARE-1) led by Takeshi Nagata on board *R/V Soya* left Tokyo on 08 November 1956. The *Umitaka-maru* of Tokyo University of Fisheries escorted her. Scientists in the summer team included meteorologists, geophysicists (seismology, geomagnetism, cosmic ray, aurora and ionospheres), marine biologists, geographers, oceanographers and surveyors. They constructed Syowa Station (69°00′S, 39°35′E) with 4 small huts and a 20 kVA generator on Ongul Island in the Lützow-Holm Bay, and officially opened it on 29 January 1957. Eleven personnel wintered over in order to conduct meteorology, upper atmospheric physics, geography and geology. Since then, Syowa has been the primary station for JARE operations both as a permanent observatory and as a supply station for field activities, although there was a three season break before the icebreaker *Fuji* was commissioned in 1966. The station has expanded in size, to more than 60 buildings and huts covering an area of 6,778 m² by the time of IPY 2007–2008.

After IGY, Japan signed the Antarctic Treaty as one of 12 original signatory countries in 1961.

The inland activities. During the early stages of JARE operations, Antarctica was still "*Terra incognita*". Satellite image maps were not available and topographic maps, if available, were rudimentary. A number of inland traverses were undertaken to conduct geographical surveys after 1957, finding previously unknown mountain ranges. In 1960, the Yamato Mountains were discovered from the air by a Belgian team and geological and geomorphological surveys were undertaken by JARE-4.

One of the great steps to the inland Antarctica was a traverse led by Masayoshi Murayama to the South Pole in 1968. After 141 days, after leaving Syowa Station, the 12-man party with four snow vehicles completed the return trip to the South Pole, covering a distance of 5,182 km. This overland traverse was only the 9th successful expedition to the South Pole. En route to South Pole and return, the expedition conducted geophysical, glaciological and meteorological studies. This traverse illustrated significant progress and expertise of Japanese inland traverse capability and development in the technology of snow vehicles.

After the success of this South Pole traverse, glaciologists and geologists proposed to extend survey operations to inland areas. One of the first such inland operations was the "Enderby Land Project" (1969–1974), the objectives of which was to survey the catchment area of the Shirase Glacier, which is the largest glacier in the vicinity of Syowa Station, to study the ice dynamics of the glacier.

In 1970 the first inland station "Mizuho" was established ~300 km away from the coast as a base for glaciological and meteorological studies. In 1985, a second inland station "Asuka" was built ~630 km west of Syowa Station primarily as a base for geoscience programs. Through these inland activities, JARE accumulated extensive experience conducting remote inland activities.

Science in JARE. JARE have been conducting extensive and diverse science programs since commencing operations in the Antarctic, both at stations and in the deep field. At Syowa Station many routine observation and monitoring studies have been conducted, including meteorology, upper atmospheric geophysics, atmospheric sciences, geodesy, gravity and geomagnetic study, biology and environmental sciences, to name a few. In inland field operations and on board ship voyages, geology, geophysics, oceanography, marine and terrestrial biology, geodesy, glaciology and other studies have been conducted. Many significant scientific contributions have been reported. For example, the study of aurora has been developed because Syowa Station is conveniently located within the aurora oval. In 1982, JARE meteorologists established the existence of the ozone hole in the upper atmosphere.

Another significant scientific contribution to earth and planetary science is the discovery of a vast numbers of meteorites in the blue ice fields of the continent (Yamaguchi, this volume). Since 1969, JARE have collected 17,100 specimens (>1.5 tonnes) including Martian and lunar meteorites. Using suitable satellite images, we can predict where meteorites are concentrated. During 1976–79, a Japan–USA joint search for meteorites program was held in the Transantarctic Mountains, continuing to the current ANSMET (the Antarctic Search for Meteorites) program by NSF. Nowadays, several other national

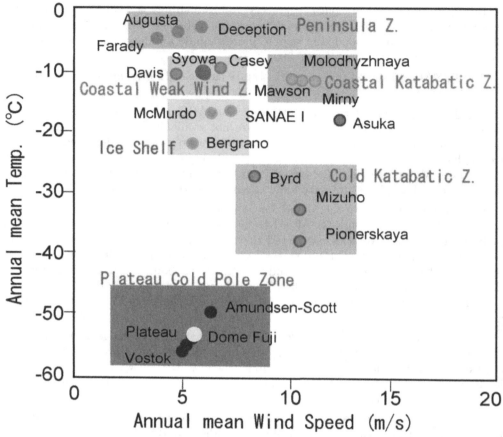

Figure 2. Climate classification of stations in Antarctica using wind velocity (m/s) *vs.* air temperature (°C).

Antarctic programs, such as China, Korea, Italy and Belgium, are conducting meteorite searches for research and have collected more than 30,000 specimens of meteorites.

3. Dome Fuji Station

The Antarctic continent is covered by a thick ice sheet. East Antarctica forms a huge dome-like shape in cross section, reaching as much as 4000 m a.s.l. at the highest point. In 1995, JARE established a new inland station, Dome Fuji Station (77°19′S, 39°42′E), which is located at ~3,810 m a.s.l. at one of ice domes of East Antarctica.

Figure 2 shows the classification of the Antarctic stations in terms of annual mean wind speed and air temperature. The inland stations experience extreme climatic conditions on the Antarctic plateau. Records obtained at Dome Fuji during 1995–1997 show that the annual mean temperature is −54.3°C and −79.7°C minimum. The annual mean air pressure of 598.4 hPa makes conditions extremely difficult for humans to live (Watanabe *et al.*, 2003).

The snow accumulation is heavy at the coastal zone of the Antarctic continent except for the steep edge where ablation of ice is predominant. However, the rate of snow

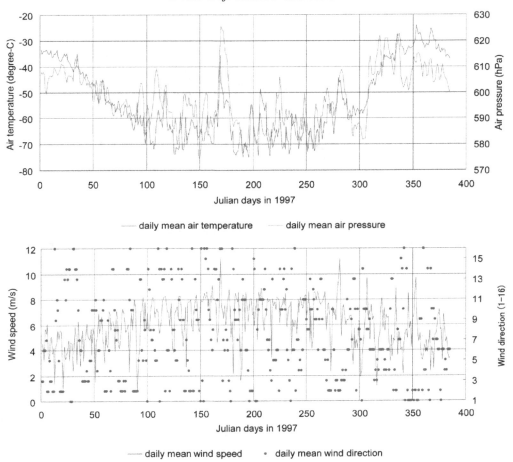

Figure 3. Meteorological data at Dome Fuji Station in 1997. (Motoyama, personal communication)

accumulation decreases inland, toward inland high plateau and, as a result, snowfall at the Dome Fuji is very rare.

Ice core drilling at Dome Fuji The ice core drilled at Dome F records the history of the snow accumulation, and with each annual layer of snow, air bubbles are trapped preserving ancient air. Therefore, an ice core is just like a time capsule of the earth's history, enabling scientists to study changes in the atmosphere back through time.

A deep ice core drilling project to obtain ice core samples of Antarctic ice sheet reaching to the bedrock was carried out in 1993–1998. During the first overwintering at the Dome Fuji, drilling commenced in 1995 and successfully collected ice core down to 2,503 m deep, an ice record that extends back 320,000 years. In 2006, second deep ice core drilling program succeeded reaching 3,035 m in depth, almost the bottom of the ice sheet. The oldest ice was estimated ~720,000 years old. Unexpectedly, the bottom of ice was melted due to geothermal energy and could not reach the basement rocks.

Through detailed analysis, glaciologists identified climate cycles of several tens of thousands years and 100 years in scale preserved in the ice core. The deep ice core preserved volcanic ash layers and even micrometeorite layers (Misawa *et al.* 2010). It is beyond the scope of this paper to describe the glaciological results in detail and there are many

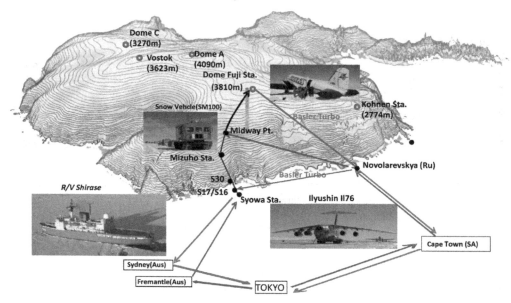

Figure 4. Access to Dome Fuji by sea, air and land.

published reports on the study of Dome Fuji ice core (*e.g.* Kawamura *et al.*, 2007, Uemura *et al.*, 2012).

4. Access to Dome Fuji

Transportation to the Antarctic is essential for Antarctic expeditions. Figure 4 illustrates an example of traverse to Dome Fuji. Normally the Icebreaker *Shirase* departs the Port of Tokyo in mid-November, and two weeks later expedition personnel depart, via air, for Fremantle, Western Australia, where the icebreaker visits en route to Syowa Station. The icebreaker *Shirase* is 12,600 tonnes displacement and 138 m in length, is operated every austral summer season and carries around 1,100 tonnes of cargo and fuel, including supplies supporting Antarctic operations. However, the fast sea ice around Syowa Station is not always in good condition. In the 2011–2012 season, the icebreaker *Shirase* could not reach Syowa Station due to extremely difficult sea ice conditions, with the sea ice reaching 8 m in thickness. It has been 18 years since the last time the sea-ice conditions were as unfavorable as this.

Transportation into the inland. The traverse party with several snow vehicles and sledges leaves from the S16 point, a coastal depot 20 km east of Syowa. As already shown, snow accumulation on the inland plateau is less than in coastal areas, in contrast, the katabatic wind (which is a cold air mass sliding down along the surface of ice sheet from higher altitudes) forms a wind sculptured rough snow surface feature called "sasturugi". Sasturugi make oversnow travel very slow and difficult and it takes roughly three weeks to travel to Dome Fuji. During the traverse, various glaciological, meteorological and geophysical observations are carried out. On the way to Dome Fuji, there are a couple of unmanned stations for meteorology, glaciology, geomagnetism and seismology.

Air transportation. Compared with other areas in the Antarctic, such as the Ross sea region and the Antarctic Peninsula area, regular air transportation has not been available in East Antarctica. However, in the last ten years an air network called DROM-LAN (Dronning Maud Land Air Network) was established by the cooperation of eleven countries, including Japan. A flight takes only 6.5 hours from Cape Town to the Russian Novolazarevskaya Station with a cargo jet, Ilyushin Il76. From Novolazarevskaya, intra-continental feeder flights are carried out by Basler Turbo BT67 ski-equipped aircraft. This aircraft is a modified version of the versatile Douglas DC3. The Basler Turbo is able to land and take off at Dome Fuji, however, avoiding altitude sickness for personnel, it is safe not to fly directly using aircraft except for emergency. Normally, JARE operations use aircraft from Novolazarevskaya to the S17 point or the midway point between Mizuho and Dome Fuji. From the landing point, travel to Dome Fuji is completed on surface with snow vehicles.

5. New Era of JARE

Frontier of Earth-Planetary Sciences. At the occasion of the 50th anniversary of JARE, the JARE Headquarters discussed the vision of a new science strategy and selected topics on "Global Warming" as the principle research projects to be considered for high-level national and international research collaboration. In addition, there are general topics proposed by many scientists including a number of routine monitoring observations, conducted by NIPR and other Japanese government agencies such as the Japan Meteorological Agency, Geographical Survey Institute and Japan Coast Guard.

In recent years, it is widely known that the environment of the inland of Antarctica is suitable for astronomical observation and Japanese astronomical community has identified Dome Fuji Station as a candidate of the future astronomical observatory (*e.g.* IAU Aymposium 288 in 2012). In addition, the Antarctic Inland Plateau is recognised as an important area for many scientific fields. NIPR and collaborators are planning to propose new projects utilising the new Dome Fuji Station, as a part of major science projects in the next decade planned by the Japan Science Council.

The new projects at and around Dome Fuji may include the following targets:

(*a*) The oldest ice in the world, which is expected to elucidate the Earth history up to one million years ago,

(*b*) 3D-observation for aurora in cooperation with Syowa Station, which will reveal the mechanism of aurora and upper atmospheric geophysics,

(*c*) Atmospheric science in the polar vortex, which will reveal the matter cycle between stratosphere and troposphere,

(*d*) Age and composition of continental crust underneath the ice sheet and its role in the evolution of the Antarctic continent,

(*e*) Sub-glacial lakes to explore for unknown micro-organisms,

(*f*) Microbiology under extreme conditions related to the global circulation of microbes,

(*g*) Biorhythms during polar night as a physiological study under extreme environment.

New Dome Fuji Station and logistics in the inland area. To make this sciences possible in inland areas, we have to develop new technology and build the new Dome Fuji Station. Currently, it is in a design phase and several plans have been proposed.

For designing the new Dome Fuji Station, the following points are considered:

(*a*) Keeping the building on the surface for more than 10 years,

(*b*) Foundation of buildings on snow surface, avoiding unequal subsidence,

(*c*) Using light weight and thermal insulated panels of buildings, for reducing the load of transportation,

(*d*) Heating used sustainable energy,

(*e*) Effective energy consumption in terms of electricity, water *etc.*

The most difficult issue is the foundation of the buildings to avoid tilting and snow drift during the lifetime of buildings. In addition, saving energy, utilising discharged grey water, and other environmentally friendly measures are requested.

Traverse technology. Cargo transportation with a limited number of personnel is essential to maintain the station. It is necessary to carry a huge amount of cargo and fuel. One of the ideas to keep stable transportation in the future is the "unmanned tractor system" to minmise the logistic burden. Testing of an original model is now being undertaken.

6. Summary

In summary, it is safe to say that:

(*a*) Japan has a long history of conducting traverses into the inland area of Antarctica since early last century. These activities fostered development of expertise, experience, logistics and technology to successfully conduct deep inland traverses.

(*b*) Dome Fuji Station is one of the inland stations which has advantages for frontier research in earth, planetary and life sciences.

(*c*) In the future, astronomical research will be a high priority at Dome Fuji Station, but will also include atmospheric and glaciological studies.

However, there are many difficulties and challenges to maintain the year-round station, such as new buildings based on innovative concepts of construction, a new transportation system and so on.

Acknowledgements

The author would like to thank to Professors Takashi Ichikawa of Tohoku University, Michael Burton and John Storey of the University of New South Wales for inviting him to the IAU symposium in Beijing. He is also indebted to Dr. Chris Carson of Geoscience Australia for comments to the draft. The author thanks Professor Hideaki Motoyama of NIPR for information on the deep ice core drilling project at Dome Fuji.

References

Dagnell, L., & Shibata, H. (Translated into English) 2011 *The Japanese South Pole Expedition 1910–12 A Record of Antarctica.* (Compiled and edited by the Shirase Antarctic Expedition Supporters' Association. Originally published in Japanese in 1913 by Nankyoku Tanken Koenkai, Tokyo), Bluntisham Books & Erskine Press, 414pp.

Kawamura, K., Parrenin, F., Lisiecki, L., Uemura, R., Vimeux, F., Severinghaus, J. P., Hutterli, M. A., Nakazawa, T., Aoki, S., Jouzel, J., Raymo, M. E., Matsumoto, K., Nakata, H., Motoyama, H., Fujita, S., Goto-Azuma, K., Fujii, Y., & Watanabe, O. 2007 *Nature*, 448, 912

Misawa, K., Kohno, M., Tomiyama, T., Noguchi, T., Nakamura, T., Nagao, K., Mikouchi, T., & Nishiizumi, K. 2010 *Earth and Planetary Science Letters*, 289, 287

Uemura, R., Masson-Delmotte, V., Jouzel, J., Landais, A., M otoyama, H., & B. Stenni, B. 2012 *Climate of the Past*, 8, 1109

Watanabe, O., Kamiyama, K., Motoyama, H., Fujii, Y., Igarashi, M., Furukawa, T., Goto-Azuma, K., Saito, T., Kanamori, S., Kanamori, N., Yoshida, N., & Uemura, R. 2003 *Memoirs of National Institute of Polar Research*, Special Issue No. 57, 1

Astrophysics from Antarctica
Proceedings IAU Symposium No. 288, 2012
M. G. Burton, X. Cui & N. F. H. Tothill, eds.

© International Astronomical Union 2013
doi:10.1017/S1743921312016833

The US Long Duration Balloon Facility at McMurdo Station

W. Vernon Jones

Astrophysics Division, DH000, Science Mission Directorate,
NASA Headquarters, Washington, DC 20546
email: w.vernon.jones@nasa.gov

Abstract. A sea change in scientific ballooning occurred with the inauguration of 8–20-day flights around Antarctica in the early 1990's. The attainment of 28–31-day flights and 35–42-day flights, respectively, in two and three circumnavigations of the continent has greatly increased the expectations of scientific users. There is a scientific need for the capability to provide similar-duration flights for investigations that cannot be done in the Polar Regions. A new super-pressure balloon is currently under development for future flights of 60–100 days at any latitude. This first new balloon in more than half a century would meet this need and allow the focus to change from increasing the durations of flights over and around Antarctica to ultra-long-duration flights from Antarctica.

Keywords. NASA Balloon Program, Antarctic Ballooning, LDB, ULDB

1. Introduction

The U.S. National Aeronautics and Space Administration (NASA) routinely launches scientific balloons from sites in Antarctica, Australia, Sweden and within the U.S. The Antarctic Long-Duration Balloon (LDB) flights are launched from a site on the Ross Ice Shelf near McMurdo, in cooperation with the U.S. National Science Foundation Office of Polar Programs (NSF/OPP). The buildings comprising the LDB launch site are shown in Fig. 1. The two large buildings on the left are the so-called payload buildings, which are available to the science teams for final integration and test of their payloads prior to launch. These are the largest buildings on the Antarctic Continent. They are on skids to facilitate their movement to nearby elevated berms of snow at the end of each launch season. They are dragged back to the launch site configuration at the beginning of the next season.

This LDB camp site is very remote from both the home base of the Columbia Scientific Balloon Facility (CSBF) in Palestine, Texas and the scientific user team laboratories, which are mainly in the Continental U.S. (Conus). Capabilities for making on-site payload repairs are minimal. The payload buildings are intended to support re-assembly and checkout of "flight-ready" payloads after their shipment "to the Ice." Until recently, LDB flight candidates were required to have completed an engineering test flight of all the flight components approximately a year in advance of their desired LDB flight. Recently, in lieu of an engineering test flight, the Balloon Program has offered the option for an extensive Thermal-Vacuum Test at the Plumbrook Facility of the Glenn Research Center. In either case, the payloads are shipped to Antarctica only after having completed successful hang-tests at CSBF to verify that they are indeed flight ready.

Figure 2 shows three payloads being readied for launch outside the two large payload buildings during the 2007-2008 Austral season, the first time three payloads were launched. These three payloads shared the two payload buildings, with the largest BLAST

Figure 1. The LDB Camp Site near McMurdo.

(Balloon Large Aperture Submillimeter Telescope) in one and both ANITA (Antarctic Impulsive Transient Antenna) and SBI (Solar Bolometric Imager) in the other. NASA is currently planning to add a third payload building, a critical need for accommodating the exceedingly large payloads wanting Antarctic LDB flights. Another, corresponding critical need is a dedicated airplane with large cargo capability for same-season recovery of the payloads at the end of their flights.

2. Impact of Antarctic LDB Flights

Scientific ballooning offers a unique capability for frequent access to near-space for instruments ranging in mass from a few kilograms to a few tons. Balloon payloads for science, applications, and new technology development have been flown for periods of 1–2 days since the 1950's. The flight times were extended to 10–20 days in the early 1990's by conducting launches in Antarctica during the austral summer. These long-duration balloons (LDB) float in the nearly circumpolar stratospheric wind vortex during the Antarctic summer. They employ zero-pressure polyethylene balloons identical to those

Figure 2. ANITA, left, hanging from the launch vehicle; BLAST, middle, hanging from the payload building; and SBI, right, standing on the Ice.

utilized for conventional 1–2 day flights, whose durations are limited because ballasting is required to minimize their altitude excursions during day-night transitions. The zero-pressure balloons used today have changed only incrementally from the large polyethylene balloons introduced in the 1950's. The order of magnitude improvement in flight duration in the polar region is possible because of the constant daylight during local summer. The nearly constant solar heating ensures nearly constant altitudes with minimal or no ballasting (Gregory & Stepp 2004).

Most LDB missions have carried suspended payloads of 2,300 – 2,800 kg, with scientific instruments of 900 – 1,400 kg, to altitudes of 37 – 41 km for one circumnavigation of the continent. In 2002, a record was set when a 0.83 MCM (million cubic meter) balloon carrying the Trans Iron Galactic Element Recorder (TIGER) payload flew in excess of 31 days in two rounds of the South Pole (Geier *et al.* 2005). In 2005 a new LDB flight record was set when a 1.11 MCM balloon carrying the Cosmic Ray Energetic and Mass (CREAM) experiment (Seo *et al.* 2008) flew for nearly 42 days while circumnavigating the continent three times. The CREAM payload has accumulated a total of 162 days of exposure with six Antarctic flights, the record for a single balloon project (Seo 2012).

Scientific ballooning is a vital infrastructure component for astronomy and astrophysics in general. Instruments carried on high-altitude balloons have produced important scientific results, and many instruments developed initially for balloon flights have been used on spacecraft for significant astrophysical observations. Ballooning seems essential for continued scientific progress and instrument development, since it is highly unlikely that all of the worthy space flight projects being studied can be funded within any plausible federal budget during the coming decade. Scientific ballooning is simultaneously an excellent environment for training graduate students and young post-doctoral scientists. Indeed, many leading astrophysicists, including 2006 Nobel laureates John Mather and George Smoot, gained invaluable early experience conducting balloon-borne science investigations (Israel *et al.* 2009).

The Antarctic LDB flights enabled by the NASA–NSF/OPP cooperative agreement have been subsequently dubbed "jewels in the crown of the NASA Balloon Program." Among the 41 LDB flights launched to date in Antarctica were a 0.201 Million Cubic Meter (MCM) SPB flown successfully for 54 days around Antarctica between December 2008 and February 2009, and a 0.402 MCM SPB flown successfully for 22 days in January 2011. See Fig. 3 for a photograph of the former balloon at its float altitude, along with a graphic of its trajectory showing the turnaround of the polar vortex.

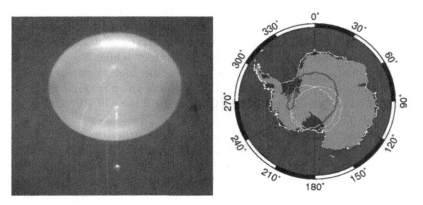

Figure 3. A photograph of the 7 MCF super-pressure balloon at its float altitude and its 54-day flight trajectory in Antarctica showing the turnaround of the polar vortex.

The NASA-NSF/OPP partnership that established the Antarctic LDB program revitalized ballooning. This new capability facilitated several high impact cosmic ray / particle astrophysics projects. Recent examples include the Advanced Thin Ionization Calorimeter (ATIC), which reported an unexpected surplus of high-energy cosmic ray electrons after two LDB flights in Antarctica (Chang *et al.* 2008). The source of these excess electrons would need to be a previously unidentified and relatively nearby cosmic object, within about 1 kilo parsec (3,260 light years) of the Sun. Annihilation of exotic particles postulated to explain dark matter is among other explanations proposed. The Balloon Experiment with a Superconducting Spectrometer (BESS) has conducted a negative search for annihilation signatures of dark matter in the antiproton channel (Abe *et al.* 2008). The electron excess in ATIC and lack of excess antiprotons in BESS provide interesting constraints on dark matter models.

Another notable example is the Cosmic Ray Energetics And Mass (CREAM) investigation, which extends direct elemental composition measurements to the highest energy practical in a balloon experiment to explore the theoretical limit of supernova shock wave acceleration (Ahn *et al.* 2010) This project has already achieved a record-breaking cumulative exposure of $\sim$ 162 days in 6 successful flights over Antarctica. Its report of hardening in the elemental spectra calls for a cosmic-ray acceleration and propagation model that is more realistic than current models based on a steady state/continuous source distribution.

The Trans-Iron Galactic Element Recorder (TIGER), a non-magnet spectrometer, has measured the elemental composition of cosmic rays heavier than iron in a search for the origin of cosmic rays (Rauch *et al.* 2009). TIGER was the first payload to make two circumnavigations of Antarctica in one flight. It produced a strong indication that cosmic rays originate and are accelerated in associations of massive stars called OB associations. A larger instrument called Super-TIGER is one of three pay-loads planned for LDB flights in December 2012. Super-TIGER is a large-area instrument being developed to measure the abundances of $30 \leqslant Z \leqslant 42$ elements with unprecedented individual-element resolution. It will test the emerging model of cosmic-ray origin in massive (i.e., OB) star associations, and models for atomic processes by which nuclei are selected for acceleration to cosmic ray energies.

The Antarctic Impulsive Transient Antenna (ANITA) is a unique neutrino experiment to constrain the origin of the highest energy particles in the universe (Gorham *et al.* 2010). It is designed to detect coherent radio Cherenkov radiation from neutrino-initiated showers in the Antarctic ice. Figure 4 illustrates the underlying technique, and Figure 5 shows a photograph of the instrument. Ultra High Energy Cosmic Rays guarantee an associated GZK (Greisen-Zatsepin-Kuzmin) neutrino flux from the interactions of extreme energy hadrons with cosmic microwave background photons. ANITA observes $\sim 10^6$ km^3 of ice from balloon altitudes of $\sim$ 110,000 feet, which is nearly optimal for the ultrahigh energy neutrinos of interest. ANITA now has highest-energy sample of radio-detected ultra high-energy cosmic rays. Its detection of these extreme energy cosmic ray events was featured on the cover of the October 8, 2010 issue of Physical Review Letters (Hoover *et al.* 2010).

Balloon-borne cosmic microwave background (CMB) experiments have had exceptionally high impact, most notably the Balloon Observations Of Millimetric Extra-galactic Radiation and Geophysics (BOOMERanG), which established that the universe is flat, i.e., that its geometry is not curved but Euclidean. This result was obtained by measuring a detailed map of the CMB temperature fluctuations. The Principal Investigators of this project were awarded the 2006 Balzan Prize for Astronomy and Physics "for their contribution to cosmology, in particular the Boomerang Antarctic balloon experiment." This

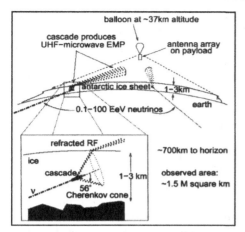

Figure 4. Schematic of ANITA measurement concept with balloon over the Antarctic ice.

Figure 5. Photograph of the ANITA payload preparing for launch in Antarctica.

prize is considered to be one of the highest awards for science, culture and humanitarian achievement, ranking close to the Nobel Prize.

The COBE and WMAP CMB Explorer missions were enabled by precursor balloon flights beginning in the 1970's (Israel *et al.* 2009). The currently operating Planck CMB satellite also relied on advances made in these balloon missions. Polarization-sensitive focal planes employing Transition Edge Sensor (TES) bolometers, polarization modulation strategies, and developing filter technologies are being employed by high priority instruments discussed at this Symposium and currently awaiting LDB flights in Antarctica to search for signatures of inflation. The IRAS, ISO and Spitzer observatories all relied on far–IR telescope and detector technologies proven during balloon flights in the 1970's and 1980's. The balloon-borne High Energy Focusing Telescope, predecessor of the NuSTAR mission launched in early 2012, utilized similar multilayer optics and CdZnTe pixel detector technologies.

3. Super Pressure Ballooning and ULDB Flights

In parallel with increasing Antarctic LDB flight durations, NASA has continued development and qualification flights leading to heavy-lift super pressure balloons capable of supporting 1,000 kg science instruments to 33 km for upwards of hundred day missions, with plans for increasing the altitude to 38 km. This goal is even more important now, in view of the National Research Council Astro2010 Decadal Study recommendation that NASA should support ultra-long duration ballooning (ULDB) for science missions in a range of disciplines. Figure 6 illustrates a comparison of the performance of super-pressure and zero-pressure balloons. The volume of zero-pressure balloons used for conventional and polar LDB flights changes as the ambient atmospheric pressure changes, causing large droop at night. By contrast a super-pressure balloon maintains nearly constant volume, which allows LDB flights in non-polar latitudes, including ULDB flights.

The current vision for scientific ballooning includes development of super-pressure balloons (SPB) designed to maintain essentially constant volume, day and night, and thus to float at nearly constant altitude without the need for dropping ballast at sunset. These sealed balloons are designed to withstand slight differential pressure. They are inflated with enough helium to fill the volume at the coldest temperatures, and they have sufficient strength to hold that helium when sunlight heats it (Cathey 2011). They would permit LDB flights of one- to two-week durations at any latitude without diurnal altitude variation. They would also permit ULDB flights that circumnavigate the globe at any latitude, with the potential for durations on the order of a hundred days. This is in contrast to conventional zero pressure balloons, which cannot keep altitude for more than about a week at mid-latitudes, even with substantial ballast drops to limit excursions due to day/night cycles. The Astro2010 decadal survey strongly supported ULDB flights, especially for cosmic microwave background and particle astrophysics research (Panel on particle astrophysics and gravitation 2010).

The Antarctic LDB flight capabilities have dramatically increased access of heavy payloads to near space for durations as long as $\sim$ 45 days using zero pressure balloons. The ULDB capability based on SPB technology can extend the flight time to as much as 100 days, even at mid-latitudes, thereby opening up the entire sky to investigators with payloads having substantial weight and power requirements. These expanded capabilities will allow investigations from balloons that previously could be done only from Explorer class missions, for example, at a fraction of the cost. Currently, ULDB is defined as a 1,000 kg science instrument suspended along with its flight support equipment from a SPB floating above 33 km for up to 100 days. Comparable flights of smaller instruments to higher altitudes around 38 km on larger SPB's are also being pursued.

Figure 7 compares the altitude variations of the 2008 SPB (introduced in Fig. 3) with two LDB payloads flown during the same season: Cosmic Ray Energetics and Mass (CREAM) and Antarctic Impulsive Transient Antenna (ANITA). The figure shows that the Super-Pressure Balloon maintains a stable altitude with little variation while the zero-pressure balloons significantly droop during a diurnal cycle. The SPB's differential pressure varied as expected due to time of day and solar and Earth IR inputs. At the end of the flight, ballast was dropped to verify the balloon's structural envelope to the maximum design pressure. The payload and portions of the balloon were recovered for post-flight testing.

The 2011 test flight of a 0.402 MCM super-pressure balloon launched in Antarctica on January 9, 2011 flew for 22 days. The flight performance matched predictions very closely. The SPB balloon carried 1,815 kg suspended payload, and it fully deployed just before reaching the target float altitude at essentially zero differential pressure. It took $\sim$3 hours to ascend to its float altitude of $\sim$ 33.9 km, and it demonstrated almost no

Figure 6. Illustration of the day-time and night-time altitude performance of a super-pressure and zero-pressure balloon.

altitude change during the 22-day flight. The balloon demonstrated stable pressure and remained at its designed float altitude for the flight duration. The average altitude variations were about ± 180 m, or ∼ 0.5%.

The first deployment verification test of a 0.525 MCM super-pressure balloon vehicle was successfully launched August 14, 2012 with a 2,270 kg suspended payload. This test flight demonstrated the balloon vehicle performance, obtained in-flight video data of the balloon inflation, measured the differential pressure at the base of the balloon, and measured the tension in a select number of tendons. It also obtained temperature and altitude data throughout the flight, demonstrated vehicle altitude stability and performance, and validated the structural envelope through pressurization. This SPB also fully deployed at very low differential pressure. All ballast was expended during the course of 2 hours at float to pressurize the balloon as much as possible. The balloon performance was judged to be excellent.

4. Future Super Pressure Balloon Flights

The next 0.525 MCF SPB flight is tentatively planned for May/June of 2013 from Kiruna, Sweden with suspended payload up to ∼2,500 kg, including a small piggyback science instrument. Further plans call for this pumpkin balloon to carry its first science payload for an extended duration flight over Antarctica, before leaving the continent for an extended ULDB flight. After completing of tests of the 0.525 MCM balloon, it is planned to scale up the SPB design for future test flights of a 0.746 MCM balloon toward meeting the project goal for development of a balloon vehicle capable of carrying 2,721 kilograms to 33.5 kilometers for 60–100 day duration missions.

The Balloon Program is planning a mid-latitude demonstration flight of the full-scale SPB by 2014, with a goal of developing a Southern Hemisphere launch site that will support LDB and ULDB science flights, while also complying with flight safety policies. In addition, once operational, NASA also plans to launch ULDB missions from Antarctica with recovery off the continent in the southern hemisphere.

5. Wallops Arc Second Pointer (WASP)

The NASA Scientific Balloon Program affords researchers the opportunity to conduct research in a near-space environment. Flight altitudes above more than 99.5% of the

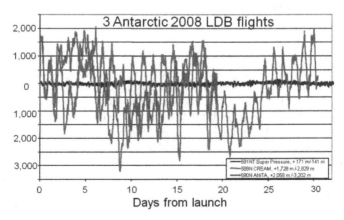

Figure 7. Comparisons of the altitude variation of the 2008 SPB test flight relative to the CREAM and ANITA flights on zero-pressure balloons.

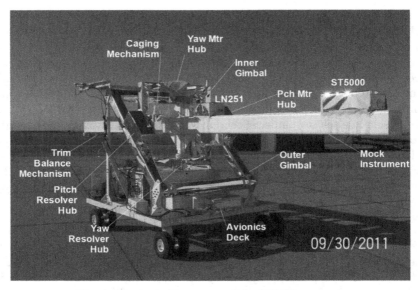

Figure 8. Artist's concept of the WASP payload.

earth's atmosphere are typical. Recognizing that there is significant interest from the science community in a reliable balloon-borne fine pointing system, the Balloon Program has developed and tested the Wallops Arc Second Pointer (WASP) for the user community. See Fig. 8. Potential user areas include extra-solar planetary finders, cosmic background exploration, astronomy at a variety of wavelengths, high-energy astrophysics, upper atmospheric science and ultra-high-energy neutrinos, etc.

The first test flight of the WASP gondola with a mock telescope occurred in October 2011. During the 5-hour flight, the WASP system was exercised at a float altitude of 32 km for ~ 2 hours. It demonstrated sub-arcsecond pointing stability with the mock telescope in a typical flight environment. The mock instrument was uncaged, and inertial target offsets were issued from the ground to demonstrate acquisition dynamics. The system was able to maintain arcsecond pointing stability during discrete ground-commanded gondola azimuth adjustments. The second flight in September 2012 provided more extensive tests. Specifically, the WASP system was in fine-pointing mode for over seven hours during daytime, and approximately three hours during nighttime using its onboard Star Tracker. The pointing performance was sub-arcsecond consistently.

6. Acknowledgements

The outstanding staff of the NASA Balloon Program Office, Columbia Scientific Balloon Facility, and Aerostar, Inc. are responsible for the advancements being made in scientific ballooning. The NASA Balloon Program Office deserves special recognition for leading this effort with its in-house studies and oversight of the design tools, film technologies, and fabrication techniques required for its development. Special recognition is due the staff of the Columbia Scientific Balloon Facility for a lifetime of contributions to ballooning, and to the staff of Aerostar, Inc. for innovative approaches to pumpkin-balloon fabrication and consistent fabrication of high quality balloons. The Antarctic LDB program, the crown jewel of ballooning, would not be possible without the crucial contribution of the U.S. National Science Foundation Office of Polar Programs and its Antarctic support contractors.

References

Abe, K. *et al.* 2008, *Phys. Lett. B*, 670, 103–108

Ahn, H. S. *et al.* 2010, *Astrophys. J.*, 714, L89–L92

Cathey, H. 2011, Internal NASA balloon program report, NASA `http://sites.wff.nasa.gov/code820/`

Chang, J. *et al.* 2008, *Nature*, 456, 362–365

Geier, S, *et al.* 2005, *Proc. 29th ICRC*, Pune, India OG1.5

Gorham, P. *et al.* 2010, *Phys. Rev. D*, 82, 022004

Gregory, D. D. & Stepp, W. E. 2004, *Adv. Space Res.*, 33, No. 10, 1688

Hoover, S. *et al.* 2010, *Phys. Rev. Lett*, 105, 151101

Israel, M. *et al.* 2009, Report of the Scientific Ballooning Roadmap Team: NASA Stratospheric Balloons. `http://sites.wff.nasa.gov/code820/`

New Worlds, New Horizons in Astronomy and Astrophysics, 2010, Panel on Particle Astrophysics and Gravitation, `http://www.nap.edu`

Rauch, B. *et al.* 2009, *Astrophys. J.*, 697, 2083–2088

Seo, E. S. *et al.* 2008, *Adv. Space Res.*, 42, 1656–1663

Seo, E. S. 2012, *Astropart. Phys.* (in press)

Astrophysics from Antarctica
Proceedings IAU Symposium No. 288, 2012   © International Astronomical Union 2013
M. G. Burton, X. Cui & N. F. H. Tothill, eds.   doi:10.1017/S1743921312016845

The French-Italian Concordia Station

Djamel Mekarnia[1] and Yves Frenot[2]

[1]Université de Nice, Observatoire de la Côte d'Azur, CNRS UMR 7293,
Bd de l'Observatoire, BP 4229, F-06304 Nice Cedex 4, France
email:mekarnia@oca.eu

[2]IPEV - Institut polaire français Paul Emile Victor, Technopôle Brest-Iroise,
BP 75, F-29280 Plouzané, France
email: y.frenot@ipev.fr

Abstract. Concordia is a French-Italian permanent station located at Dome C, Antarctica. The station provides accommodation for up to 16 people over winter and more than 70 scientists and technicians during the austral summer. The scientific projects implemented at Concordia are strictly dependent on the characteristics of the site: *a)* the presence of a 3 300 m thick ice cap that allows access to the planet's climate archives and the reconstruction of glacial-interglacial cycles over more than 800 000 years; *b)* a particularly stable pure and dry atmosphere ideal for astronomy observations and for research on the chemical composition of the atmosphere; *c)* a distant location from coastal perturbations favourable to magnetic and seismological observatories to complement a poor world data network in the southern hemisphere; and *d)* a small totally isolated group of people confined to the station over a long winter, offering an opportunity for a range of medical and psychological studies useful to prepare long duration deep space missions. We will address the main characteristics of this station and its interest for science.

1. Introduction

The East Antarctic plateau at Dome C is one of the coldest and among the most remote places on Earth. Dome C is 1 100 km from the coast at a height of 3 233 m above sea, surrounded by thousands of kilometers of solid ice. Temperatures hardly rise above $-25°C$ in summer and can fall below $-80°C$ in winter with record of $-84.6°C$ reached in 2010. Dome C, one of the high points of the Plateau, was originally selected for glaciological reasons as one of the two sites of the EPICA deep ice-core drilling project that reached solid ground at a depth of 3 270 m in December 2004. Ten countries were associated in this program which provided a climate reconstruction of about 800 000 years. These results were at the basis of the IPCC (Intergovernmental Panel on Climate Change) discussions during the last years and provided evidence of the responsibility of Humans in the recent increase in the CO_2 or methane concentration in the atmosphere.

Concordia is jointly funded, staffed and operated by the French polar institute, IPEV and the Italian national program, PNRA, through the Italian national agency, ENEA, under a cooperative agreement signed by IPEV and PNRA in 1993. IPEV is a French governmental support agency for the scientific researches in the northern and southern polar regions, including the sub-antarctic islands (Crozet, Kerguelen and Amsterdam), with a staff of 50 permanent and up to 200 contract employees. The main IPEV partners are the French Ministry of Research and the CNRS (French National Centre for Scientific Research). ENEA is the Italian national agency for new technologies, energy and sustainable economic development.

Concordia opened in 1997 with the establishment of a summer camp. The year round facility station was built in five summer seasons, and Concordia became the third permanent continental station in February 2005, besides Vostok (Russia) and Amundsen-Scott

Figure 1. Main buildings of the Concordia station.

(USA) stations. As one of the newest stations in Antarctica, Concordia uses an efficient energy and waste management systems.

With mostly katabatic winds of less than 20 m/s, almost no absolute humidity, no light or air pollution, very low atmospheric turbulence and an inversion layer reaching up to just 30 m above ground, Dome C, which was selected for glaciology, is shifting it interests to atmospheric sciences and astronomy and astrophysics. The station hosts a large number of international scientific programs with participation of the researchers from mostly European countries.

2. The Station

Concordia station consists of two main identical polygonal buildings which are interlinked by enclosed walkways (Fig. 1). The two large three-storeyed buildings which provide the station's main living and working quarters, are divided for quiet and loud activities, respectively. The *calm* building houses laboratories, lodging, communications, and medical facilities, while the *noisy* building houses the kitchen, the dining area, the gym, a video room, and some storage and technical plants. Each housing unit is 18.5 m in diameter and has 250 sqm. A third structure, made up of stacked and interlocked containers, houses power station and mechanical facilities.

During the summer, up to 34 people can be housed in the main station buildings, while the overflow (up to 50) population can be housed in the summer camp. During the winter, only the main buildings are opened (the summer camp becoming an emergency camp) and capable of providing home to up to 16 people winter over crew.

The station, which has a fully equipped medical facility, including an operating room and a dental suite, provides an absolute comfort in a place where external temperature can drop as low as $-84°$ C.

3. Logistics

Access to Concordia station is limited to the austral summer due to the extreme weather conditions. For more than nine months per year, the station functions fully autonomously and has access to the outside world only through communication systems. There are two different ways to access the station. Coming from Hobart, Australia, by ship (L'Astrolabe) to the French coastal station Dumont d'Urville, or from Christchurch (New-Zealand) by intercontinental plane (Hercules aircraft) to the Italian station Mario-

Zucchelli at Terra-Nova bay. For transportation of people, Twin Otter or Basler aircrafts are used, starting from either Dumont d'Urville or Mario Zucchelli stations.

L'Astrolabe's cargo is transferred from Dumont d'Urville, which is located on an island, to Cap Prud'homme station, a small annex station on the continent, 5 km far from Dumont d'Urville, by helicopter, by pontoon or by sledges, depending on the sea-ice and weather conditions.

Selected light cargo is transported to Concordia by plane while heavy equipment is brought to Concordia using ground traverses. Cap Prud'homme serves as the continental launching point for these inland traverses. A typical traverse convoy consists of two or three snow grading machines leading eight to ten tractors towing the cargo sledges, and covering 1 200 km in 8-10 days depending on weather conditions. Approximately 150 tons of cargo is brought on each traverse, two thirds of which is fuel. IPEV is the leader in the development and implementation of such inland traverses in Antarctica. Even if less rapid than aircraft, this concept has many advantages: it is cheaper, it is not weather-dependent (geolocation system is needed to allow navigation in bad weather conditions), it requests less maintenance, and is safer for the environment.

4. Power station

The power station is made up of 3 Diesel generators, adapted to the particular conditions of the air in Dome C. Each generator can deliver at least 125 kW at full load. The system is a co-generator one that lets the recovering of waste heat for a total amount of around 80%. The waste heat is used to heat the station year round without additional heaters.

Concordia station was designed for power production and fuel consumption efficiency. Maximum consumption in the summer season averages 150 kW. Concordia's fuel as well as kerosene, petrol and other fuels for vehicles arrives by tractor traverse and is stored in tanks inside specially constructed containers. A spatially removed supply of fuel is present at the summer/emergency camp.

For safety reasons, there is also an emergency Diesel generator inside the noisy building.

5. Telecommunications

Dome C is still in the line of sight of existing geostationary satellites, then Concordia station has a permanent VSAT 512 kbps connection using a 3.8 m antenna. The satellite connection is used mainly for text emails and small data volume transfer. The station has also Iridium and Inmarsat satellite systems allowing a bandwidth of 64 kbps, a VoIp conferencing room and medical teleconference capabilities.

All around the station up to about 1 km, scientific instruments and warmed shelters are connected by fiber optics to a switch located into the main buildings, where the network runs at 1 Gbps.

6. Water-recycling

Fresh water is obtained by melting snow collected from a designated clean area. To protect the Antarctic environment, waste water is appropriately treated. It is separated into two different units: the *grey* and the *black* water units. The grey water treatment unit (GWTU) was designed together with the European Space Agency (ESA). It converts waste water to potable water which is used only for personal hygiene applications, even if it can be safely consumed. A four-step treatment process including ultrafiltration, nano filtration and two stages of reverse osmosis is realized. Treated water from the black water system is taken up by the grey water treatment unit and recycled. Solid waste from the black water system is containerized and removed from the continent as hazardous waste.

7. Scientific activities

The objective of Concordia is to operate, year-round, as an international research facility. Beside the glaciology programs, the remote inland location of the station provides an excellent site for astronomy and astrophysics, geophysics (seismology and geomagnetism), atmospheric sciences and human biology and medicine.

Concordia's science facilities are located in area surrounding the main station up to one kilometer away (Fig. 2). During summer, experiments can be performed up to a few kilometers (5 to 20) from the main station. If necessary, summer science traverses can be organized to perform scientific studies (glaciology, seismology, ...) up to Vostok station located at 560 km from Concordia.

The main scientific activities conducted at Concordia are the following :

• Beside, the EPICA ice-core, which was completed in December 2004, reaching a depth of 3 270.2 m, 5 m above bedrock, and extended the record of climate variability to an age estimated to be around 800 000 years old, Concordia remains an active site for glaciology with a large number of scientific experiments conducted on this field year round.

• During winter-over, Concordia station shares many stressor characteristics with long duration deep space missions, particularly extreme isolation and confinement. For this reason, ESA is interested in Concordia as a model of space vessel, simulating a human mission to Mars by the year 2030. The space agency accordingly supports several experiments in the field of space medicine and psychology, involving volunteers from the winter crew.

During the winter, the crew are without possibility of evacuation or deliveries for 9 months and live for a period of 3 months in total darkness, at altitude almost equivalent to 4 000 m at the equator. Experiments conducted at Concordia during winter-over are

Figure 2. Panoramic view of Concordia station showing the main scientific installations.

linked to the physiological and psychological strains on the crew, particularly for the study of chronic hypobaric hypoxia, the stress secondary to confinement and isolation, the circadian rhythm and sleep disruption and the individual and group psychology.

• Concordia station is an opportunity for long term observatories in meteorology, seismology and earth magnetism, to supplement the worldwide Earth observatories network in a region where its coverage remains very sparse.
Since 1998, Concordia has housed a permanent seismological observatory. The seismology station was constructed progressively over six summer campaigns, at the same time as the scientific base itself. The seismology program at Concordia has two main goals: the set-up and operation of a broad-band "observatory-quality" permanent seismic station, and the deployment of a temporary seismometer array. Both aspects of the program aim to contribute to the study of both Earth structure and earthquakes.

• Concordia permanent magnetic observatory was opened at the beginning of 2005. The observatory is equipped with standard instruments for continuous three-component field variation records and absolute measurements. The observatory, located above more than 3 000 m of layered ice, is unaffected by crustal field contaminations and coastal tidal effects. Furthermore, it lays inside the Southern polar cup, close to the South corrected geomagnetic pole. In these regions, the Earth's magnetic field is stronger and the external field influences are enhanced.

• Several programs are implemented in the field of atmospheric sciences, including air/snow interactions, as well as in the specific domain of ozone dynamics. Among them, the *Concordiasi* project, a French-US initiative for climate/meteorology over Antarctica at a global scale is a good example of a large international study (Hertzog *et al.* 2010) (Fig. 3). Many agencies or scientific organizations were involved: Météo-France, CNES, CNRS/INSU, NSF, NCAR, University of Wyoming, Purdue University, University of Colorado, the Alfred Wegener Institute, the Met Office and ECMWF. *Concordiasi* has benefited from logistic or financial support of the operational polar agencies IPEV, PNRA, USAP and BAS, and from BSRN (Baseline Surface Radiation Network) mea-

Figure 3. Concordiasi: Launch preparation from McMurdo station *(CNES Image)*.

surements at Concordia. *Concordiasi* was part of the THORPEX-IPY cluster within the International Polar Year effort.

Three field experiments were conducted, two which have occurred during the austral autumn 2008 and austral spring 2009 in Antarctica and a third one in austral spring 2010. In September and October 2009, 18 drifting balloons were released to acquire measurements in the stratosphere. Twelve of them released 600 parachuted dropsondes to establish vertical profiles of the troposphere. Additional in-situ measurements include radio-soundings at the Concordia station and at Dumont d'Urville, and high altitude balloons able to drop dropsondes, were launched on demand under a parachute to measure atmospheric parameters on their way down over Antarctica in 2010. Some of the balloons carried experimental instruments measuring ozone and particles at flight level. Others, also, have been improved for carrying GPS receivers to perform radio-occultation measurements.

8. Astronomy and Astrophysics

Concordia station had received world-wide attention from the astronomical community when it became known that the seeing conditions on Dome C are likely the best on the entire planet at a height of approximately 30 m above ice surface (Lawrence *et al.* 2004, Agabi *et al.* 2006, Aristidi *et al.* 2003, 2005a, 2005b). Extensive site quantification were conducted since the first winter-over in 2005, confirming an excellent seeing above a thin boundary layer (Aristidi *et al.* 2009, Fossat *et al.* 2010), a very low scintillation (Kenyon *et al.* 2006) a low sky brightness (Kenyon & Storey 2006) and a high duty cycle (Mosser & Aristidi 2007, Crouzet *et al.* 2010). In addition, as expected by Dempsey *et al.* (2005), photometric observations conducted at Concordia are not affected by Aurorae.

Thanks to the cold, dry, and calm atmosphere, Concordia offers the best conditions on Earth to investigate astronomical objects at high angular resolution, particularly in

Figure 4. ASTEP 400 telescope at Concordia.

the near thermal IR and submillimeter-wave ranges. The high latitude location allows 3-month continuous night-time photometric observations, which makes it an ideal site for long-duration observations such as those required to study the periodic variability of celestial objects and to detect and characterize exoplanets. In several well-identified domains (niches), the Antarctic plateau may even compete with space missions, but at a much lower cost, and with the invaluable bonus of using the most advanced technologies.

The main astronomy experiments conducted at Concordia are the following :

• IRAIT – the International Robotic Antarctic Infrared Telescope – is a Cassegrain 80 cm telescope having an alt-azimuth mount and two stable Nasmyth focal stations (Tosti *et al.* 2006). The telescope was installed at Concordia in 2008. It is provided with a wobbling secondary mirror to perform focusing, dithering (for near-infrared observations) and fast chopping (for mid-infrared observations). A plane tertiary mirror can alternatively feed the two Nasmyth foci. One of the Nasmyth focal stations will be equipped with AMICA (Antarctic Multiband Infrared Camera), a NIR/MIR double armed camera operating in the near- and mid-infrared spectral regions (Dolci *et al.* 2010), the other focus is foreseen for the CAMISTIC bolometer (Minier *et al.* 2007). First observations are expected for 2013.

• COCHISE is a Cassegrain 2.6 m millimetric telescope, with a wobbling secondary mirror and a field of view of few arc-minutes (Sabbatini *et al.* 2011). This telescope was installed in 2007 as a pathfinder for future Antarctic millimetric telescopes.

• QUBIC (formerly BRAIN) is a ground-based project of observational cosmology dedicated to measuring the Cosmic Microwave Background (CMB) polarization anisotropies (Piatt *et al.* 2012). QUBIC (Q&U Bolometric Interferometer for Cosmology) is an international collaboration involving universities and laboratories in France, Italy, the U.K. and the U.S.A. The QUBIC instrument is based on bolometric interferometry, an innovative technology that is a promising alternative to direct imaging. The first module of the final instrument should be installed at Concordia in 2014.

• The ASTEP (Antarctica Search for Transiting Exo-Planet) program is dedicated to exoplanet transit search (Fig. 4). A pathfinder instrument called ASTEP South, a 10 cm refractor telescope, was installed in 2008 in order to evaluate the photometric capacities of the site (Crouzet *et al.* 2010) and is in operation since its installation. The main instrument, ASTEP 400, was then built in order to achieve the major objective of exoplanet detection.

ASTEP 400 is an optical 40 cm telescope with a field of view of $1° × 1°$ (Daban *et al.* 2010). The achieved photometric sensitivity is about 1 mmag for the brightest stars of the field. The optical design of the instrument guarantees high homogeneity of the PSF sizes in the field of view. The use of carbon fibers in the telescope structure guarantees high stability. The focal optics and the detectors are enclosed in a thermally regulated box which withstands extremely low temperatures. The telescope designed to run at $-80°C$ was set up at Concordia during the Southern summer 2009-2010. It began its nightly observations in March 2010.

References

Agabi, A., Aristidi, E., Azouit, M., Fossat, E., Martin, F., Sadibekova, T., Vernin, J., & Ziad, A., 2006, *PASP*, 118, 344

Aristidi, E., Agabi, A., Vernin, J., Azouit, M., Martin, F., Ziad, A., & Fossat, E., 2003 *A&A*, 406, 19

Aristidi, E., Agabi, K., Azouit, M., Fossat, E., Vernin, J., Travouillon, T., Lawrence, J. S., Meyer, C., Storey, J. W. V., Halter, B., Roth, W. L., & Waldern V., 2005a *A&A*, 430, 739

Aristidi, E., Agabi, A., Fossat, E., Azouit, M., Martin, F., Sadibekova, T., Travouillon, T., Vernin, J., & Ziad, A., 2005b *A&A*, 444, 651

Aristidi, E., Fossat, E., Agabi, A., Mékarnia, D., Jeanneaux, F., Bondoux, E., Challita, Z., Ziad, A., Vernin, J., & Trinquet, H., 2009 *A&A*, 499, 955

Crouzet, N., Guillot, T., Agabi, A., Rivet, J.-P., Bondoux, E., Challita, Z., Fantei-Caujolle, Y., Fressin, F., Mékarnia, D., Schmider, F.-X, *et al.*, 2010 *A&A*, 511,36

Daban, J.-B., Gouvret, C., Guillot, T., Agabi, A., Crouzet, N., Rivet, J.-P., Mékarnia, D., Abe, L., Bondoux, E., Fante-Caujolle, Y., *et al.*, 2010 *Proc. SPIE*, 7733, 151

Dolci, M., *et al.* 2010 *SPIE Proc.*, 7735, 121

Dempsey, J. T., Storey, J. W. V., & Phillips, A., 2005, *PASA*, 22, 91

Fossat, E., Aristidi, E., Agabi, A., Bondoux, E., Challita, Z., Jeanneaux, F., & Mékarnia, D., 2010 *A&A*, 517, 69

Hertzog, A. *et al.*, 2010 *Cosp.*, 38, 4057.

Kenyon, S. L., Lawrence, J. S., Ashley, M. C. B., Storey, J. W. V., Tokovinin, A., & Fossat, E., 2006 *PASP*, 118, 924

Kenyon, S. L., Storey, J. W. V., 2006 *PASP*, 118, 489

Lawrence, J. S., Ashley, M. C. B., Tokovinin, A., & Travouillon, T., 2004, *Nature*, 431, 278

Minier, V., Durand, G., Lagage, P.-O., & Talvard, M., 2007 *HiA*,14, 709

Mosser, B. & Aristidi, E., 2007 *PASP*, 119, 127

Piatt *et al.*, 2012 *Journal of Low Temperature Physics*, 167, 872

Sabbatini, L., Cavaliere, F., DallOglio, G., Miriametro, A., & Pizzo, L., Mancini D., Torrioli G., 2011 *Experimental Astronomy*, 31, 199

Tosti, G. *et al.*, 2006 *Proc. SPIE*, 6267

Astrophysics from Antarctica
Proceedings IAU Symposium No. 288, 2012
M. G. Burton, X. Cui & N. F. H. Tothill, eds.

© International Astronomical Union 2013
doi:10.1017/S1743921312016857

Winterover scientists in Antarctic Astrophysics

N. F. H. Tothill[1] and C. L. Martin[2]

[1]University of Western Sydney, Locked Bag 1797, Penrith 2751 NSW, Australia
email: n.tothill@uws.edu.au

[2]Physics and Astronomy Dept., Oberlin College, 110 N. Professor St., Oberlin, OH 44074, USA

Abstract. Astronomy in Antarctica is largely carried out in winter, and so winterover scientists are required to run the instruments. A winterover appointment is a unique opportunity for a scientist, but brings challenges for both the scientist and the larger instrument team. We give a brief review of how winterovers work and their experiences. Although recent projects have required less support from winterover scientists, we believe that they will be a feature of Antarctic astronomy and astrophysics into the future.

Keywords. Winterover Scientists, Instrumentation

1. Introduction — who/what are Winterover Scientists?

Winterover scientists have been on the Antarctic continent for more than half a century — the first South Pole winterover crew (1957) included the astronomer Arlo Landolt — but they remain somewhat anomalous in Antarctic science. Antarctic science (glaciology, ecology, geology...) is largely carried out in the summer months by visiting scientists. Year-round research stations usually have skeleton winter crews to maintain the infrastructure for summer operations and sometimes to carry out upgrades or repairs. Some science is year-round (usually monitoring programmes such as meteorology, atmospheric sampling, seismic monitoring...). Antarctic astronomy, however, is largely done in winter: Optical/IR observations need twilight or darkness, and FIR/submm/mm-wave observations require the very low levels of precipitable water vapour (pwv) found only in winter. Therefore, most Antarctic instruments are maintained and upgraded over summer, and carry out science during winter. The winterover scientists who run these telescopes constitute an important resource for their scientific projects.

2. The winterover's view

A winterover appointment is a unique opportunity for a scientist. The challenges are obvious: communications are limited, and physical isolation is absolute; the climate is harsh, with cold temperatures and no daylight. The opportunities, however, are greater than the challenges: The opportunity to work independently, to understand the components, subsystems and systems engineering of a research instrument, and the responsibility of managing such an instrument to maximise the scientific return. In an era of large projects, and larger teams, the winterover scientist's experience is unusual.

3. The project leader's view

From the perspective of the project leader or manager, to whom the winterover is responsible, the effectiveness of winterover personnel in Antarctica follows a quite standard track: The winterover arrives in Summer (usually at the start of the summer season) and quickly learns their way around the project, becoming highly effective. The summer season is usually very busy, and the winterover may be quite tired at the end — but their real job is only just beginning. The effectiveness of the winterover is probably at a peak in the early part of the winter: There are few distractions, they have learned the system, and the environment is still quite benign, with daylight/twilight and fairly warm conditions. For many instruments, this is also a time with a significant workload, as calibration, telescope pointing etc. are carried out to ready the instrument fully for the winter. As the winter goes on, the effectiveness of the winterover steadily declines: The lack of daylight takes a gradual toll on cognitive abilities, and people get more and more tired. This can be a particular problem when significant workload occurs towards the end of the winter. For example, the lowest pwv conditions generally happen in August–September, and so this is when commissioning of THz instrumentation was done at the South Pole. Thus, a peak in the demand on the winterover can coincide with the time when they are at their least effective. The project leader should try to anticipate the likely workload on the winterover and manage that workload, bearing in mind that the winterover is a finite resource.

4. A year in the life of a winterover

We (the authors) both spent winters at the South Pole Station. This section is therefore specific to the South Pole, but should give some idea of winterover life at other stations in Antarctica, and even the Arctic, *mutatis mutandis.*

4.1. *Summer*

Figure 1. On the way to the South Pole Station in an LC-130.

Figure 2. 2003 December 13: A solar halo.

Figure 3. Upgrading old instruments and fitting new instruments: lifting SPIFI onto AST/RO.

The most common pattern for a winterover year is to arrive on station near the start of the Antarctic summer (Fig. 1), so as to spend the summer learning the instrument and getting used to the working environment. The South Pole Station summer (see Fig. 2) starts in late October, and continues to the middle of February. It is used for instrument maintenance and upgrading, including the installation of new instruments (Figs. 3, 4). The fast pace of maintenance and upgrade gives the winterover an excellent, albeit tiring, opportunity to learn the instrument systems. The peak of this activity is in January, when the weather is warmest and logistics are easiest.

Figure 4. Upgrading old instruments and fitting new instruments: Installing SPIFI.

4.2. *Winter*

The winter season runs for 8–9 months from February to October. In the first phase, both station and instruments are prepared for the winter. This time is generally less favourable for science observations, due to continued daylight (optical/IR) and still-high pwv (sub/mm-wave), so is largely used for calibration, e.g. rebuilding telescope pointing models after summer maintenance.

Figure 5. March: The sun sets near the equinox.

Once the sun is fully set (Fig. 5), the long period of twilight starts. During this time, the weather is getting steadily colder and the pwv is steadily decreasing. This is the start of the core science operations time (Fig. 6), and uptime is at a premium. In the case of equipment failures, the emphasis is on getting the instrument back into operation (Fig. 7), rather than finding a permanent solution. Ideally, the instrument settles into a long winter without incident (Figs. 8, 9).

Figure 6. April: Delivery of a dewar of liquid helium in the twilight.

Scientific priorities may require complex technical activities in the middle of winter, such as commissioning and testing of THz instruments at the South Pole (Figs. 10, 11).

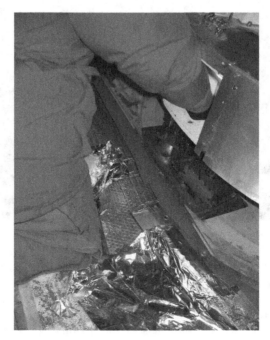

Figure 7. Telescope repairs in winter.

Figure 8. June: Midwinter greetings are sent throughout the continent.

Figure 9. Aurora Australis over AST/RO at the South Pole.

As the sun rises, the pwv remains very low, and observations continue all the way into the start of summer. Winterovers usually leave as soon as possible after the station opens, following a short handover to their replacements.

5. Do we still need winterovers?

Over the last few years, winterover scientists have been less involved in Antarctic astrophysics. There are two trends at work here:

(a) Many of the operational instruments (e.g. IceCube, SPT...) are designed to work in 'survey mode' rather than 'observatory mode', in which the science programme for the winter is entirely predetermined, and the main job of the winterover is equipment maintenance, rather than science operations — even observation scheduling may not need to be carried out on-site.

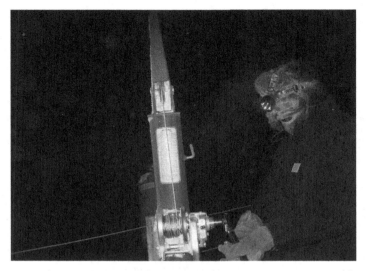

Figure 10. August: Using a crane in winter — compare to summer (Fig. 3).

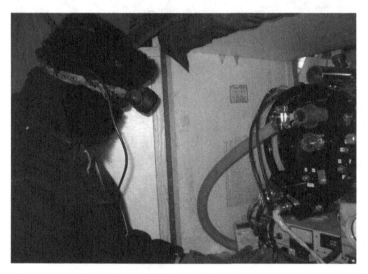

Figure 11. August: Operating SPIFI in winter — compare to summer (Fig. 4).

(*b*) The development of the PLATO observatories (e.g. Yang *et al.* 2009, Lawrence, Ashley & Storey 2012) to support site-testing and scientific instruments at remote sites, which require no on-site personnel over the winter, and only short servicing missions in summer.

These trends are likely to continue — survey telescopes producing datasets are becoming more and more important to astronomy, and the automation of instruments is likely to increase. However, fully-autonomous instrumentation is probably some way off (see, e.g., Strassmeier *et al.* (2007), Tothill *et al.* (2008) for discussion of instrument automation), and remote operation (similar to that of current survey instruments) is likely to require high-bandwidth communication, of the kind available at the South Pole. As new instruments test the limits of what can be done with current communication bandwidth, (and

Figure 12. September: The sun rises on the winter's snowdrifts.

Figure 13. October: The first cargo plane of the summer arrives.

some will be built in locations without high-bandwidth communications available) the role of the winterover scientist may become far more important.

References

Lawrence, J. S., Ashley, M. C. B., & Storey, J. W. V. 2012, *these proceedings*
Strassmeier, K., *et al.* 2007, *Astr. Nachr.*, 328, 451
Tothill, N. F. H., Martin, C. L., Kulesa, C. A., & Briguglio, R. 2008, *Astr. Nachr.*, 329, 326
Yang, H., *et al.* 2009, *PASP*, 121, 174

Astrophysics from Antarctica
Proceedings IAU Symposium No. 288, 2012
M. G. Burton, X. Cui & N. F. H. Tothill, eds.

© International Astronomical Union 2013
doi:10.1017/S1743921312016869

Astronomy from 80 Degrees North on Ellesmere Island, Canada

Eric Steinbring

National Research Council of Canada,
5071 West Saanich Rd, Victoria, BC V9E 2E7, Canada
email: `Eric.Steinbring@nrc-cnrc.gc.ca`

Abstract. Site testing carried out on Ellesmere Island over recent years has shown that mountainous coastal terrain there can provide high clear-sky fractions in the long dark season, with low precipitable water-vapour column and prospects for excellent seeing. This presents new possibilities for time-domain and survey-mode science in the northern hemisphere, allowing uninterrupted high-precision photometry in the optical/near-infrared, but also gains in the submillimetre/millimetre. Efforts underway at the Eureka research station, at 80 degrees latitude, are reviewed. This location provides year-round access to a nearby site being developed as a pathfinder observatory. A program of variable-star and transient searches involving a wide-field imaging system has begun, with some early results. Plans include extrasolar-planet hunting via transit surveys, and future directions are discussed.

Keywords. arctic, site testing, telescopes, optical, submillimeter

1. Introduction

Ellesmere Island, in the territory of Nunavut, Canada, stretches beyond 82°N latitude at its far northern tip, near the military base of Alert. Some mountains here are the highest within 10 degrees of the Pole, topping 2,600 m at Barbeau Peak. Many coastal summits reach from 1,000 m to 1,900 m, above a strong atmospheric thermal inversion that persists through the winter. Site testing on three of the four highest peaks within 100 km of the northwestern coast began in 2006. These locations are indicated by red crosses on the map in Figure 1. They are accessible by helicopter from the Eureka research base in summer, which allowed the installation of small autonomous weather stations and cameras for remote monitoring through winter (Steinbring *et al.* 2008; Wallace *et al.* 2008). Although the initial stations were power-limited – relying on wind-turbines and a fuel-cell – they were able to confirm the excellent weather conditions expected at these sites based on meteorology and satellite analysis: stable temperature near −30°C, clear-sky fractions approaching 70% and long periods of calm winds (Steinbring *et al.* 2010). And importantly, this established the feasibility of access and potential for developing and maintaining more advanced astronomical instrumentation at the highest-elevation/highest-latitude mountains attainable (Steinbring *et al.* 2012b).

Comparison can also be made to a lower-elevation coastal location nearer to Eureka, the site of the Polar Environment Atmospheric Research Laboratory (PEARL; 80°N, 600 m). This manned facility, although less ideal than the remote sites, allows realistic demonstration of a variety of scientific programs at a High Arctic astronomical observatory. This review article summarizes the current status of that work near Eureka. Section 2 provides an outline of logistics and the currently-known astronomical sky properties, which is followed in Section 3 by a discussion of scientific programs being undertaken and possible future directions.

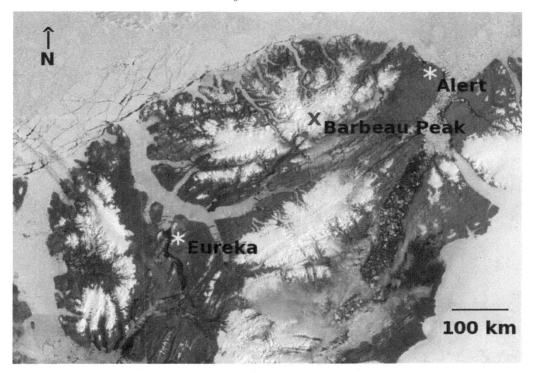

Figure 1. A map of northern Ellesmere Island; coastal mountains reach above 1,000 m, the tallest, Barbeau Peak is over 2,600 m. Study sites are marked with red crosses; PEARL is nearest to Eureka. (Colour image available in the online version.)

2. Site Testing and Development Near Eureka

Eureka is a base of Environment Canada, the weather service of the Canadian federal government. The PEARL facility, to the northwest of Eureka at the end of a 15 km road, is shown in Figure 2. Although it was originally designed for atmospheric studies, particularly the monitoring of ozone using Laser Detection and Ranging (LIDAR), it also offers a good platform for small astronomical instruments. Power is supplied from diesel generators in Eureka, and the building provides a large, flat roof. Below this is warm laboratory space, dormitory areas and a kitchen. Broadband internet and technical support have been provided through the Canadian Network for the Detection of Atmospheric Change (CANDAC).

2.1. *Logistics*

The key advantages of Eureka and PEARL are existing infrastructure and the relative ease of access to high-elevation terrain nearby the ice-locked coast. Eureka has a 5,000 foot all-season runway which provides year-round access via commercial charters: small turboprop aircraft, but also including large transports, e.g. Boeing 737. In late summer, when ice conditions allow, a yearly re-supply vessel provides fuel and large cargo for the base facility. The weather station maintains a rotating staff of typically eight people: a manager, three weather technicians, a cook, a mechanic, a heavy-equipment operator, and a handyperson, although in summer the number of additional visitors can swell to 20 or 30. PEARL can be accessed from here by truck in all seasons, and observers either go up to "the summit" periodically or stay for more extended periods.

Figure 2. The PEARL facilty near Eureka; the observing platform is accessed with the external stairway at left. A 360 degree panoramic view (above) is from the observing platform itself. The orientation of both images is to the south, aligned with the mast of the rooftop meteorological station. Photos: Thomas Pfrommer.

2.2. *Clear-Sky Fraction*

Each hour at the Eureka weather station, a visual estimate of cloud cover is obtained by a technician as part of the standard meteorological record. There are over 50 years of continuous data which show that much of the time in winter is clear or with thin, uniform attenuation due to suspended ice crystals. More recently, a detailed study of cloud cover has been carried out using an all-sky-viewing camera installed on the PEARL roof, discussed in Steinbring *et al.* (2012a).

The PEARL All-Sky Imager (PASI) was designed for atmospheric studies, particularly buoyancy wave phenomenon, not astronomy. However, the central wavelength of one of its narrow-band filters is similar to V. Near-continuous photometry of Polaris over two winters provides a good estimate of sky clarity, after subtraction of variation in this well-known Cepheid. About 84% of the time clouds, primarily in the form of ice crystals, have an attenuation less than 2 mag, up to 68% of the time is clear (below 0.5 mag) which can persist for over 100 hours at a time; and 48% of that time is truly photometric (no cloud at all).

Interestingly, the PASI data are also consistent with the expectations from LIDAR and RADAR observations that the density of ice crystals decreases with elevation, under calm conditions. If correct, some higher terrain near Eureka should have clearer conditions than those experienced at PEARL (Steinbring *et al.* 2012a).

Figure 3. An observer checking on DIMM and MASS/DIMM site-testing telescopes deployed on the PEARL observing platform. Behind is the ice-covered mast of the meteorological station, beyond and to the right is the constellation Orion. Photo: Pierre Fogal.

2.3. *Sub-Millimetre Opacity*

The precipitable water vapour column at sea level in winter for Eureka is very low, with a median under 2 mm, so it should not be a surprise that millimetre-wave opacity is good. This was verified using a 225 GHz tipping radiometer, deployed on the PEARL observing platform for three months in the winter/spring of 2011. The main result was that the opacity was $\tau = 0.1$, and stable with temperature and humidity (Asada *et al.* 2012). This is also discussed in more detail by Matsushita *et al.* (2012) in this Proceedings.

2.4. *Sky Brightness and Aurorae*

Sky brightness has been measured from photometry with PASI and using a commercial single-diode device. These agree that without a moon, sky brightness reaches a minimum for Sun elevations just below -12 degrees of 21 mag per square-arcsec, as expected for the latitude of Eureka (Steinbring *et al.* 2012a; Sivanandam *et al.* 2012). Eureka is within the auroral hole; when aurorae are seen, they are typically along the southern horizon. There is no evidence of strong aurorae in the PASI data, although that would not be anticipated for the filter bandpass. Initial measurements at J-band also suggest near-infrared skies are as dark as the best mid-latitude sites (Sivanandam *et al.* 2012), with further observations planned in upcoming winters.

2.5. *Seeing*

Ground-layer turbulence measurements sensitive up to 100 m have been made from the PEARL rooftop using an autonomous lunar scintillometer: the Arctic Turbulence Profiler (ATP). These show a thin boundary layer, with a median contribution to total seeing

Figure 4. An AWCam deployed on the PEARL observatory platform. Photo: Wayne Ngan.

of 0.5 arcsec at an elevation of 15 m (Hickson *et al.* 2010), comparable to the best mid-latitude sites.

Because northern Ellesmere is usually within the Arctic polar vortex, with the jet stream well to the south, the expectation is that Eureka will also benefit from excellent free-seeing – analogous to atmospheric conditions already demonstrated on the Antarctic glacial plateau (Lawrence *et al.* 2004). To confirm this, Differential Image Motion Monitor (DIMM) and Multi-Aperture Scintillation Sensor (MASS) observations have been obtained during two campaigns in winter 2011 and 2012 from the PEARL roof. Seeing of 0.6 arcsec to 0.9 arcsec is typical, consistent with the ATP results plus weak free-atmosphere seeing, although there is still not sufficient data for solid statistics. Obtaining data over a longer period, using an autonomous system on a standard 6-m tall tower away from the building should be conclusive, and is planned.

3. Initial Science and Future Directions

Astronomical studies requiring steady, uninterrupted observations are well suited to a near-PEARL site. The cold and dry conditions lead to long periods of low opacity in winter darkness, and at these latitudes high-elevation targets maintain essentially constant airmass. By taking advantage of the ease of accessibility afforded by the location, efforts can focus on open-shutter efficiency. Taken together, this can be a particularly useful path towards wide-field surveys and time-domain scientific programs, such as high-cadence surveillance of variable or transient objects in the optical. But other science can also benefit, from the UV through to the millimetre.

So far, the main scientific pursuit being considered near PEARL is the detection of extra-solar planets via transits. One way to exploit the persistence of relatively uniform and thin sky transparency is to monitor large areas of sky. A first test of this approach was obtained with the Arctic Wide-field Cameras (AWCams) deployed in winter 2012, observing a fixed field centred on the celestial North Pole. One of these instruments is shown in Figure 4. The details of the design can be found in Law *et al.* (2012) and first results will appear in the literature soon: millimag-level photometry of many bright stars near Polaris, including known eclipsing binary W UMi. A logical extension of the AWCams called the Compound Arctic Telescope Survey (CATS) is to expand the field by deploying multiple unit cameras, with a rotating mount counteracting star trails and allowing coverage of the entire sky down to airmass of about 2. In complement, a larger aperture telescope monitoring a smaller field may realize the benefits of good seeing. The 0.5-m aperture Dunlap Arctic Telescope (DIAT), also discussed in Law *et al.* (2012), is planned for deployment on the ground nearby PEARL. Again, high-cadence millimag photometry would be the focus of a survey of M dwarf stars, with an advantage of the long twilight at this latitude possibly allowing multiple transit detections during a single winter season.

A larger, future optical/near-infrared telescope might take full advantage of the thin boundary layer and expected weak free-atmospheric seeing. A concept similar to that proposed by Rene Racine over 20 years ago for the site now occupied by Gemini North was outlined in a White Paper by Carlberg, Hickson, & Steinbring (2010), employing low-order adaptive optics to achieve wide-field diffraction-limited performance. This 1m to 4m-class facility is endorsed in the 2010 Canadian Long Range Plan for Astronomy (www.casca.ca/lrp2010/), which recommends a design study once site testing is completed.

References

Asada, K., Martin-Cocher, P. L., Chen C.-P., Matsushita, S., Chen, M.-T., Inoue, M., Huang, Y.-D., Inoue, M., Ho, P. T. P., Paine, S. N., & Steinbring, E., 2012, *SPIE Conf. Series*, in press

Carlberg, R., Hickson, P., & Steinbring, E., 2010, White Paper *Super Seeing Sites in the Canadian High Arctic*

Hickson, P., Carlberg, R., Gagne, R., Pfrommer, T., Racine, R., Schoeck, M., Steinbring, E., & Travouillon, T., 2010, *SPIE Conf. Series*, 7733, 53

Law, N., Sivanandam, S., Murowinski, R., Carlberg, R., Ahmadi, P., Steinbring, E., & Graham, J., 2012, *SPIE Conf. Series*, in press

Lawrence, J. S. *et al.* 2004, *Nature*, 431, 238

Matsushita, S., Chen, M.-T., Martin-Cocher, P., Asada, K., Chen, C.-P., Inoue, M., Paine, S., Turner, D., & Steinbring, E., 2012, *These Proceedings*

Sivanandam, S., Tekatch, A., Welch, D., Abraham, B., Graham, J., & Steinbring, E., 2012, *SPIE Conf. Series*, in press

Steinbring, E., Leckie, B., Welle, P., Hardy, T., Cole, B., Bayne, D., Croll, B., Walker, D. E., Carlberg, R. G., Fahlman, G. G., Wallace, B., & Hickson, P., 2008, *SPIE*, 7012, 1

Steinbring, E., Carlberg, R., Croll, B., Fahlman, G., Hickson, P., Leckie, B., Pfrommer, T., & Schoeck, M., 2010, *PASP*, 122, 1092

Steinbring, E., Ward, W., & Drummond, J. R., 2012, *PASP*, 124, 185

Steinbring, E., Leckie, B., Hardy, T., Caputa, K., & Fletcher, M., 2012, *SPIE Conf. Series*, in press

Wallace, B., Steinbring, E., Fahlman, G., Leckie, B., Hardy, T., Fletcher, M., Pennington, M., Caputa, K., Carlberg, R., Croll, B., Bayne, D., Cole, B., Hickson, P., Pfrommer, T., & Thorsteinson, S., 2008, *AMOS Conf. Series*, Ed. S. Ryan, 7, The Maui Economic Development Board

Astrophysics from Antarctica
Proceedings IAU Symposium No. 288, 2012
M. G. Burton, X. Cui & N. F. H. Tothill, eds.

© International Astronomical Union 2013
doi:10.1017/S1743921312016870

Sub-mm VLBI from the Arctic — Imaging Black Holes

Makoto Inoue[1]
and the Greenland Telescope team[1,2,3,4]

[1] Academia Sinica Institute of Astronomy and Astrophysics,
P.O. Box 23-141, Taipei, 10617 Taiwan, R.O.C.
email: inoue@asiaa.sinica.edu.tw

[2] Smithsonian Astrophysical Observatory,
60 Garden Street, Cambridge, MA 02138, USA
[3] MIT Haystack Observatory,
Off Route 40, Westford, MA 01886-1299, USA
[4] National Radio Astronomy Observatory,
520 Edgemont Road, Charlottesville, VA 22903-4608, USA

Abstract. We are deploying a new station for sub-millimeter Very Long Baseline Interferometry (VLBI) to obtain shadow images of Supermassive Black Hole (SMBH). Sub-mm VLBI is thought to be the only way so far to get the direct image of SMBH by its shadow, thanks to the superb angular resolution and high transparency against dense plasma around SMBH. At the Summit Station on Greenland, we have started monitoring the opacity at sub-mm region. The Summit Station subtends long baselines with the Atacama Large Milimeter/submillimeter Array (ALMA) in Chile and Submillimeter Array (SMA) in Hawaii. In parallel, we started retrofitting the ALMA North America prototype telescope (renamed as Greenland Telescope: GLT) for the cold environment.

Keywords. Supermassive Black Hole (SMBH), Sub-millimeter telescope, Very Long Baseline Interferometry (VLBI)

1. Introduction

An experiment in the sub-millimeter (submm) Very Long Baseline Interferometry (VLBI) has shown the capability for direct imaging of Sumermassive Black Hole (SMBH) (Doelaman *et al.* 2007). The angular resolution reaches the expected size of several tens of micro arcsec (μas) for some of SMBHs. Submm observations also allow us to see the immediate vicinity of SMBH, penetrating into the dense plasma around it. Supported by imaging simulations (e.g., Luminet 1979, Falcke *et al.* 2000, and Dexter *et al.* 2012) and good mass estimations of SMBHs (Ghez *et al.* 2008, and Gebhardt *et al.* 2011), we are now confident to be able to observe and resolve the shadow image of SMBHs: Sgr A* and M87 being the best targets based on their apparent angular sizes.

The image of SMBH can be seen as a shadow against bright accreting materials, and the shadow size is expected to be about 5 times of the Schwarzschild radius for non rotating black hole, enlarged by a lensing effect of the strong gravity. The shadow size depends on the black hole spin (Luminet 1979). This means we could measure directly the key parameters of black hole, *i.e.*, the mass and spin of SMBH by its shadow image with submm VLBI, in addition to observing other physical conditions under strong gravity field, such as the energy transfer in the accretion flows and the launching mechanism of relativistic jets.

However, the number of the submm VLBI telescopes is very limited, and the distribution of baselines, or the *uv* coverage, is not well arranged. Under such circumstances, development of new submm VLBI sites is a key to promote the SMBH science. Academia Sinica Institute of Astronomy and Astrophysics (ASIAA) has awarded a new submm telescope with CfA and started a site development for the telescope to establish a new submm VLBI station. We will describe below the science target and the siting situation.

2. Science Target

As the Schwarzschild radius r_s is proportional to the SMBH mass, the apparent size of the shadow depends on the mass and distance. Sgr A* at the center of our Galaxy shows the largest apparent diameter of 52 μas because of its distance. The second largest one is in M87, showing a comparable apparent diameter of 40 μas to Sgr A*. The apparent diameter of other galaxies is at most 10 μas or smaller.

For SMBHs of Sgr A* and M87, we are now able to resolve the black hole shadow with ground-based submm VLBI. In terms of the apparent size of the black hole shadow, Sgr A* would be the best target. However, it becomes clear that Sgr A* shows rapid time variations within a day even at mm and submm wavelengths (Miyazaki *et al.* 2004, and Fish *et al.* 2011). This is probably due to the small mass of the black hole. The rapid variations make it difficult to generate synthesized images of Sgr A*. On the other hand, the mass and size of the SMBH of M87 are 10^3 times larger than that of Sgr A*, and much longer time scale is expected for M87. The SMBH mass of M87 indicates a dynamical timescale of about 5 hours, and a typical orbital timescale at the Innermost Stable Circular Orbit (ISCO) is 2-18 days, depending on the SMBH spin. Therefore, unlike Sgr A*, the structure of M87 may not change during an entire day for the image synthesis. In this point, M87 is better candidate for the VLBI imaging synthesis technique by Earth rotation. It would be possible to produce a sequence of images of dynamical events in M87, occurring over many days. M87 is thus a promising target to image the black hole shadow and bright accreting materials rotating around it.

Further, as shown in Figure 1, the tracing back of the jet radius is one of the crucial tests to identify the launching point of the jet, and put constrains on the formation mechanism. If the parabolic shape of the jet extends towards the innermost region, *i.e.*, up to $2r_s$, the jet must originate within the ISCO of the accretion disk around the SMBH or the SMBH itself. Submm VLBI could make observations of the jet radius to reveal the launching point of the relativistic jet (Asada & Nakamura 2012).

3. Submm VLBI Network

As seen in Figure 2, the baselines between Greenland, Hawaii, and Atacama in Chile form a big triangle for M87, extending 9,000 km long, which achieves the angular resolution of 20 μas at 350 GHz. At 230 GHz, other submm telescopes like IRAM 30-m (Pico Veleta, Spain), PdBI (France), SMT (Arizona, USA), LMT (Mexico), CARMA (California, USA), etc., will provide a good *uv* coverage to make plausible quality images of M87. The Summit Station on Greenland has good baselines between both European and American submm telescopes. The North-South baselines subtended by the Summit Station and other telescopes provide good opportunity to investigate the structure of M87 jet which is running toward almost West (see Figure 1).

ASIAA is the only Institute which has close relations to both the SMA and ALMA. The project for phase-up ALMA is in progress with international collaboration, and the phased array of the SMA, JCMT and CSO is already under development by the Haystack

Observatory and the SAO. This triangle is then expected to provide high sensitivity and high angular resolution at 350 GHz and even higher frequencies. At 230 GHz, many other submm telescopes will collaborate with this big triangle to form a good submm VLBI network.

4. Retrofitting Telescope

Since April 2011, the ALMA-North America (NA) prototype telescope has been re-tested and studied to retrofit for the cold environment on Greenland. After retrofitting, the telescope, renamed Greenland Telescope (GLT), will be shipped to Greenland in 2014. In parallel, construction works for the foundation and infrastructures have started, collaborating with a company who operates the summit facilities.

For the submm VLBI receivers, 230, 350 and 650-GHz receivers are planned. The GLT is also expected to be a forerunner of Cerro Chajnantor Atacama Telescope (CCAT) in science and as a test bench of receiver developments for single-dish observations. Heterodyne multi-feed receivers and multi-pixel bolometers up to 1.5 THz have been discussed.

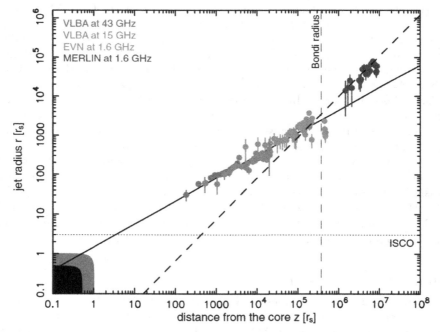

Figure 1. Distribution of the radius of the jet as a function of the deprojected distance r_s from the core. Images were obtained by previous VLBA measurements at 43 GHz (red circles) and at 15 GHz (orange circles), EVN measurements at 1.6 GHz (green circles), and MERLIN measurements at 1.6 GHz (blue circles). The jet is described by two different shapes. The solid line indicates a parabolic structure with a power-law index $\alpha \sim 1.7$, while the dashed line indicates a conical structure with $\alpha \sim 1.0$. HST-1 is located around 5×10^5 r_s, near the Bondi radius. The black area near the core shows the size of the minor axis of the event horizon of the spinning black hole with maximum spin. The gray area indicates the size of the major axis of the event horizon of the spinning black hole with maximum spin, which also corresponds to the size of the event horizon of the Schwarzschild black hole. The horizontal dotted line indicates the ISCO size of the accretion disk for the Schwarzschild black hole (Asada & Nakamura 2012).

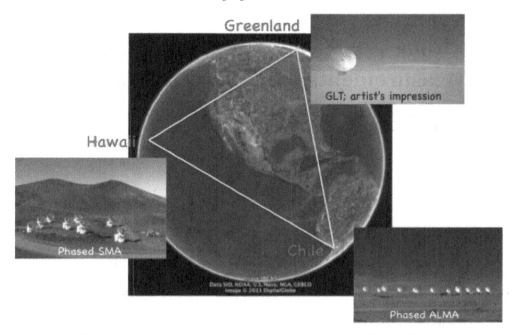

Figure 2. A big triangle formed by Summit Station on Greenland, the SMA in Hawaii and ALMA in Chile, as seen from the direction of M87. This network provides a 20-μas resolution at 350 GHz.

5. Summary

Direct imaging of SMBHs has been anticipated by the recent development of submm VLBI observatories. New sites for a submm VLBI network could provide good image quality, and a good site has been identified at the Summit Station of the ice cap on Greenland. ASIAA has been monitoring the sky opacity at the Summit Station, and the preliminary monitoring shows satisfactory opacity condition at submm region (see Matsushita *et al.* 2012). The ALMA-NA prototype telescope was awarded to ASIAA/CfA. Retrofitting the telescope, now called GLT, is under way for the Greenland Summit Station. The new submm VLBI network by the big triangle with GLT, ALMA, SMA and other submm telescopes, will be a powerful tool for imaging the SMBH shadows and related studies. Multi-feed and multi-pixel receivers are also planned at up to the THz region for single-dish observations.

References

Asada, K. & Nakamura, M. 2012, *ApJ*, 745, L28
Dexter, J. *et al.* 2012, *MNRAS*, 421, 1517
Doeleman, S. *et al.* 2008, *Nature*, 455, 78
Falcke, H., F. Melia, F., & Agol, E. 2000, *ApJ*, 528, L13
Fish, V. *et al.* 2011, *ApJ*, 727, L36
Gebhardt, K. *et al.* 2011, *ApJ*, 729, 119
Ghez, A. M. *et al.* 2008, *ApJ*, 689, 1044
Luminet, J.-P. 1979, *A&Ap*, 75, 228
Matsushita, S. *et al.* 2012, this proceedings
Miyazaki, A. *et al.* 2004, *ApJ*, 611, L97

Astrophysics from Antarctica
Proceedings IAU Symposium No. 288, 2012
M. G. Burton, X. Cui & N. F. H. Tothill, eds.

© International Astronomical Union 2013
doi:10.1017/S1743921312016882

225 GHz Atmospheric Opacity Measurements from Two Arctic Sites

S. Matsushita[1], Ming-Tang Chen[1], P. Martin-Cocher[1], K. Asada[1], C.-P. Chen[1], M. Inoue[1], S. Paine[2], D. Turner[3] and E. Steinbring[4]

[1] Institute of Astronomy and Astrophysics, Academia Sinica,
P.O.Box 23-141, Taipei 10617, Taiwan, R.O.C.
email: satoki@asiaa.sinica.edu.tw

[2] Smithsonian Astrophysical Observatory,
160 Concord Ave., Cambridge, MA 02138, USA

[3] National Severe Storms Laboratory,
120 David L. Boren Boulevard, Norman, OK, 73072, USA

[4] National Research Council, Herzberg Inst of Astrophysics,
5071 W Saanich Rd, CA Victoria BC V9E 2E7, Canada

Abstract. We report the latest results of 225 GHz atmospheric opacity measurements from two Arctic sites; one on high coastal terrain near the Eureka weather station, on Ellesmere Island, Canada, and the other at the Summit Station near the peak of the Greenland icecap. This is a campaign to search for a site to deploy a new telescope for submillimeter Very Long Baseline Interferometry and THz astronomy in the northern hemisphere. Since 2011, we have obtained 3 months of winter data near Eureka, and about one year of data at Summit Station. The results indicate that these sites offer a highly transparent atmosphere for observations in submillimeter wavelengths. Summit Station is particularly excellent, and its zenith opacity at 225 GHz is statistically similar to the Atacama Large Milllimeter/submillimeter Array site in Chile. In winter, the opacity at Summit Station is even comparable to that observed at the South Pole.

Keywords. Arctic sites, 225 GHz opacity, site testing

1. Introduction

The success of Doeleman *et al.* (2008) in obtaining a scatter-free size estimate of submillimeter emission in Sagittarius A* (Sgr A*) using Very Long Baseline Interferometry (VLBI) promises a new window for direct imaging of supermassive black holes (SMBHs). Although Sgr A* is the nearest known and biggest (in apparent size), its mass is relatively small among SMBH. This means a short timescale of variability, leading to undesirable smoothing of that signal during integration. On the other hand, the second largest source in apparent size is the SMBH in M87 (Virgo A), which has a large mass. A further scientifically interesting point is that M87 has a strong jet activity. But to image the SMBH in M87 with submm-VLBI requires a longer baseline in the northern hemisphere than currently available (see Inoue *et al.* 2012 in this proceeding) and so we began a search for a new submm telescope site.

2. Site Selection

For the site selection, we set criteria as follows: (1) annual precipitable water vapor (PWV) of less than 3 mm for good submm opacity; (2) longest-possible baseline with existing telescopes, for best imaging resolution; (3) observable sky together with the

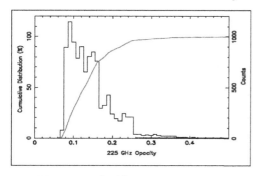 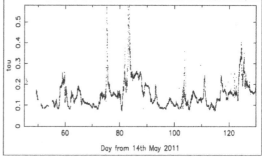

Figure 1. *(Left)* Time variation plot of 225 GHz opacity at the PEARL site. *(Right)* Histogram and cumulative plot of the opacity.

Atacama Large Millimeter/submillimeter Array (ALMA) to achieve the highest possible sensitivity; (4) accessibility to the site.

Based on the first two criteria, there are three potential broad regions of interest; western China and Tibet, the highest mountains of southern Alaska, or the high Arctic polar desert, including northern Canada and Greenland. The Western China and Tibet region does not have common sky with ALMA, so it does not meet criterion (3), and the tallest peaks in Alaska (e.g., Mount McKinley) are excluded due to criterion (4). The Eureka research base on Ellesmere Island, Canada and Summit on the Greenland icecap meet all four criteria, and we considered these two sites for further study.

3. 225 GHz Tipping Radiometer

For the site survey, we purchased a 225 GHz tipping radiometer from Radiometer Physics GmbH. The reason for the choice of this frequency is that there are many site survey results from all over the world, including the summit of Mauna Kea, the ALMA (Chajnantor) site and South Pole.

For the opacity measurements, we use the tipping method: We observe five angles (90°, 42°, 30°, 24°, and 19.2°, corresponding to $\sec(z)$ of 1.0, 1.5, 2.0, 2.5, and 3.0) with 4 second integration at each angle, for a duration of 75 second per tipping measurement. For each measurement, the opacity is derived from the instrument output voltage as a function of zenith angle. Measurements were obtained every 10 minutes.

We first tested at the Academia Sinica, Institute of Astronomy and Astrophysics (ASIAA) in Taipei, Taiwan, and then repeated on Mauna Kea, Hawaii to check the consistency of our measurements with the 225 GHz tipping radiometer at the Caltech Submillimeter Observatory (CSO). Simultaneous opacity measurements were performed between 31 December 2010 and 11 January 2011, and are consistent with each other (linear regression coefficient = 1.04).

4. Results

Identical observations were then carried out at the high Arctic sites, allowing direct comparison to those worldwide.

4.1. *Eureka, Ellesmere Island, Canada*

Eureka is a manned weather station on Ellesmere Island, Canada. It has a 5000-foot all-season airstrip. Yearly resupply by ship occurs in late summer. A road allows access to the Polar Environment Atmospheric Research Laboratory (PEARL), located on a 610 m-high ridge at 80.05° N, 86.42° W. Although this site is not as high as other good

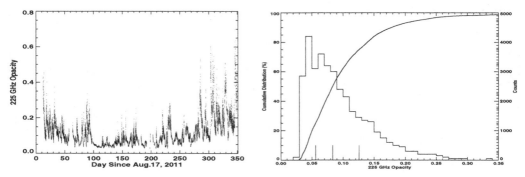

Figure 2. *(Left)* Time variation plot of 225 GHz opacity at the Summit Camp. *(Right)* Histogram and cumulative plot of the opacity.

submm sites, it is expected to have low opacity conditions due to the low temperatures, typically between −20°C and −40°C in winter (Steinbring *et al.* 2010). The radiometer was deployed on the rooftop observing platform of PEARL, and measured the 225 GHz opacity between 14 February and 10 May 2011. Tipping direction was south, providing 10,522 data points.

Fig. 1 shows the time variation, histogram, and cumulative plots. The lowest 225 GHz opacity measured was 0.07, the 25% quartile was 0.11, the median (50% quartile) was 0.14, and the most frequent opacity was 0.09 during our measurements. These statistics indicate that submm-VLBI at this frequency is feasible at this site.

4.2. *Summit Camp, Greenland*

Summit Camp is a research station near the peak of the Greenland ice sheet, located at 72.57° N, 38.46° W, with an elevation of 3,200 m. Access is by C-130 air transport in summer and by Twin Otter in winter, or traverse from Thule. Very good opacity conditions are expected for this site, based on the low winter temperatures, between −40°C and −60°C with minimum temperature of −72°C (Vaarby-Laursen 2010), and the high altitude. We put our radiometer on the roof of the Mobile Science Facility (MSF), and started measuring the opacity on 17 August 2011 (and still continue to do so). Here we show data up to 31 July 2012. Tipping directions were both south and north, and provide 36,555 data points.

Fig. 2 shows the time variation, histogram, and cumulative plots. The lowest 225 GHz opacity measured was 0.027, the 25% quartile was 0.056, the median was 0.083, and the most frequent opacity was 0.04 during our measurements. These statistics are excellent, strongly indicating that submm-VLBI at these frequencies, or even higher, is feasible at this site. THz astronomy is also worth considering.

4.3. *Comparison with Other Sites*

To illustrate how a high Arctic submm site can compare to the best worldwide, we compared Summit Camp statistics with those of the ALMA site (elevation = 5,050 m) and South Pole (2,800 m). The opacity data for the ALMA site have been taken from Radford & Chamberlin (2000) and Radford (2011), measured between April 1995 and April 2006, and the data for South Pole from Chamberlin & Bally (1994) and Chamberlin & Bally (1995), measured between January and December 1992.

Winter in the northern hemisphere is defined as between the beginning of November and the end of April, with summer May through October. The opposite is taken to be the case for the southern hemisphere; winter May through October, summer November

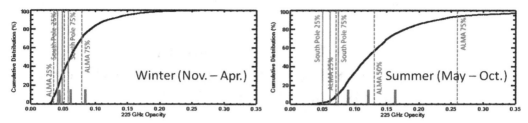

Figure 3. Opacity quartiles comparison between the Summit Camp (histogram and cumulative plots with quartiles as short and thick solid lines), the ALMA site (dashed vertical lines), and South Pole (thin solid vertical lines). Left plots are winter season, and right summer.

through April. We put the quartiles for the three sites on the cumulative plots of the Summit Camp in Fig. 3.

In winter season, 225 GHz opacities of all three sites are less than 0.1 even at the 75% quartile. There is small difference between the sites, but the opacity statistics of Summit Camp are rather similar to that of the ALMA site. The opacity statistics of South Pole do not vary much from winter to summer, but the ALMA site shows much greater variation. Summit Camp falls somewhere in between. It is worth noting, however, that 2011/12 represents a record warm period for Greenland, which may have affected opacity statistics. Long-term monitoring is needed to carefully judge the quality of Summit Camp as a submm site.

5. Summary

We have presented a program of site testing for a submm telescope site in the northern hemisphere. Both Eureka, Ellesmere Island, and Summit Camp, Greenland offer the potential for new submm VLBI observations. Based on the best 225 GHz opacity measurements, we selected Summit Camp, and efforts are currently underway to retrofit an antenna for the extremes of the site. Opacity measurements are still ongoing to collect long-term opacity variation data. This will reveal whether the record warmth in Greenland has affected the opacity statistics or not, and the true fraction of time that Summit Camp reaches the quality of South Pole. In addition, atmospheric characterisation using various instrumentations (cloud radars and lidars, radiosondes, microwave radiometers, precipitation measurements) are ongoing at Summit Camp by atmospheric researchers (e.g., Shupe *et al.* 2012). We are closely collaborating with them to estimate the atmospheric conditions more accurately, and to construct accurate atmospheric models for future submm/THz astronomy at this site.

References

Chamberlin, R. A. & Bally, J. 1994, *App. Opt.* 33, 1095
Chamberlin, R. A. & Bally, J. 1995, *Int. J. IR MM Waves* 16, 907
Doeleman, S. S., *et al.* 2008, *Nature* 455, 78
Inoue, M., *et al.* 2012, these Proceedings
Radford, S. J. E. 2011, *RevMexAA (SC)* 41, 87
Radford, S. J. E. & Chamberlin, R. A. 2000, *ALMA Memo* 334
Shupe, M. D., *et al.* 2012, *Bull. Amer. Meteo. Soc.* in press
Steinbring, E., Carlberg, R., Croll, B., Fahlman, G., Hickson, P., Ivanescu, L., Leckie, B., Pfrommer, T., & Schoeck, M. 2010, *PASP* 122, 1092
Vaarby-Laursen, E., *DMI Tech. Rep.* 10-09

Astrophysics from Antarctica
Proceedings IAU Symposium No. 288, 2012
M. G. Burton, X. Cui & N. F. H. Tothill, eds.

© International Astronomical Union 2013
doi:10.1017/S1743921312016894

Precision CMB measurements with long-duration stratospheric balloons: activities in the Arctic

P. de Bernardis[1], S. Masi[1]
for the OLIMPO and LSPE teams

[1]Dipartimento di Fisica, Sapienza Universitá di Roma, P.le A. Moro 2 00185 Roma, Italy
email: paolo.debernardis@roma1.infn.it

Abstract. We report on the activities preparing long duration stratospheric flights, suitable for CMB (Cosmic Microwave Background) measurements, in the Arctic region. We focus on pathfinder flights, and on two forthcoming experiments to be flown from Longyearbyen (Svalbard islands): the OLIMPO Sunyaev-Zeldovich spectrometer, and the Large-Scale Polarization Explorer (LSPE).

Keywords. Cosmic Microwave Background, Arctic, Stratospheric Balloons

1. Introduction

In 1998 the BOOMERanG experiment started the use of long-duration balloon (LDB) flights in the Antarctic stratosphere to perform precision Cosmic Microwave Background (CMB) anisotropy measurements, producing detailed maps of the CMB where causal horizons in the cosmic photosphere are fully resolved (B98, see e.g. de Bernardis *et al.* 2000). This was followed by the successful Antarctic flight of BOOMERanG-polarization (B03, see e.g. Montroy *et al.* 2006). Nowadays the EBEX experiment (see Reichborn-Kjennerud *et al.* 2010) is going to be flown on a similar long duration flight, promising unprecedented sensitivity in the measurement of CMB polarization, as well as the SPI-DER experiment (see Filippini *et al.* 2010), targeting at larger angular scales.

The Arctic region can be used to perform similar LDB flights. Developing these flights will double the availability of long-duration flight opportunities every year, reducing the waiting time for users, and will enable access to the northern hemisphere, which cannot be observed during Antarctic flights. Longyearbyen, in the Svalbard islands, is at a latitude of 78°N (roughly at the antipodes of McMurdo) and can provide all the required infrastructure. A number of small-payload flights and one heavy-lift payload have been successfully launched from the Longyearbyen airport, demonstrating the feasibility of long duration flights in the Arctic (Peterzen *et al.* 2008). In addition, we have launched small payloads from the Dirigibile Italia station in Ny-Alesund (also in the Svalbard islands) during the Arctic winter, demonstrating the eastward trajectory (see below). In the following we shortly report on the Arctic flights, and then focus on two CMB payloads we plan to launch from Longyearbyen: OLIMPO and LSPE.

2. Long Duration Flights in the Arctic

The first LDB flight fully circumnavigating the north pole at high latitudes was launched from Longyearbyen in the summer of 2006, by an international team composed of personnel from the Italian Space Agency, Andoya Rocket Range, Sapienza Universitá

Figure 1. Top: Launch of a path-finder balloon from Ny-Alesund (Svalbard, Jan. 2012); bottom: Ground-track of a path-finder balloon launched from Ny-Alesund (Svalbard, Jan. 2011)

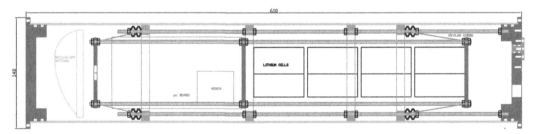

Figure 2. Sketch of the light-weight payload used to study stratospheric circulation in the Arctic during the polar winter. Dimensions in mm.

di Roma, Istituto Nazionale di Geofisica, ISTAR. The ground track of that flight can be found in Peterzen *et al.* 2008. In 2009 the same team launched the heavy-lift ($>$ 1 ton) SORA experiment, with a 800,000 m^3 balloon, also from Longyearbyen. The ground track for that flight is shown in Flamini & Pirrotta 2011.

A team composed of personnel from La Sapienza, ISTAR and CNR, launched in Jan. 2011 and Jan. 2012 path-finder balloons (3,000 and 9,000 m^3) from the Dirigibile Italia base in Ny-Alesund (Svalbard), to demonstrate the east-bound trajectories during the polar winter. The ground track of one of these flights is shown in Fig. 1.

The balloon lifted a lightweight (5 kg) payload consisting of a GPS receiver, temperature and pressure sensors, an Iridium SBD communication system, a radiation-hard

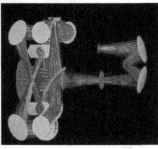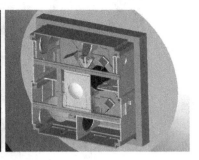

Figure 3. Left: The OLIMPO payload with solar panels, ground-shields and sun-shields removed. **Center:** Optical configuration of the lossless Martin-Pupplett DFTS for OLIMPO (on the left in the figure) and existing reimaging optics inside the cryostat (on the right in the figure); **Right:** Rendering of the mechanical setup for the DFTS: the center lens and the following wedge can be shifted on command, allowing the beam to by-pass the interferometer, for photometric measurements.

microcontroller and a battery pack. The main challenge in the preparation of a payload for a long duration flight in the polar night is thermal insulation. The system has to operate in an environment with a pressure of a few mbar and a temperature of $\sim -80°$C, supplied only by batteries. In order to minimize the weight of the system, we used lithium cells, which maximize the stored energy to weight ratio, and insulated very carefully the battery pack and the system electronics to avoid the use of electrical heaters. The system is surrounded by mylar superinsulation, suspended by kevlar cords, and inserted in a vacuum vessel to minimize heat dispersion. A sketch of the payload is shown in Fig. 2. The system has been able to operate in the harsh conditions of the polar winter stratosphere with just 1W of power.

Although a full winter circumnavigation has not been achieved yet, the east-bound trajectory has been demonstrated. We plan to repeat the path-finders launch campaign in the coming winter seasons, to improve the statistics and to find the best launch period during the winter.

3. OLIMPO

The OLIMPO (Masi *et al.* 2008) balloon payload is a 2.6 m millimeter-wave telescope aimed at measuring the Sunyaev-Zeldovich (SZ, Sunyaev & Zeldovich 1972) effect in clusters of galaxies. It works at high frequencies (up to 500 GHz), which cannot be observed from the ground, and with angular resolution higher than the recent Planck survey (Planck Collaboration, 2011a). At $\sim$ 450 GHz (near the positive maximum of the SZ spectral brightness), the angular resolution of OLIMPO is comparable to the resolution of the 10 m class mm-wave telescopes operating at $\sim$ 140 GHz in Antarctica (see Carlstrom *et al.* 2011) and in Atacama (see Swetz *et al.* 2011 and Schwan *et al.* 2011). The OLIMPO payload has been recently upgraded with a lossless, differential, low-resolution (6 GHz), Fourier transform spectrometer (DFTS), which was placed as a plug-in between the rear part of the primary mirror and the window of the cryostat (see Fig. 3).

This instrument can measure 33 independent frequency bands, located below, across and above the null of the SZ spectrum (see Table 1). The sensitivity (NEPD, noise equivalent power density) has been computed per detector, per beam, and assuming photon

Table 1. Main features of the OLIMPO spectroscopic instrument

Band (GHz)	Sub-bands	FWHM (′)	Array Elements	Background (pW)	NEPD ($fW \sqrt{s}/GHz$)
135–160	4	4.0	19	6	5
180–250	12	3.0	19	20	12
330–365	6	1.7	23	7	9
450–515	11	1.2	23	14	14

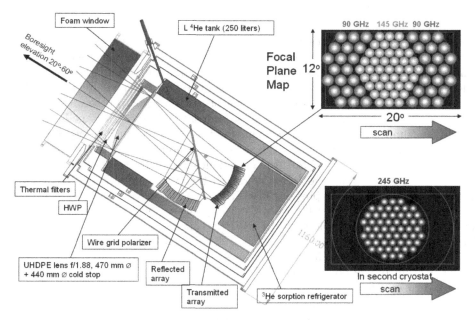

Figure 4. Left: The SWIPE instrument on the LSPE.

noise limited detectors. The DFTS is intrinsically rejecting the large common mode signal from the telescope, the residual atmosphere and the CMB monopole. The effectiveness of this instrument in removing degeneracies between cluster (and foreground) parameters has been discussed in de Bernardis *et al.* 2012. OLIMPO will carry-out an ambitious observation program, including spectroscopic observations of the SZ in 40 known clusters, plus a $10° \times 10°$ blind deep-sky survey. In addition, OLIMPO represents a first step in the development of space-based differential spectrometers for the mm and sub-mm bands, which will eventually lead to a DFTS instrument for the 10 m space telescope on the Millimetron mission. The flight of OLIMPO is scheduled for June 2013.

4. LSPE

The Large-Scale Polarization Explorer (see LSPE collaboration 2012) is a balloon-borne mission polarimeter aimed at measuring the polarization of the CMB at large angular scales, to constrain the curl component of CMB polarization (B-modes) produced by cosmic inflation. The target is to constrain the ratio of tensor to scalar perturbations amplitudes down to $r = 0.03$, at 99.7% confidence. This will be achieved measuring multi-color wide maps of the mm sky, in order to beat cosmic variance and to monitor the polarized foreground generated in our Galaxy by synchrotron emission and interstellar dust emission. The mission is optimized for large angular scales, with coarse angular

Table 2. Main features of the LSPE polarimeters

Band (GHz)	FWHM (deg)	Array Elements	NET_{array} $(\mu K \sqrt{s})$
43	1.0	49	33
90	0.5	7	104
95	1.8	80	1.9
145	1.5	86	1.8
245	1.2	110	1.9

resolution (around 1.5° degrees FWHM), and wide sky coverage (25% of the sky). To achieve this, the payload will fly on a circumpolar LDB in the polar winter. Using the Earth as a giant solar shield, the instrument will spin in azimuth, observing a large fraction (>25%) of the northern sky. The payload will host two instruments. STRIP (the STRatospheric Italian Polarimeter), an array of coherent polarimeters using cryogenic HEMT amplifiers, will survey the sky at 43 and 90 GHz (see Bersanelli *et al.* 2012). SWIPE (the Short Wavelength Instrument for the Polarization Explorer), an array of bolometric polarimeters, will survey the same sky region in three bands at 95, 145 and 245 GHz (see de Bernardis *et al.* 2012b). The resulting wide frequency coverage will allow optimal control of the polarized foregrounds, with comparable angular resolution at all frequencies. While the main purpose of STRIP is to monitor low-frequency polarized foregrounds, SWIPE provides high sensitivity by means of arrays of large-throughput bolometers. In fact, each bolometer of SWIPE collects > 20 modes of the incoming radiation, boosting by a factor > 4 the sensitivity with respect to single-mode detectors. Polarization modulation is achieved by means of a rotating half-wave-plate and a static polarizer in front of the bolometers. The main characteristics of LSPE are summarized in Table 2. In Fig. 4 we show the setup of the SWIPE polarimeters.

The flight of LSPE is scheduled for January 2015.

Acknowledgements: The OLIMPO and LSPE projects have been funded by the Italian Space Agency. The balloon campaigns at Ny-Alesund have been supported by Programma Nazionale Ricerche in Antartide - CNR contract 2010-A3.02. We gratefully acknowledge the support of the CNR team at the Dirigibile Italia base in Ny-Alesund. The OLIMPO collaboration includes teams from University of Rome La Sapienza, IFAC-CNR Firenze, INGV Roma, University of Milano Bicocca, Dept. of Physics and Astron. Univ. of Cardiff, NIST Boulder, CEA Saclay, INPE Brasil, Agenzia Spaziale Italiana. See Masi *et al.* 2008 and de Bernardis *et al.* 2012 for the full list of collaborators. The LSPE collaboration includes teams from University of Rome La Sapienza, University of Milano, University of Milano Bicocca, IFAC-CNR Firenze, IASF-INAF Bologna, University of Firenze, Cavendish Laboratory Cambridge, University of Trieste, Jodrell Bank CFA, University of Manchester, IEIIT-CNR Torino, Istituto Nazionale di Geofisica e Vulcanologia Roma, IASF-INAF Milano, Agenzia Spaziale Italiana, OAT-INAF Trieste. See LSPE collaboration 2012 for the full list of collaborators.

References

Bersanelli, M., *et al.*, 2012, in *Proceedings of the SPIE Astronomical Telescopes + Instrumentation 2012 Conference - Ground-based and Airborne Instrumentation for Astronomy IV*, astro-ph/1208.0164

Carlstrom, J. E., *et al.*, 2011 *Publ. Astron. Soc. of the Pacific* 123, 568

de Bernardis, P., *et al.*, 2000, *Nature*, 404, 955

de Bernardis, P., *et al.*, 2012, *Astronomy & Astrophysics*, 538, A86 de Bernardis, P., *et al.*, 2012b, in *Proceedings of the SPIE Astronomical Telescopes + Instrumentation 2012 Conference - Ground-based and Airborne Instrumentation for Astronomy IV*, astro-ph/1208.0282

Filippini, J.P., *et al.*, in: *Proceedings for SPIE Millimeter, Submillimeter, and Far-Infrared Detectors and Instrumentation for Astronomy V* astro-ph/1106.2158

Flamini, E. & Pirrotta, S., 2011, *Mem. Soc. Astron. It. Supp.* 16, 145

LSPE Collaboration, Amico G. *et al.*, 2012, *Proceedings of the SPIE Astronomical Telescopes + Instrumentation 2012 Conference - Ground-based and Airborne Instrumentation for Astronomy IV*, astro-ph/1208.0281

Masi, S., *et al.*, 2008, *Mem. Soc. Astron. It. Supp.*, 79, 887

Montroy, T., *et al.*, 2006, *Ap.J.*, 647, 813

Peterzen, S., *et al.* 2008, *Mem. Soc. Astron. It.* 79, 792

Planck Collaboration, Ade, P. A. R., Aghanim, N., & Arnaud, M. e. a., 2011a, *Astronomy & Astrophysics* 536, A8

Reichborn-Kjennerud, B., *et al.* in: *Proceedings for SPIE Millimeter, Submillimeter, and Far-Infrared Detectors and Instrumentation for Astronomy V* astro-ph/1007.3672

Schwan, D., *et al.* 2011, *Review of Scientific Instruments* 82, 091301

Sunyaev, R. A. and Zeldovich, Y. B., 1972, *Comments on Astrophysics and Space Physics* 4, 173

Swetz, D. S., *et al.*, 2011, *Ap.J.Suppl.*, 194, 41

Astrophysics from Antarctica
Proceedings IAU Symposium No. 288, 2012
M. G. Burton, X. Cui & N. F. H. Tothill, eds.

© International Astronomical Union 2013
doi:10.1017/S1743921312016900

Present and Future Observations of the Earthshine from Antarctica

Danielle Briot[1], Luc Arnold[2] and Stéphane Jacquemoud[3]

[1] Observatoire de Paris, 61 avenue de l'Observatoire, 75014 Paris, France
email: `danielle.briot@obspm.fr`

[2] Observatoire de Haute-Provence, 04870 Saint-Michel l'Observatoire, France
email: `luc.arnold@oamp.fr`

[3] Université Paris Diderot / Institut de Physique du Globe de Paris
39 rue Hélène Brion, 75013 Paris, France
email: `jacquemoud@ipgp.fr`

Abstract. It is likely that images of Earth-like planets will be obtained in the next years. The first images will actually come down to single dots, in which biomarkers can be searched. Taking the Earth as a example of planet providing life, Earthshine observations showed that the spectral signature of photosynthetic pigments and atmospheric biogenic molecules was detectable, suggesting that, in principle, life on other planets could be detected on a global scale, if it is widely spread and distinguishable from known abiotic spectral signatures. As for the Earth, we already showed that the Vegetation Red Edge which is related to chlorophyll absorption features was larger when continents, versus oceans, were facing the Moon. It proved that an elementary mapping of a planet was even possible. In the frame of the LUCAS (LUmière Cendrée en Antarctique par Spectroscopie) project, the Earthshine has been measured in the Concordia Research Station (Dome C, Antarctica) long enough to observe variations corresponding to different parts of the Earth facing the Moon. An extension of this project, called LUCAS II, would allow long-term observations to detect seasonal variations in the vegetation signal. These data, together with precise measurements of the Earth's albedo, will help to validate a model of global and spectral albedo of our planet.

Keywords. Astrobiology, Extraterrestrial life, Earth, Moon, Earthshine, Vegetation

1. Looking for life in extrasolar planets

To date, about one thousand exoplanets have been detected and we know a few thousand others are candidates [see the *Extrasolar Planets Encyclopaedia* published daily on the Internet by Jean Schneider (http://exoplanet.eu/)]. The CoRoT and Kepler satellites have sped up the discovery of new planets. In December 2011, the first Earth-like planet in the habitable zone of a Sun-like star, namely Kepler 22b, has been revealed. Since then, about ten near-Earth sized candidates orbiting in the habitable zone of their host star have been detected. We expect future extremely large instruments and space missions to provide soon the first low-resolution image of an extrasolar planet. Such images will appear as a dot so that detecting life on an unresolved Earth-like extrasolar planet is the next challenge. The spectrum of the light reflected by the planet, when normalized to the parent star spectrum, will gives a reflectance spectrum that mixes information about its surrounding atmosphere and its ground colour. What would the spectrum of an unresolved Earth-like planet look like? Life on an extrasolar planet, if any, probably substantially differs from that on Earth. However, our planet is the only one that is known to provide life. A way to answer this question would consist of measuring the global and spectral albedo of the Earth from a very long distance, typically several parsecs. This could be achieved using a space craft traveling through the Solar System and looking back at the Earth, like Voyager 1 in 1990 or Mars Express in 2003.

2. Earthshine observations

According to Jean Schneider (1998, *private communication*), the global spectrum of the Earth can be obtained in another way by measuring the spectral reflectance of the Moon Earthshine, i.e. the light back-scattered by the non-sunlit Moon. A spectrum of the Earthshine directly gives the disk-averaged spectrum of the Earth. Because of the lunar surface roughness, any place of the Earthshine reflects the whole enlighted part of the Earth facing the Moon. As soon as 1912, Arcichovsky (1912) suggested to look at the chlorophyll absorption features in the Earthshine spectrum, to calibrate this photo-synthetic pigment in the spectrum of other planets. However, such a detection was im-possible at that time. This approach was then completely forgotten and independently rediscovered in 1998. In the Earth reflected spectrum, we shall look for the signature of atmospheric molecules (biogenic products) O_2, O_3, CH_4, H_2O, CO_2 and biological activity. Up to no detection of animal life on Earth in observational conditions of an extrasolar planet has been reported. For instance, the Great Barrier Reef (2,600 km), which is considered as the world's largest structure composed of living entities, cannot be detected at long distance from its spectral signature. Contrary to Earth vegetation. The best example is the Amazonian forest that covers about 7×10^6 km^2. However, one should keep in mind that in another physical environment, vegetation may take forms and colours that are unfamiliar to us (e.g. Kiang (2008)). We decided to take the Earth as a model. The reflectance of plant canopies is about 9 times higher in the near in-frared (900 nm) than in the red (660 nm) part of the electromagnetic spectrum. This produces a sharp edge around 700 nm, the so-called Vegetation Red Edge (VRE). Since 2002 (Arnold *et al.* (2002), Woolf *et al.* (2002)), several observations of the VRE have been made of the VRE using the Earthshine. Although VRE is only a few percents, this signature that is typical of vegetation is detectable in an integrated (or disk-averaged) Earth spectrum. Arnold *et al.* (2002) and Hamdani *et al.* (2006) showed that it was the lowest when an ocean was facing the Moon (1.3% for the Pacific Ocean) and the highest when continents were present (4%). In addition the red side (600–1000 nm) of the Earth reflectance spectrum shows the presence of O_2 and H_2O absorption bands, while the blue side (320–620 nm) shows the Huggins and Chappuis ozone (O_3) absorption bands (Hamdani *et al.* (2006)).

3. Conditions of Earthshine observations

As it is well known, at mid- or low-latitude, Earthshine observations are possible during twilight, i.e. just after the sunset or just before the sunrise. Observations are consequently possible only during a short period of time. Roughly speaking, only two enlighted parts of Earth face the Moon during one observation: either the part located at the West of the telescope for evening observations (beginning of the lunar cycle), or the part of Earth located at the East of the telescope for morning observations (last days of the lunar cycle). From an idea of Jean Schneider (2002, *private communication*), if observations are made from a site located at high latitudes, the Earthshine can be monitored during several hours, about 8 times a year. Near the poles, they can even last 24 hours, i.e. a total nychthemeron. During such a long period, due to the Earth's rotation, different landscapes alternately face the Moon. The Dome C in Antarctica actually offers a perfect spot for that kind of observations. The French–Italian Concordia station (latitude 75°S, altitude 3,250m) which is open to winter-over teams since 2005, provides exceptional conditions for astronomical observations, specially in the infrared. Information about observational conditions obtained after 10 years of site testing can be found in Fossat (2011).

4. The LUCAS program

After inquiring about the possibility of observing the Earthshine from Concordia, we designed the LUCAS (LUmière Cendrée en Antarctique par Spectroscopie) experiment, which consists in an instrument dedicated to Earthshine spectroscopic observations in extreme conditions. The first observing campaign has been unsuccessful due to extreme temperatures and atmospheric conditions (problems of thermal insulation). The feedback we got from this campaign was very important to detect, analyze and correct the instrumental weaknesses. During the following winterover campaign, we obtained Moon spectra during all the observational sequences from the winter (June) solstice to the end of observing time of the year 2009. Observations could be carried out during up to 8 contiguous hours. Such a long observational time is impossible at mid- or low-latitude!

5. The future of Earthshine observations from Antarctica: LUCAS II

Our goal is to continue the Earthshine observations during at least four years to study the Earth global and spectral albedo. It will allow us to determine its variations throughout the year and to detect the interannual variability. Using the PROSAIL radiative transfer model, we expect to detect seasonal variations of the vegetation signal as a function of the Leaf Area Index (LAI). PROSAIL is the short name of two coupled models that are very popular in the remote sensing community: PROSPECT, a leaf optical properties model (Jacquemoud & Baret (1990)), and SAIL, a canopy reflectance model (Verhoef (1984)). It could be possible to estimate the average amount of vegetation present on Earth. The interaction with the atmosphere can be taken into account using a radiative transfer model through the atmosphere. The LUCAS II program will reuse part of the instrumentation deployed during LUCAS I. Results will be correlated to satellite measurements and to other measurements of the Earth albedo, in the context of climate change studies. It would be also very interesting to include similar observations made in Northern high latitude in this program, in order to double the number of Earthshine observations. In Antarctica, observations from around the March equinox up to the June solstice will correspond to the last days of the lunar Cycle, i.e. from the last quarter to a few days before the new moon; observations from the June solstice to approximatively the September equinox will correspond to the first days of the lunar cycle, i.e. from a few days after the new moon to the last quarter. Northern high latitudes observations would occur as long, but in opposite configurations.

When the Earthshine cannot be observed, LUCAS II could be used to study stars with rapid variations, like Be stars, the period of which ranges from hours to days. These rapid variations are attributed to non-radial pulsations. Continuous observations over a long period would allow us to detect them. The interest of such observations from Antarctica is exposed in Briot (2005).

6. Conclusion

LUCAS is one of the first astronomy research program with spectroscopic observations at Dome C. We measured the Earth vegetation spectrum during several hours, detecting variations during the Earth's rotation, as it will be possible in the future for extrasolar planets. The LUCAS II program will extend the observations of the Earth-shine from Antarctica during several years, providing very important information on the Earth albedo and variation of vegetation remote sensing. Concerning observations of Earthshine, some geophysical applications also exist, according to the recommendations made by the NASA Navigator Program: "Continued observations of Earthshine are needed to discern diurnal, seasonal, and interannual variations."

The main collaborators of this study are Jérome Berthier (Observatoire de Paris-Meudon), Patrick Rocher (Observatoire de Paris-Meudon), Jean Schneider (Observatoire de Paris-Meudon) and the winterover observers in Antarctica: Karim Agabi, Eric Aristidi, Erick Bondoux, Zalpha Challita, Denis Petermann and Cyprien Pouzenc.

References

Arcichovsky V. M. 1912, *Ann. Inst. Polytech. Don Tsar. Alexis (Novotcherkassk)*, 1(17), 195

Arnold, L., Gillet, S., Lardière, O., Riaud, P., & Schneider, J., 2002, *A&A*, 392, 231

Briot, D., 2005, *Dome C Astronomy and Astrophysics Meeting*, EAS Publications Series 14, 275

Fossat E., 2012, *arXiv:1101.3210*, http://arxiv.org/abs/1101.3210

Hamdani, S., Arnold, L., Foellmi, C., Berthier, J., Billeres, M., Briot, D., François, P., Riaud, P., & Schneider, J., 2006, *A&A* 460, 617

Jacquemoud, S., & Baret, F., 1990, *Remote Sensing of Environment*, 34, 75

Kiang, N. Y., 2008, *Scientific American*, 298, April 1, 12

Verhoef, W., 1984, *Remote Sensing of Environment*, 16, 125

Woolf, N. J., Smith, P. S., Traub, W. A., & Jucks, K. W., 2002, *ApJ*, 574, 430

Astrophysics from Antarctica
Proceedings IAU Symposium No. 288, 2012
M. G. Burton, X. Cui & N. F. H. Tothill, eds.

© International Astronomical Union 2013
doi:10.1017/S1743921312016912

Time domain astronomy from Dome C: results from ASTEP

J.-P. Rivet[1], L. Abe[1], K. Agabi[1], M. Barbieri[1], N. Crouzet[2], I. Goncalves[1], T. Guillot[1], D. Mekarnia[1], J. Szulagyi[1], J.-B. Daban[1], C. Gouvret[1], Y. Fantei-Caujolle[1], F.-X. Schmider[1], T. Furth[3], A. Erikson[3], H. Rauer[3], F. Fressin[4], A. Alapini[5], F. Pont[5] and S. Aigrain[6]

[1] Laboratoire Lagrange, UMR7293, Université de Nice Sophia-Antipolis, CNRS, Observatoire de la Côte d'Azur, BP 4229, F-06304 Nice Cedex 4 France.

[2] Space Telescope Science Institute, Baltimore, MD 21218, USA

[3] DLR Institute for Planetary Research, 12489 Berlin, Germany

[4] Harvard-Smithsonian Center for Astrophysics, Cambridge, MA 02138, USA

[5] School of Physics, University of Exeter, Stocker Road, Exeter EX4 4QL, United-Kingdom

[6] Department of Physics, University of Oxford, Oxford OX1 3RH, United Kingdom

Abstract. ASTEP (Antarctic Search for Transiting Exo Planets) is a research program funded mainly by French ANR grants and by the French Polar Institute (IPEV), dedicated to the photometric study of exoplanetary transits from Antarctica.

The preliminary "pathfinder" instrument ASTEP–South is described in another communication (Crouzet et al., these proceedings), and we focus in this presentation on the main instrument of the ASTEP program : "ASTEP–400", a 40 cm robotized and thermally-controlled photometric telescope operated from the French-Italian Concordia station (Dome C, Antarctica).

ASTEP–400 has been installed at Concordia during the 2009-2010 summer campaign. Since, the telescope has been operated in nominal conditions during 2010 and 2011 winters, and the 2012 winterover is presently in progress. Data from the first two winter campaigns are available and processed. We give a description of the ASTEP–400 telescope from the mechanical, optical and thermal point of view. Control and software issues are also addressed. We end with a discussion of some astronomical results obtained with ASTEP–400.

Keywords. Transiting exoplanets, photometry, Dome C

1. Introduction

Since the first discovery, in 1999, of an exoplanet (HD209458b) transiting across its host star (Charbonneau et al. 2000, Henry et al. 1999), more than 282 transiting exoplanets have been discovered around 230 stars. The transit method has revealed to be a powerful method to detect new exoplanets, and also to characterize already known ones, as a complement to the radial velocity method.

Several ground based instruments are at least partially dedicated to transiting exoplanet studies. For example, HATNet is a network of 6 robotic 11 cm wide field telescopes spread around the word, dedicated to variable stars and transiting exoplanets search (Bakos et al. 2007). It has discovered so far 41 transiting exoplanets. The WASP project (Christian et al. 2006) has discovered 79 with a pair of wide angle small photometric telescopes (SuperWASP I and II) located in La Palma and at Sutherland in South Africa respectively. Other ground-based transiting exoplanets surveys (MEarth project, OGLE, BEST, ...) have also contributed to the 282 discoveries. More recently, space borne instruments have joined the race (CoRoT: Baglin *et al.* 2006; Kepler: Borucki *et al.* 2010), yielding an important increase in the exoplanet detection rate and in photometric accuracy.

The periods of the 282 known exoplanets range from 0.45 to more than 300 days. Thus, long time series of photometric data are required for exoplanet discovery and qualification. Moreover, the photometric drop during the transit is very low. Even for a Jupiter-like planet orbiting a solar type star, the depth of the drop in the lightcurve hardly reaches 1%, and much lower depths are frequent. Thus high accuracy photometric data are required, with sub-millimagnitude noise (white and red). Consequently, observations from a good photometric site in polar regions would be of interest to discover new exoplanet candidates, but also to perform follow-up observations of known systems, for a better characterization.

The Dome C site (Antarctic Plateau; 75°06′ South, 123°19′ East, altitude: 3250 m) has been thoroughly tested as far as atmospheric turbulence parameters (seeing, isoplanatic angle and outer scale) are concerned (Agabi *et al.* 2006; Ziad *et al.* 2008; Aristidi *et al.* 2009). However, little was known about the photometric quality of the site.

The aim of the ASTEP program is twofold: first, assessing the photometric capabilities of Dome C, and second, studying transiting exoplanets (discovery and follow-up). ASTEP is a two-instrument program: ASTEP–South and ASTEP–400, both installed at the French–Italian Concordia station (Dome C).

ASTEP–South is a small "pathfinder" instrument observing the South pole region since 2008. It is a commercial 10 cm refractor on a commercial mount. No motor is needed, since the target is the south celestial pole. The field of view covered by the 4096 × 4096 pixels camera is 3.88° × 3.88°. The data from the first season of observation (2008) has revealed the good photometric quality of the Dome C site (Crouzet *et al.* 2010). Results from the 2008 to 2011 seasons are discussed at length in the presentation by N. Crouzet in this symposium (Crouzet *et al.* 2012), and thus, fall outside of the scope of the present contribution.

ASTEP–400 is a 400 mm semi-robotic custom telescope on a commercial equatorial mount, in operation at the Concordia station since 2010. The instrument itself and the results from its first winter observation campaign are described here.

This paper is organized as follows. We first recall basic facts about the ASTEP program, and give a detailed description of the ASTEP–400 instrument, data, and reduction pipelines. Then, we discuss the results of the 2010 and 2011 winter campaigns (exoplanets candidates found, follow-up of WASP-19b transit). We end up with some conclusions about the photometric accuracy of ASTEP–400.

2. The ASTEP program

2.1. *The context*

The ASTEP program (Antarctic Search for Transiting Exo-Planets) has been proposed in 2005 by F. Fressin, during his PhD at the *Observatoire de la Cote d'Azur* (France). The funding, by an ANR grant, was obtained in 2006, and the "pathfinder" instrument ASTEP–South was installed at the Concordia station in late 2007, ready for operation in the 2008 winter campaign (1,500 hours of data delivered in 2008).

The main instrument, ASTEP–400, was tested in France from March to October 2009, and installed at Concordia in December 2009, and ready for operation at the beginning of the 2010 austral winter (daytime first light late December 2009; nighttime first light on

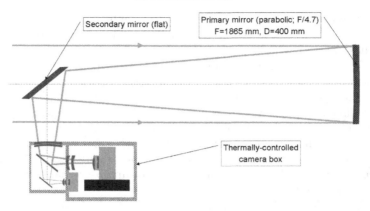

Figure 1. ASTEP–400 optical scheme.

March 25th, 2010). Presently (September 2012), the third winter season of ASTEP–400 and the fifth for ASTEP–South is about to end.

The ASTEP collaboration was initiated by the University of Nice-Sophia-Antipolis and the *Observatoire de la Cote d'Azur* (France). It includes the *Deutschen Zentrum für Luft- und Raumfahrt* (Germany), the Exeter University (UK), the *Observatoire de Haute Provence* and the *Observatoire de Marseille* (France).

2.2. *The ASTEP-400 instrument*

ASTEP–400 is a custom 400 mm Newtonian telescope with coma corrector, designed to withstand both summer and winter temperatures at Dome C, that is, temperatures ranging from $-20°$C to $-75°$C ($-4°$F to $-103°$F), with rapid variations. Figure 1 sketches the ray path of the optical design.

The custom five-lens comma corrector yields a $1° \times 1°$ field of view on a 4096×4096 pixels FLI *Proline* camera. The plate scale is $0.88''$ per pixel (1 pixel $= 9\mu$m). The PSF (Point Spread Function) is stable and homogeneous from the centre to the corner of the CCD, with a full width at half-maximum (FWHM) below 3 pixels when atmospheric turbulence is low (the entrance pupil is at less than 2 m above the ice surface, which is far from optimal for the seeing conditions).

To achieve a good thermal and mechanical stability, the structure of the telescope is an 8-th order double Serrurier assembly with carbon fiber legs, connected to an aluminium alloy main frame through Invar sleeves. To protect the optics both against stray light and ice dust, the Serrurier structure is covered by a two-layer cloth envelope.

Even with this careful design and manufacturing, the thermal expansion during rapid temperature variations may yield minor but unacceptable defocus (0.15 mm for a temperature variation of 30°C according to a finite elements simulation with NASTRAN). Thus, the science camera stands on a motorized, computer-controlled translation stage.

For accurate guiding, the incoming beam is split by a dichroic plate, which reflects the light with $\lambda > 600$ nm to the science camera, and transmits the light with $\lambda < 600$ nm to an auxiliary guiding camera, a SBIG *ST402*. In normal observing conditions, the resulting guiding accuracy is about $0.5''$ or better.

The science and guiding cameras, the motorized translation stage, and the electronic interfaces for the camera signals cannot withstand the Antarctic temperatures. Thus, a multi-zone thermally controlled casing has been designed to maintain the cameras and their surroundings at temperatures compatible with safe operations. The thermal control is done by electric heating resistors, PT100 temperature sensors and commer-

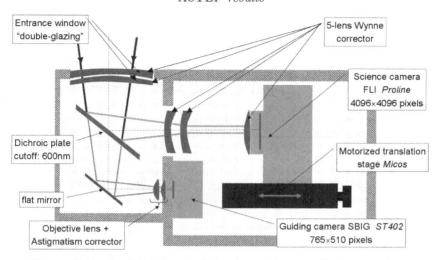

Figure 2. A simplified sketch of the thermally-controlled camera box.

cial computer-supervised controllers. This "camera box" (see Figure 2) receives the light from the Newtonian telescope's flat secondary mirror through the two first lenses of the 5-lens Wynne coma corrector, which act as a "double glazing" insulation window. The first compartment, which contains only optical components (dichroic plate, M3 mirror) is stabilized around −20°C (−4°F). The second compartment, which contains the cameras and electronic devices, is held around −10°C (+14°F), and the cameras' front panel is held at +5°C (+41°F), since the mechanical shutters cannot withstand too low temperatures. All the thermal controllers are daisy-chained on a RS485 bus, which communicates with the main control computer through a RS485/RS232 adapter. So, all temperatures can be monitored, and the set points of the controllers can be adjusted by software.

In Antarctic winter conditions, frost can form and/or ice dust can fall on the primary mirror of the telescope, and, to a lesser extent, on the secondary mirror. During the first winter season (2010), this problem required direct human intervention to manually defrost the mirror, or to remove the ice dust with a soft brush. To reduce direct human intervention as much as possible, we have introduced a defrosting mechanism for the primary and secondary mirrors during the 2010-2011 summer campaign. This system consists of custom designed film resistors, attached to the rear surfaces of the primary and secondary mirrors. Temperature probes are glued on the side of these mirrors, and a pair of additional commercial temperature controllers are added to drive the heaters. Since these controllers are also daisy-chained with all other controllers, they communicate with the mainframe computer driving the telescope. So, the mirror temperatures can be monitored easily, and the power supplied to the heaters can be adjusted by the winter crew, from the control room.

This defrosting system can operate in two different modes: a "preventive" mode and a "curative" mode. In the preventive mode, a small percentage of the maximum power is supplied to the heaters, so as to maintain the mirror surface a few degrees above ambient temperature (less than 4° for the primary, and less than 2° for the secondary). This mode can run during observations without hampering the image quality. Figure 3 shows typical temperature curves of the primary and secondary mirrors, for the preventive mode. Under special wether conditions frost can appear, or, more frequently, ice dust grains can fall on the optics. Then, the curative mode is activated, and 100% of the nominal power of the heaters is supplied. This removes the frost or ice dust. Of course, no image can be

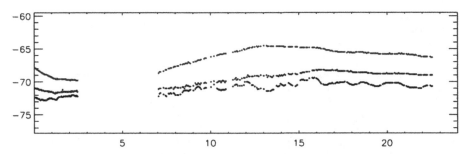

Figure 3. Temperatures (Celsius) of the primary mirror (upper curve), of the secondary mirror (middle curve) and outside temperature (lower curve), as a function of the UT time, on July 17th, 2012.

taken during this curative phase. This system was used during the 2011 and 2012 winter campaigns.

The telescope is semi-robotic and fully computer-controlled, so that little intervention is required from the winter-over crew. Only regular inspection of the primary mirror may be needed. Through a single graphic user interface, the winter crew member in charge with ASTEP–400 can monitor the relevant parameters of the telescope, perform any modifications, send orders to the mount motors, or run pre-determined observation scripts.

2.3. *The data and the reduction strategy*

Each frame produced by th FLI *Proline* camera is 32 MiB. A typical observation "night" produces some 400 frames, including calibration frames. This corresponds to more than 12 GiB of data. The 2010 winter season produced around 6 TiB. Since the internet connection at Concordia is not sufficient for bulky data transfer, all these data are transferred to France on hard disks in one of the summer crew members' personal luggage, at the end of the following summer campaign. Thus, data from the winter campaign number n are available in France for processing on February of year $n + 1$, typically.

Four types of frames are produced by the science camera: science frames, bias (offset) frames, dark frames, and "sky-flat" frames. Indeed, it is not possible to do either twilight flats or dome flats. So, the photometric gain calibration is performed with "sky-flat" frames. To get these frames, we take 200 frames during the few hours where the sky background rises as the Sun comes close to the horizon. Between each of these frames, a slight random shift is applied, to both the declination and right ascension axes of the telescope. Star images, which are visible on individual frames, are eliminated by taking the median of all these individual frames.

The ASTEP team uses two distinct data processing pipelines: a fast preliminary pipeline running daily at the Concordia station, and a more thorough one, running in France, on more powerful computers, when the data are available (February of Year $n+1$ for the winter season of Year n).

The fast pipeline is a custom IDL code making extensive use of "Daophot" routines. It first performs standard frame calibrations (bias, dark, flat). Then, it extracts the photometry of all stars in each frame by aperture photometry, to obtain raw light curves for all the stars brighter than the 16th magnitude (in R band). This yields light curves for several thousands of stars in a typical field. To reject transparency variations, a global calibration is performed on these light curves: the pipeline selects 4, 8 or 16 "reference" stars with stable fluxes, and corrects the light curves of other stars accordingly. The reference stars are chosen so as to minimize the resulting photometric noise on the whole

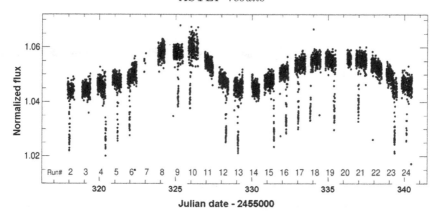

Figure 4. WASP19 lightcurve from May 1st to May 23rd, 2010. The 10.5 day stellar variation has not been removed.

field. Finally, the BLS transit detection algorithm (Box Least Square, Kovacs *et al.* 2002) is applied to the calibrated light curves, to extract the lightcurves with interesting features (exoplanets transits, eclipsing binaries, variable stars).

A more thorough custom pipeline has been written, to obtain more refined lightcurves. Being much more time-consuming, this IDL pipeline cannot run on the computers at Concordia station, but only in France. In addition to the standard frame calibrations, this pipeline also corrects for the so-called "sky concentration" effect (Andersen *et al.* 1995; Bellini *et al.* 2009). Then, the OIS method (Optimal Image Subtraction, Alard & Lupton 1998, Miller *et al.* 2008) is applied, and the raw light curves for all the measurable stars in the field are extracted. These raw lightcurves are calibrated against reference stars, chosen to minimize the residual photometric noise, as for the fast pipeline.

On lightcurves corresponding to exoplanet transit candidates, stellar variation is removed to flatten the lightcurve. Then, the lightcurve is phase-folded and the primary and secondary transits are fitted with the analytic transit model by Mandel & Agol (2002) with the Levenberg-Marquardt algorithm, to obtain the transit parameters.

3. Results from the 2010 winter campaign

3.1. *Photometric accuracy*

To assess the photometric quality of the lightcurves from ASTEP–400, we have studied a well-known transiting exoplanet : WASP-19b (Hebb *et al.* 2010) This exoplanet transits across its host star ($Mv = 12.3$) with a very short period : 0.78884 day. Each individual frame had an exposure time of 130 s. The lightcurve extends from May 1st to May 23rd, 2010 (see Figure 4).

The stellar variability is estimated through a median filtering of the original lightcurve, and the resulting filtered contribution (period 10.5 ± 0.2 days, consistent with Hebb *et al.* 2010) is subtracted out. This leads to a flattened light curve, which is phase-folded, binned over 0.001 phase units, and fitted with the analytic transit model of Mandel & Agol (2002). Figure 5 shows the folded lightcurve (binned and un-binned), and the best fitting transit curve for primary and secondary transit.

The rms photometric accuracy is estimated on out-of-transit data. This leads to 3 millimagnitudes for the un-binned data, and to 0.5 millimagnitudes or less, for binned data. This is comparable to using telescopes 2 times bigger at other good temperate astronomical sites.

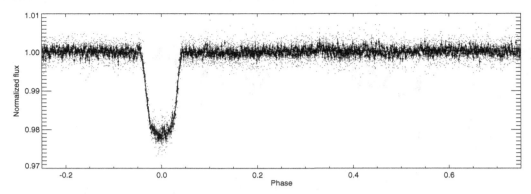

Figure 5. WASP19 lightcurve from May 1st to May 23rd, 2010. The stellar variation has been removed and the curve is phase-folded. The small grey dots correspond to individual frames. The large black dots with error bars correspond to data binned over 0.001 phase unit, and the solid curve is the best fit of the primary and secondary transit by the analytic transit model of Mandel & Agol (2002).

To estimate the contribution of red noise for the 1 hour time scale (typical for transit durations), we have applied the prescription in Pont *et al.* (2006) on each of the 23 "nights" during which WASP-19 has been observed with ASTEP–400. This yields the following statistics: the median value of the 1 hour time scale red noise is 0.7 millimag. The best and worst values are respectively 0.3 millimag and 2 millimag. In this sample of 23 "nights", one third are less than ideal, with red noise around 1 millimag; some nights are very good, with red noise below 0.4 millimag, and most nights are "average", with red noise around 0.7 millimag. These statistics are typical of very good ground-based photometric sites.

3.2. *Exoplanet candidates*

Only the data from 2010 and 2011 winter seasons have been processed so far. In these data, the BLS algorithm has found 93 lightcurves displaying transit-like structures. Among those, 8 targets are potential exoplanet candidates, with transit depths below 3% and periods ranging from 0.68 to 4.4 days. Follow-up observations are in progress, to obtain the stellar type of the host star, and to estimate the radial velocity.

4. Conclusion

ASTEP is a two-instrument program dedicated to exoplanet discovery and follow-up from Dome C (Antarctic Plateau), and to site qualification for long time scale high duty cycle photometry.

ASTEP–South, a small "pathfinder" instrument, has been in operation at Dome C since 2008, and its fifth winter season is currently in progress.

ASTEP–400, a 400 mm custom Newtonian telescope on a commercial mount, was installed at Dome C in less than 2 months during the 2009-2010 summer campaign. This instrument has been in operation since the 2010 austral winter, and is presently in its third winter season.

The data from the first two winter seasons have led to lightcurves with white noise around 3 millimag r.m.s. for un-binned data, and less than 0.5 millimag r.m.s. for binned data on a $M_v = 12.3$ star with 130 seconds exposures. This is comparable to a telescope twice as big in a good temperate astronomical site.

The red noise statistics (worst: 2 millimag; median: 0.7 millimag; best: 0.3 millimag) is comparable to very good ground-based astronomical sites.

In the 2010 and 2011 data, 93 lightcurves displaying transit-like structures were detected by the BLS algorithm, among which 8 are promising exoplanet candidates. Follow-up observations are in progress to ascertain their status.

Acknowledgements

The ASTEP project has been funded by the French agencies ANR (Agence National pour la Recherche), INSU (Institut National des Sciences de l'Univers), PPF–OPERA (Programme Pluri-Formation "Objectif Planètes Extrasolaires et Recherche en Antarctique") and PNP (Programme National de Planétologie). The authors are grateful to the French Polar Institute Paul-Emile Victor (IPEV) for their valuable logistic assistance.

References

Agabi K. *et al.* 2006, *PASP*, 118, 344

Alard C. & Lupton R. H. 1998, *ApJ*, 503, 325

Andersen M. I., Freyhammer L., & Storm J. 1995, "Gain calibration of array detectors by shifted and rotated exposures", *ESO Conference and Workshop Proceedings, Proceedings of an ESO/ST-ECF workshop on calibrating and understanding HST and ESO instruments, 25-28 April 1995* P. Benvenuti Ed., p. 87

Aristidi E. *et al.* 2009, *A&A*, 499, 955

Baglin A. *et al.* 2006, *36th COSPAR Scientic Assembly*, 36, 3749

Bakos G. A. *et al.* 2007, *ApJ*, 656, 552

Bellini A. *et al.* 2009, *A&A*, 493, 959

Borucki W. J. *et al.*, 2010, *Science*, 327(5968), 977

Charbonneau, D., Brown T., Latham D., & Mayor M. 2000, *ApJ*, 529, L45

Christian D. J. *et al.* 2006, *Astron. Nachr.*, 327, 800

Crouzet, N., *et al.* 2010, *A&A*, 511, A36

Crouzet N. *et al.* 2012, "An analysis of 4 years of data from ASTEP South", These proceedings.

Charbonneau, D., Brown T., Latham D., & Mayor M. 2000, *ApJ*, 529, L45

Hebb, L. *et al.* 2010, *ApJ*, 708, 224

Henry, G., Marcy, G., Butler, R. P., & Vogt, S. S. 1999, *IAU Circ.* 7307

Kovacs G., Zucker S., & Mazeh T. 2002, *A&A*, 391, 369

Mandel K. & Agol E., 2002, *ApJ*, 580, 171

Miller J. P., Pennypacker C. R., & White G. L. 2008, *PASP*, 120, 449

Pont F., Zucker S., & Queloz D. 2006, *MNRAS*, 373, 231

Ziad A. *et al.* 2008, *A&A*, 491, 917

Astrophysics from Antarctica
Proceedings IAU Symposium No. 288, 2012
M. G. Burton, X. Cui & N. F. H. Tothill, eds.

© International Astronomical Union 2013
doi:10.1017/S1743921312016924

ASTEP South: a first photometric analysis

N. Crouzet[1], T. Guillot[2], D. Mékarnia[2], J. Szulágyi[2,3], L. Abe[2],
A. Agabi[2], Y. Fanteï-Caujolle[2], I. Gonçalves[2], M. Barbieri[2],
F.-X. Schmider[2], J.-P. Rivet[2], E. Bondoux[4], Z. Challita[4],
C. Pouzenc[4], F. Fressin[5], F. Valbousquet[6], A. Blazit[2],
S. Bonhomme[2], J.-B. Daban[2], C. Gouvret[2], D. Bayliss[7],
G. Zhou[7] and the ASTEP team

[1] Space Telescope Science Institute, 3700 San Martin Drive, Baltimore, MD 21218, USA
email: crouzet@stsci.edu

[2] Laboratoire Lagrange, UMR 7293 UNS-CNRS-OCA, Boulevard de l'Observatoire, BP 4229,
06304 Nice Cedex 4, France

[3] Konkoly Observatory, Research Centre for Astronomy and Earth Sciences, Hungarian
Academy of Sciences, Konkoly Thege Miklós út 15-17, H-1121 Budapest, Hungary

[4] Concordia Station, Dome C, Antarctica

[5] Harvard-Smithsonian Center for Astrophysics, 60 Garden Street, Cambridge, MA 02138, USA

[6] Optique et Vision, 6 bis avenue de l'Estérel, BP 69, 06162 Juan-Les-Pins, France

[7] Research School of Astronomy & Astrophysics, The Australian National University, Cotter
Road, Weston Creek, ACT 2611, Australia

Abstract. The ASTEP project aims at detecting and characterizing transiting planets from
Dome C, Antarctica, and qualifying this site for photometry in the visible. The first phase of
the project, ASTEP South, is a fixed 10 cm diameter instrument pointing continuously towards
the celestial South Pole. Observations were made almost continuously during 4 winters, from
2008 to 2011. The point-to-point RMS of 1-day photometric lightcurves can be explained by
a combination of expected statistical noises, dominated by the photon noise up to magnitude
14. This RMS is large, from 2.5 mmag at $R = 8$ to 6% at $R = 14$, because of the small size
of ASTEP South and the short exposure time (30 s). Statistical noises should be considerably
reduced using the large amount of collected data. A 9.9-day period eclipsing binary is detected,
with a magnitude $R = 9.85$. The 2-season lightcurve folded in phase and binned into 1,000
points has a RMS of 1.09 mmag, for an expected photon noise of 0.29 mmag. The use of the 4
seasons of data with a better detrending algorithm should yield a sub-millimagnitude precision
for this folded lightcurve. Radial velocity follow-up observations reveal a F-M binary system.
The detection of this 9.9-day period system with a small instrument such as ASTEP South and
the precision of the folded lightcurve show the quality of Dome C for continuous photometric
observations, and its potential for the detection of planets with orbital periods longer than those
usually detected from the ground.

Keywords. techniques: photometric, methods: data analysis, site testing, (stars:) binaries:
eclipsing

1. Introduction

Dome C offers exceptional conditions for astronomy thanks to a 3-month continuous
night during the Antarctic winter and a very dry atmosphere. This site is located at
$75°06'$ S, $123°21'$ E at an altitude of 3,233 meters on a summit of the high Antarctic
plateau, 1,100 km away from the coast. Winter site testing for astronomy revealed a
very clear sky, excellent seeing above a thin boundary layer, very low wind-speeds (Aristidi *et al.* 2003, Aristidi *et al.* 2005, Lawrence *et al.* 2004, Ashley *et al.* 2005, Aristidi

et al. 2009, Giordano *et al.* 2012, Fossat *et al.* 2010), very low scintillation (Kenyon *et al.* 2006) and a high duty cycle (Mosser & Aristidi 2007, Crouzet *et al.* 2010). Time-series observations such as those implied by the detection of transiting exoplanets should benefit from these atmospherical conditions and the good phase coverage (Pont & Bouchy 2005). The ASTEP project (Antarctic Search for Transiting ExoPlanets) aims at determining the quality of Dome C as a site for future photometric surveys and to detect transiting planets (Fressin *et al.* 2005). The main instrument is a 40 cm Newtonian telescope entirely designed and built to perform high precision photometry from Dome C. The design is presented in Daban *et al.* (2010) and the performance is detailed in Rivet *et al.* (2012). A 10 cm instrument pointing continuously towards the celestial South Pole, ASTEP South, was first installed at Dome C (Crouzet *et al.* 2010). Both instruments use facilities provided by the Franco-Italian Concordia station at Dome C, and are installed on the AstroConcordia platform at ground level. Other photometric instruments are also observing from Antarctica. Among them is CSTAR, an instrument very similar to ASTEP South and located at Dome A (Yuan *et al.* 2008, Zhou *et al.* 2010a). Variable stars in the South pole field were detected and classified using CSTAR, and a first planetary candidate was reported (Zhou *et al.* 2010b, Wang *et al.* 2011). Here, we present a photometric analysis of the ASTEP South data. First, we briefly describe the instrumental setup. Then, we detail the lightcurve extraction and show its performance over 1 day. Finally, we present our independent detection of the planetary candidate reported by Wang *et al.* (2011), as well as follow-up observations.

2. Observations

ASTEP South consists of a 10 cm refractor, a front-illuminated 4096×4096-pixel CCD camera, and a simple mount in a thermalized enclosure. The refractor is a commercial TeleVue NP101. The camera is in the ProLines series by Finger Lake Instrumentation, equipped with a KAF-16801E CCD by Kodak (see Crouzet *et al.* 2007 for the choice of the camera). The overall transmission (600 to 900 nm) is equivalent to that of a large R band. The enclosure is thermalized to $-20°$C and closed by a double glass window on the optical path to avoid temperature fluctuations. The instrument is shown in Figure 1. ASTEP South is completely fixed and points towards the celestial South pole continuously. The observed field of view is $3.88° \times 3.88°$, leading to a pixel size of $3.41''$ on the sky. This field contains around 8000 stars to magnitude $R = 15$. This observation setup leads to stars moving circularly on the CCD with a 1-sidereal day period, and to an elongated PSF (Point Spread Function). The exposure time is 30 s with 10 s between each exposure. The PSF is defocused to a 2-pixel FWHM (Full Width Half Maximum), but varies in particular with the seeing at the ground-level where the instrument is placed. ASTEP South observed during 4 winters: from June to the end of the 2008 winter collecting ~1,500 hours of data, and all 2009, 2010, and 2011 winters collecting ~2,500 hours of data per winter.

3. Photometric analysis

Due to the rotation of the stars on the CCD, the PSFs are elongated up to 4.5 pixels at the edges of the field, with an elongation direction varying in time. We designed a specific photometric algorithm to take into account this particular feature: an elongated photometric aperture is created for each star in each image. The aim is to reduce the number of pixels compared to standard circular apertures in order to minimize the sky background and read-out noise. The aperture elongation and orientation are calculated

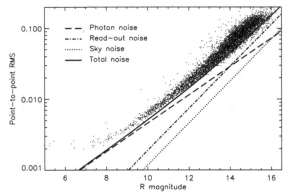

Figure 1. ASTEP South at
Dome C, Antarctica.

Figure 2. Point-to-point RMS of the June 21, 2008
lightcurves as a function of instrumental magnitude.
The RMS is calculated over the 30 s exposure time.
The photon noise is represented by a dashed line, the
read-out noise by a dash-dot line, the sky noise by a
dotted line, and their quadratic sum by a plain line.

according to the position of the star on the image. The size is optimized empirically for
each star. To this end, we perform photometry on a particular day (June 21, 2008) using
a large number of aperture sizes. The size yielding the lowest point-to-point RMS is kept.
As expected, bright stars end up with a larger aperture than faint stars. The aperture
size of each star then remains constant during the process. In particular, it does not vary
with the seeing; we found that a greater average seeing does not necessarily yield a larger
aperture when performing the optimization.

Figure 2 shows the point-to-point RMS over 30 s for the 1-day lightcurves of June 21,
2008. Each flux measurement is compared only to its neighbor, the long-term variations
are not taken into account. In each lightcurve, we remove outliers departing from the
mean by more than 3.5 times the standard deviation. This represents 6% of the data on
average. We obtain a point-to-point RMS of 2.5 mmag at $R = 8$, 6 mmag at $R = 10$,
1.8% at $R = 12$, and 6% at $R = 14$. This point-to-point RMS is compatible with the
expected photon noise, read-out noise and sky background noise, which are high due
to the small size of ASTEP South and the short exposure time. However, these statistical noises should be considerably reduced by binning the huge amount of collected data.

4. First planet candidate: an eclipsing F-M binary

Lightcurves from the 2008 season are detrended using TFA (Kovács *et al.* 2005) with
the method described in Szulágyi *et al.* (2009), and periodic signals are searched with
BLS (Kovács *et al.* 2002). A first transit candidate is found around an F dwarf star
of coordinates RA=18:30:56.777, Dec=−88:43:17.01 (J2000) and magnitude $V = 10.12$
(instrumental magnitude $R = 9.85$). This candidate was also identified by Wang *et al.*
(2011) with the CSTAR instrument at Dome A. We then gather the data from the 2008
and 2009 seasons. We use only the good images, where at least 1/5 of the stars are
detected in the field of view (compared to the typical number of stars detected under
favorable observing conditions). This represents 73.5% of the data, or 227,633 images.
The 2-season lightcurve of this candidate is calibrated with reference stars. A residual
variation with a period of 1 sidereal day appears due to imperfect flat-fielding. To correct

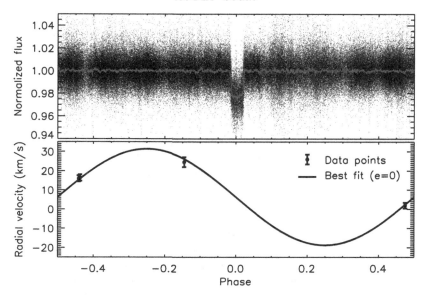

Figure 3. Top: 2-season lightcurve of the star RA=18:30:56.777, Dec=−88:43:17.01 (J2000), folded at a period $P = 9.927$ days (black dots), and binned into 1000 points (red line). The instrumental magnitude is $R = 9.85$. The binned lightcurve has RMS of 1.09 mmag. Bottom: Radial velocity measurements (filled circles), and best fit assuming no eccentricity (plain line). This system is an eclipsing F-M binary.

for this, we fold the lightcurve over 1 sidereal day, bin it into 100 points with a median filter, and divide the folded lightcurve by the binned lightcurve. This correction is applied for each season independently. Because residual trends remain, we apply a low-frequency variation correction on the lightcurve of each day by fitting and subtracting a 2nd order polynomial (the in-transit data points are excluded from the fit). The RMS over 30 s for this 2-season lightcurve is 1.1%. This is about twice the 1-day point-to-point RMS at that magnitude, indicating that trends still remain.

Twenty transits are present in the data that we analyzed: 7 in 2008 and 13 in 2009. Most of them are partial transits because of their long duration (∼10 hours). The period P is calculated using BLS; we find $P = 9.927 \pm 0.003$ days. The transit time reference is $HJD_0 = 2455060.10 \pm 0.01$. We then fold the lightcurve at the period P and bin it into 1,000 points (Figure 3). The transit depth ρ is calculated using this binned lightcurve; we find $\rho = 2.25 \pm 0.18\%$. For comparison, Wang *et al.* (2011) derive $P = 9.916$ days and $\rho = 17$ mmag (1.55%) from a smaller number of transits and using data from 2008 only. Our binned lightcurve has a RMS of 1.09 mmag for this 9.85 magnitude star. This precision shows the quality of Dome C for continuous photometric observations. The theoretical photon noise limit is however 0.29 mmag. The use of the 4 seasons of data as well as a better detrending should thus yield a sub-millimagnitude precision for this star, which would be unprecedented for a 10 cm instrument.

Follow-up radial velocity observations were conducted at the ANU 2.3 m telescope at Siding Spring Observatory, Australia, in January 2012. Four data points were taken, 2 being almost at the same phase. These measurements show an amplitude too large to be caused by a planet (Figure 3). This system is therefore an eclipsing binary. No more observations were made. Our spectra confirm the F-dwarf nature of the primary star. Assuming no eccentricity ($e = 0$), we find a radial velocity semi-amplitude $K = 25.1 \pm 2.8$ km/s. Assuming a mass $M_{prim} = 1.3$ M$_\odot$ for the primary as a standard mass for

F-dwarfs, we derive a mass $M_{sec} = 0.35 \pm 0.05$ M$_\odot$ for the secondary. This system is therefore an eclipsing F-M binary. The detection of this 9.9-day period system is however encouraging for the detection of planets with orbital periods longer than those usually detected from the ground.

5. Conclusion

ASTEP South has been observing the South pole field almost continuously during winter since 2008. Our analysis of the 2008 and 2009 seasons shows a point-to-point RMS that can be explained mostly by a combination of photon noise and read-out noise. Due to the short exposure time and the small size of ASTEP South, this RMS is high, but should be considerably reduced by the large amount of collected data when searching for periodic signals. A 9.9-day period eclipsing F-M binary is found. After combining data from the 2 first seasons, a precision of 1.09 mmag is obtained for this 9.85 magnitude system. A sub-millimagnitude precision should be reached with a better detrending and the use of the 4 seasons of data. This detection and analysis show the quality of Dome C for continuous photometric observations, even with a small instrument such as ASTEP South, and is encouraging in the search for planets with orbital periods longer than those usually detected from the ground. In addition to the search for planetary transits, we will also analyze interesting variable stars present in the field of view.

References

Aristidi, E., Agabi, A., Vernin, J., *et al.*, 2003, *A&A*, 406, L19
Aristidi, E., Agabi, K., Azouit, M., *et al.* 2005, *A&A*, 430, 739
Aristidi, E., Fossat, E., Agabi, A., *et al.* 2009, *A&A*, 499, 955
Ashley, M. C. B., Lawrence, J. S., Storey, J. W. V., & Tokovinin, A. 2005, *EAS Publications Series*, 14, 19
Crouzet, N., Guillot, T., Fressin, F., & Blazit, A., the ASTEP team 2007, *AN*, 328, 805
Crouzet, N., Guillot, T., Agabi, K., *et al.* 2010, *A&A*, 511:A36
Daban, J.-B., Gouvret, C., Guillot, T., *et al.* 2010, *SPIE Conference Series*, 7733
Fossat, E., Aristidi, E., Agabi, A., *et al.* 2010, *A&A*, 517, A69
Fressin, F., Guillot, T., Bouchy, F., *et al.* 2005, *EAS Publications Series*, 14, 309
Giordano, C., Vernin, J., Chadid, M., *et al.* 2012, *PASP*, 124, 494
Kenyon, S. L., Lawrence, J. S., Ashley, M. C. B., *et al.* 2006, *PASP*, 118, 924
Kovács, G., Bakos, G., & Noyes, R. W. 2005, *MNRAS*, 356, 557
Kovács, G., Zucker, S., & Mazeh, T. 2002, *A&A*, 391, 369
Lawrence, J. S., Ashley, M. C. B., Tokovinin, A., & Travouillon, T. 2004, *Nature*, 431, 278
Mosser, B. & Aristidi, E. 2007, *PASP*, 119, 127
Pont, F. & Bouchy, F. 2005, *EAS Publications Series*, 14, 155
Rivet, J.-P., *et al.* 2012, *IAUS288 Proceedings*, these proceedings
Wang, L., Macri, L. M., Krisciunas, K., *et al.* 2011, *AJ*, 142, 155
Szulágyi, J., Kovács, G., & Welch, D. L. 2009, *A&A*, 500, 917
Yuan, X., Cui, X., Liu, G., *et al.* 2008, *Proc. SPIE*, 7012, 70124G
Zhou, X., Wu, Z., Jiang, Z., *et al.* 2010a, *Res. Astron. Astrophys.*, 10, 279
Zhou, X., Fan, Z., Jiang, Z., *et al.* 2010b, *PASP*, 122, 347

Astrophysics from Antarctica
Proceedings IAU Symposium No. 288, 2012
M. G. Burton, X. Cui & N. F. H. Tothill, eds.

© International Astronomical Union 2013
doi:10.1017/S1743921312016936

Progress and Results from the Chinese Small Telescope ARray (CSTAR)

Xu Zhou[1,5], M. C. B. Ashley[4], Xiangqun Cui[2,5], Longlong Feng[3,5], Xuefei Gong[2,5], Jingyao Hu[1,5], Zhaoji Jiang[1,5], C. A. Kulesa[8], J. S. Lawrence[4,6], Genrong Liu[2], D. M. Luong-Van[4], Jun Ma[1], Lucas M. Macri[1,4], Zeyang Meng[15], A. M. Moore[9], Weijia Qin[7], Zhaohui Shang[10], J. W. V. Storey[4], Bo Sun,[7], T. Travouillon[9], C. K. Walker[8], Jiali Wang[1,5], Lifan Wang[3,5], Lingzhi Wang[1,14], Songhu Wang[15], Jianghua Wu[1], Zhenyu Wu[1], Lirong Xia[2], Jun Yan[1,5], Ji Yang[3], Huigen Yang[7], Yongqiang Yao[1], Xiangyan Yuan[2,5], D. York[11], Hui Zhang[15], Zhanhai Zhang[7], Jilin Zhou[15], Zhenxi Zhu[3,5] and Hu Zou[1]

[1] National Astronomical Observatories, Chinese Academy of Sciences, Beijing, 100012, China;
zhouxu@bao.ac.cn

[2] Nanjing Institute of Astronomical Optics and Technology, Nanjing 210042, China

[3] Purple Mountain Observatory, Nanjing 210008, China

[4] School of Physics, University of New South Wales, Sydney NSW 2052, Australia

[5] Chinese Center for Antarctic Astronomy

[6] Macquarie University & Anglo-Australian Observatory

[7] Polar Research Institute of China, Pudong, Shanghai 200136, China

[8] Steward Observatory, University of Arizona, Tucson, AZ 85721 USA

[9] Department of Astronomy, California Institute of Technology, Pasadena CA 91125, USA

[10] Tianjin Normal University, Tianjin 300074, China

[11] Department of Astronomy and Astrophysics and Enrico Fermi Institute, University of Chicago, Chicago IL 60637, USA

[12] Graduate University of Chinese Academy of Sciences, Beijing 100049, China

[13] Department of Astronomy, Beijing Normal University, Beijing, 100875, China

[14] Mitchell Institute for Fundamental Physics & Astronomy, Department of Physics & Astronomy, Texas A&M University, College Station TX 77843, USA

[15] Department of Astronomy & Key Laboratory of Modern Astronomy and Astrophysics in Ministry of Education, Nanjing University, Nanjing 210093, China

Abstract. In 2008 January the 24th Chinese expedition team successfully deployed the Chinese Small Telescope ARray (CSTAR) to Dome A, the highest point on the Antarctic plateau. CSTAR consists of four 14.5 cm optical telescopes, each with a different filter (g, r, i and open) and has a $4.5° \times 4.5°$ field of view (FOV). Based on the CSTAR data, initial statistics of astronomical observational site quality and light curves of variable objects were obtained. To reach higher photometric quality, we are continuing to work to overcome the effects of uneven cirrus cloud cirrus, optical "ghosts" and intra-pixel sensitivity. The snow surface stability is also tested for further astronomical observational instrument and for glaciology studies.

Due to the extremely cold, dry, calm, thin atmosphere and the absence of light and air pollution, we can obtain a low infrared background, new observation windows, and high quality photometric images from the Antarctic Plateau (Burton 2010). Over the long polar night, the observation of high quality, long-baseline uninterrupted time-series photometry has great significance to a large range of astrophysical problems, such as the search for transiting exoplanets and the study of variable stars. This kind of observation can be achieved most effectively by ambitious space-based programs such as CoRoT (Boisnard & Auvergne 2006) and Kepler (Borucki *et al.* 2010). However, the Antarctic plateau can offer a comparable alternative with significantly lower costs.

The successful site testing and astronomical observations at Dome C (Crouzet *et al.* 2010) encourage astronomers to think about the observational conditions of Dome A. Dome A, located at latitude 80°37′ S and longitude 77°53′ E, with surface elevation about 4,093 m, is the highest astronomical site in Antarctica. It is also considered to be the potential coldest place in the world (Saunders *et al.* 2009). After comparing with other Antarctic sites in weather, atmosphere, sky brightness condition, Saunders *et al.* (2009) concluded that Dome A might be the best of the existing astronomical sites on Earth.

The Chinese Small Telescope ARray (CSTAR) was developed for site testing and installed at Dome A in 2008 January. Like other optical telescopes in Antarctica, CSTAR undertook both site testing and science research tasks. CSTAR includes 4 small telescopes with apertures of 14.5 cm in diameter. Three of the telescopes are covered with SDSS *g*, *r* and *i* band filters and one has no filter. CSTAR is mounted on a tripod, with fixed pointing at the south polar sky field. Andor 1K×1K frame transfer CCD cameras are used. There is no any mechanical moving part in the whole observational system. Parts of CSTAR have operated well for the last 5 years (from 2008). One telescope of CSTAR was brought back in 2010 and the whole instrument was dismounted in 2012 January for system optimization.

With high-cadence observations from 2008 to 2011, about 2.5 TB of image data were obtained. Fig. 1 to Fig. 3 show the daily image count of observation seasons from 2008 to 2010.

Based on the data from 2008, Zhou *et al.* (2010a) released the first version of the CSTAR point source catalog of over 10,000 stars, updated recently to a magnitude limit of 14 in *i* band (S. H. Wang *et al.* 2012). The CSTAR data have already been successfully used to test the site characteristics of Dome A (Zou *et al.* 2010), for detection of the variable stars (Wang *et al.* 2011), and to study the stability of the snow surface of Dome A.

By comparing the catalog of each image with the catalog of mean magnitude and mean position of stars as reference, the statistics of the sky brightness, atmospheric transparency and appearances of the aurora are obtained. As a result for the overall night sky, a median brightness of 19.8 mag arcsec^{-2} is found across all lunar phases (left panel in Fig. 4). A median of 20.5 mag arcsec^{-2} is found for moonless clear nights (right panel of Fig. 4). Dome A has a darker sky background than the other Antarctic astronomical sites, even allowing for calibration and other uncertainties of up to several tenths of a magnitude (Zou *et al.* 2010).

In the atmospheric extinction distribution of 2008, as shown in Fig. 5, 90% of the images show an extinction of less than 0.7 mag, 80% less than 0.4 mag and more than half of the time, the extinction is less than 0.1 mag. The monthly variation of observational qualities shows that the weather is relatively bad in the second half of the 2008 and 2010 winters.

Because the variation of the sky transparency depends on the cloud coverage, it is well-correlated to the sky brightness. The sky brightness can also be well calculated with the parameters of Sun and Moon elevation, Moon phase and atmospheric extinction:

$$F_{sky} = a(F_{sun} + F_{moon})E + bE + c$$

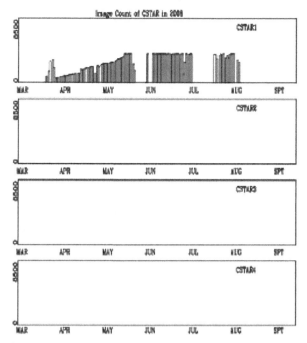

Figure 1. The daily image count of observations in 2008 from the 4 different CSTAR telescopes.

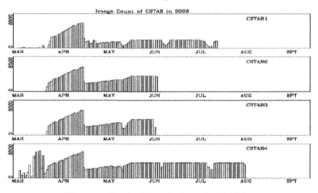

Figure 2. As Fig. 1, but for 2009.

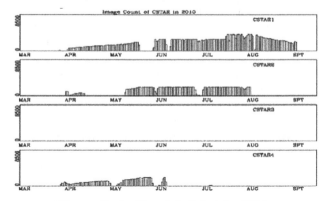

Figure 3. As Figs. 1 & 2, but for 2010.

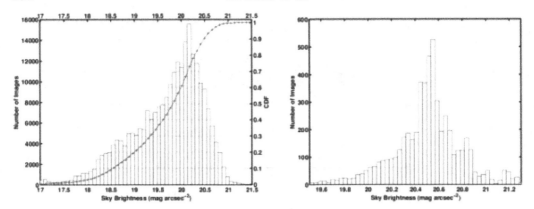

Figure 4. Left panel: Histogram and cumulative distribution function (CDF) of the *i*-band sky brightness distribution at Dome A during 2008. Right panel: Same information for the subset of images taken on moonless clear nights in 2008 June.

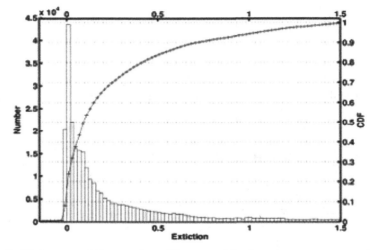

Figure 5. Histogram of the number of images with given transparency variation (in magnitudes) relative to the reference image. The curved line marked '+' is the CDF.

Here, $F_{sun} + F_{moon}$ are the solar and lunar contributions and E is transparency. a, b and c are parameters obtained from fitting.

The above relation is obtained from images without auroral pollution. The real sky brightness may be brighter than estimated. Outliers outside 3 times the rms should be considered as relatively strong aurorae. About 2% of images are affected by aurorae (Zou *et al.* 2010).

Wang *et al.* (2011) have done independent photometry on the CSTAR images to search for variable stars. In the inner circle of the CCD FOV, they found 157 variable sources from images taken in 2008, about 5 times as many as previous surveys with similar magnitude limits. Fig. 6 shows an example light curve of a contact binary. Furthermore, they analysed the data in the full CSTAR sky coverage field and found 113 new variable candidates from the CSTAR data of 2010 (L. Z. Wang *et al.* 2012).

The ongoing instrument plans are to change the CSTAR mount to that of a normal telescope, adjusting optics and improving the hardware and software of CCD control system. We plan to continue scientific research on object variation, searching for extra-solar

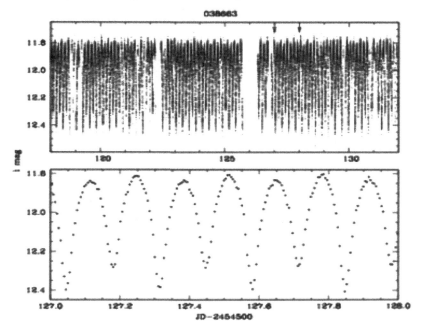

Figure 6. Time-series light curve of a contact binary in the CSTAR FOV. Only a small fraction of the complete CSTAR data is shown. Top panel: Light curve sampled at 20s intervals. Bottom panel: a portion of the top light curve (bounded by the arrows) binned into 450s intervals.

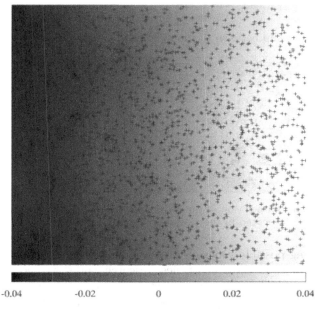

Figure 7. The cloud (extinction difference in magnitude) structure of an image (A68F2318) taken in poor photometric conditions. The grayscale denotes the structure of the cloud. '+' indicate positions of the selected stars used to calculate magnitude differences between the published catalog and the reference catalog; black is the thinnest cloud and white is the thickest cloud.

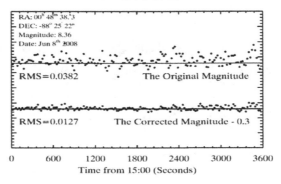

Figure 8. Comparison of the light curves from a series of images obtained under poor pho-tometric conditions. The solid line denotes the mean magnitude of the star in the reference catalog.

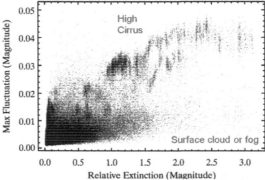

Figure 9. The relationship between the unevenness of extinction and the value of extinction at Dome A.

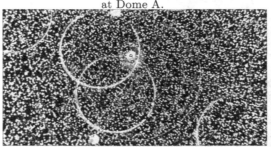

Figure 10. The calibrated position of the object detected in all the images where the objects located in the circles are mostly a "ghost".

planets and detecting possible supernovae, novae, and orphan afterglows of gamma ray bursts from the data. To fit the need of these research works, we need to do further data reduction work to get higher photometric precision, such as inhomogeneous extinction over the field of view, effect of optical "ghosts" and intra-pixel effects, etc.

Passing cirrus cloud usually appears in the Dome A area. Ideally, if the atmospheric extinction is uniform across the FOV or the cloud is absent, we can obtain the true flux of all sources by identical calibrating to standards. Because CSTAR gives about 20 square degrees FOV, we cannot assume the cloud is uniform over all the image fields. A flux calibration with a single value in an image will cause deviations when it is affected by cirrus clouds. Due to the varying extinction over the FOV, a different calibration

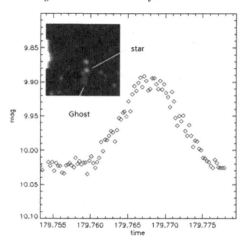

Figure 11. An example of the light curve of a star when a "ghost" passes by, from previous catalog of CSTAR data release.

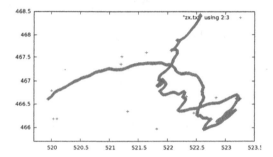

Figure 12. The position of the south celestial pole in the CSTAR field during May 12, 2008.

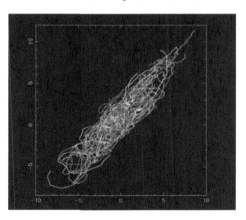

Figure 13. The total south celestial pole positions in all CSTAR images from 2008. We change the color every 24 hours.

is required for each star in the same image. In order to reduce this kind of systematic error, we first made a reference catalog from over ten thousand good photometric stars in the published catalog (Zhou *et al.* 2010a). The reference catalog contains the mean magnitude and mean position of all the stars in the CSTAR FOV. Then, we compared an ensemble of the detected magnitudes of the bright and unsaturated stars in the published catalog (Zhou *et al.* 2010a) with their magnitudes in the reference catalog. With this distribution of magnitude differences (which yields the cloud structure, as shown in Fig. 7), we updated the published catalog. In the new photometric catalog of 2008, the catalogs of about 20% of images are significantly improved. Fig. 8 shows the effect of correction. And from the relationship between the unevenness of extinction and the value of extinction at Dome A (Fig. 9), the atmospheric extinction at Dome A is considered to be caused by both high cirrus and fog near the snow surface.

In the optical system of CSTAR, many obvious "ghosts" appear in the image. These are mirror images of bright stars after multiple reflections in different optical surfaces. After careful study of the "ghost" behaviour, Meng *et al.* (2012) found that the "ghosts" make daily circular motions in the image. They pass or pass by the positions of many stars in the image. The effect of the ghost changes the measurement magnitude of the stars. The magnitude difference between the "ghosts" and its source objects are about 6 magnitudes and have a fixed symmetry point in the image. Then the influence of the "ghost" on the magnitude of all the real objects can be well calculated.

For further installation and operation of telescopes on the Antarctic plateau, the stability of the ground, floating on thousands of metres of snow and ice, needs to be considered. The observation system of CSTAR is a totally mechanically fixed system. It is firmly fixed on the snow and keeps pointing at the south celestial polar area. So the CSTAR is a unique system for checking ground stability. During the data reduction of the CSTAR, the variation of the telescope pointing is very large. The daily pointing movement can be up to 3 pixels or more than $45''$, as shown in Fig. 12. And the overall movement area during whole observation season of 2008 can be as large as 20 pixels or $5'$, as shown in Fig. 13. Further analysis of this issue will be not only very important for Antarctic astronomical observation but also very important for glaciology studies.

The author thanks the support of National Natural Science Foundation No. 11073032 (PI: Xu Zhou)

References

Bakos, G. Á., Noyes, R. W., Kovács, G., *et al.* 2007, *ApJ*, 656, 552
Boisnard, L. & Auvergne, M. 2006, *ESA Special Publication 1306*, 19
Borucki, W. J., Koch, D., Basri, G., *et al.* 2010, *Science*, 327, 977
Burton, M. G. 2010, *Astronomy & Astrophysics Review*, 18, 417
Crouzet, N., Guillot, T., Agabi, A., *et al.* 2010, *A&A*, 511, A36
Kenyon, S. L., Lawrence, J. S., Ashley, M. C. B., *et al.* 2006, *PASP*, 118, 924
Lawrence, J. S., Ashley, M. C. B., Hengst, S., *et al.* 2009, *Review of Scientific Instruments*, 80, 064501
Meng, Z. Y., *et al.*, 2012, in preparation
Pollacco, D. L., Skillen, I., Collier Cameron, A., *et al.* 2006, *PASP*, 118, 1407
Saunders, W., Lawrence, J. S., Storey, J. W. V., *et al.* 2009, *PASP*, 121, 976
Strassmeier, K. G., Briguglio, R., Granzer, T., *et al.* 2008, *A&A*, 490, 287
Taylor, M. 1990, *AJ*, 100, 1264
Wang, L., Macri, L. M., Krisciunas, K., *et al.* 2011, *AJ*, 142, 155
Wang, L. Z., *et al.*, 2012, *AJ*, submitted
Wang S. H., *et al.*, 2012, *PASP*, accepted
Yang, H., Allen, G., Ashley, M. C. B., *et al.* 2009, *PASP*, 121, 174
Yuan, X., Cui, X., Liu, G., *et al.* 2008, *Proc SPIE*, 7012E, 152
Zhou, X., Fan, Z., Jiang, Z., *et al.* 2010, *PASP*, 122, 347
Zhou, X., Wu, Z.-Y., Jiang, Z.-J., *et al.* 2010, *Research in Astronomy and Astrophysics*, 10, 279
Zou, H., Zhou, X., Jiang, Z., *et al.* 2010, *AJ*, 140, 602

Astrophysics from Antarctica
Proceedings IAU Symposium No. 288, 2012
M. G. Burton, X. Cui & N. F. H. Tothill, eds.

© International Astronomical Union 2013
doi:10.1017/S1743921312016948

Next Generation Deep 2μm Survey

Jeremy Mould

[1] Centre for Astrophysics and Supercomputing, Swinburne University, Hawthorn 3122,
Australia
[2] ARC Centre of Excellence for All-sky Astrophysics (CAASTRO)
email: jmould@swin.edu.au

Abstract. There is a major opportunity for the KDUST 2.5 m telescope to carry out the next generation IR survey. A resolution of 0.2 arcsec is obtainable from Dome A over a wide field. This opens a unique discovery space during the 2015–2025 decade.

A next generation 2μm survey will feed JWST with serendipitous targets for spectroscopy, including spectra and images of the first galaxies.

Keywords. infrared, survey, galaxies

1. Introduction: the state of the art of infrared surveys

UKIDSS† has surveyed 7,500 square degrees of the Northern sky, extending over both high and low Galactic latitudes, in the JHK bandpasses to K=18.3. This is three magnitudes deeper than 2MASS. UKIDSS has provided a near-infrared SDSS and a panoramic atlas of the Galactic plane. UKIDSS is actually five surveys, including two deep extragalactic elements, one covering 35 sq deg to K = 21, and the other reaching K = 23 over 0.77 sq deg.

VIKING-VISTA‡ is a kilo-degree infrared galaxy survey. The VIKING survey will image 1500 sq deg in Z, Y, J, H, and Ks to a limiting magnitude 1.4 mag beyond the UKIDSS Large Area Survey. It will furnish very accurate photometric redshifts, especially at z > 1, an important step in weak lensing analysis and observation of baryon acoustic oscillations. Other science drivers include the hunt for high redshift quasars, galaxy clusters, and the study of galaxy stellar masses.

Mould (2011) offers a summary of the prospects for improving on these surveys using the KDUST 2.5 m telescope.

2. KDUST camera architecture

The simplest option for a focal plane array is a Teledyne HgCdTe 2048^2. A better option is 4096^2 or 2×2 (8.5 arcmin field). ANU has delivered two such cameras to the Gemini Observatory (McGregor *et al.* 2004, McGregor *et al.* 1999). The KDUST focal plane scale is appropriate without change. JHK and Kdark filters would be required.

Plan B is for a Sofradir SATURN SW HgCdTe SWIR. However, these detectors have 150 electrons read noise and would require long exposures to overcome readout noise. Nevertheless they are feasible Plan B detectors for broadband survey work. Mosaicing many detectors is also acceptable for survey work, and, after mosaicing the focal plane, plan A and plan B detectors are fairly similar in cost.

† www.ukidss.org
‡ www.astro-wise/projects/VIKING

3. The Antarctic advantage

Above the ground layer turbulence one obtains almost diffraction limited images over a wide field with low $2\mu m$ background. This combination is only available from the Antarctic plateau, high altitude balloons and space. The competition, then, is space. We confine ourselves to WFIRST, since the ESA Euclid mission observes at H band, but not at K.

Advantages of WFIRST

- Top ranked in ASTRO 2010 (Blandford 2009)
- Broader band possible, e.g. 1.6–$3.6\mu m$.
- No clouds

Disadvantages of WFIRST

- 3 year mission lifetime
- Earliest launch 2025
- Order of magnitude higher cost

Provided the US NRO supplies a 2.5 m mirror, the following Astro2010-era disadvantages are no longer in effect.

- Smaller aperture, 1.5 m
- Lower resolution
- 200 nJy limit vs 70 nJy with KDUST

4. Science case

An excellent science case for a 2.5 m Antarctic telescope is presented by Burton *et al.* (2005). A further science case is that of WFIRST (Green *et al.* 2012).

- Kuiper Belt census and properties
- Cluster and Star-Forming-Region IMFs to planetary mass
- The H_2 kink in star cluster CMDs
- The most distant Star-Forming-Regions in the Milky Way
- Quasars as a Reference Frame for Proper Motion Studies
- Proper Motions and parallaxes of disk and bulge Stars
- Cool white dwarfs as Galactic chronometers
- Planetary transits
- Evolution of massive Galaxies: formation of red sequence galaxies
- Finding and weighing distant, high mass clusters of galaxies
- Obscured quasars
- Strongly lensed quasars
- High-redshift quasars and Reionization
- Faint end of the quasar luminosity function
- Probing Epoch of Reionization with Lyman α emitters
- Shapes of galaxy haloes from gravitational flexion

To focus on one of these areas, it is interesting to note the discovery space in the investigation of the epoch of reionization:

- $1\mu m$ band dropouts at $z = 1.1/0.09 - 1 = 11$
- J band dropouts at $z = 1.4/0.09 - 1 = 14$
- Galaxies with 10^8 year old stellar pops at $z = 6$
- Pair production SNe (massive stars) at $M_K = -23$
- Activity from the progenitors of supermassive black holes

Figure 1. The proposed survey will reach to within a magnitude of the NICMOS deep field (Thompson *et al.* 2007) with similar resolution, but cover many steradians. The field shown here is a few arc minutes. *Image courtesy NASA, Hubble Space Telescope, University of Arizona.*

- Dark stars, see Ilie *et al.* (2012)
- Young globular clusters with 10^6 year free fall times and M/L approaching 10^{-4}.
- Rare bright objects requiring wide field survey, then JWST, TMT, EELT or GMT spectra.

5. The next steps

The first question is whether this project is compatible with the KDUST 2.5 m (Cui 2010, Zhao *et al.* 2011). Assuming it is, we need to finalize the IR camera configuration, find IR camera partners, such as U. Tasmania, Swinburne University, UNSW, AAO/Macquarie University, Texas A&M, ANU and University of Melbourne. We then need to flowdown the science to camera requirements.

To maximize advantage over VISTA, the speed of a survey to a given magnitude (inverse of the number of years to complete 1 sr) is a factor of $\sim$9. The goal is to increase this and get a full order of magnitude (or better). Perhaps we should move from K to Kdark, when we have accurate measurements of the relevant backgrounds. We could consider adding a reimager to KDUST and undersample a bit. Alternatively a slightly faster secondary on KDUST could be entertained. For Sofradir chips the minimum exposure time is larger to overcome readout noise. For a background of 0.1 mJy/sq arcsec, the photon rate is half a photon per sec. This requires $> 2,000$ sec exposures for photon noise to double the Sofradir readout noise. (70% QE assumed.)

A construction and operations schedule tentatively would be:
- January 2015 ARC LIEF funding, followed by Preliminary Design Review
- 2016 Texas A & M purchases Teledyne arrays; ANU purchases dewar and filters
- 2016 Integrate and test focal plane at ANU or AAO
- January 2017 Integrate telescope/ camera in Fremantle
- 2018-2021 operations (within the international antarctic science region) at Kunlun Station
- 2022 return of focal plane to the USA.

This schedule is set by the time to manufacture and test the KDUST telescope in China. If it slipped a year or two, so could the instrument schedule, although we do have the precedent of GSAOI, where the camera was ready years before the adaptive optics. The Centre for All-Sky Astrophysics in Australia and the proposed Joint Australia–China research centre would provide a very appropriate context for this collaboration. At Dome A astronomy can have another world class astrophysics enterprise in Antarctica yielding major results.

Acknowledgements

I would like to thank our Chinese colleagues for hosting a workshop at the Institute of High Energy Physics of the Chinese Academy of Sciences in Beijing in November 2011, where a number of these ideas were developed. Survey astronomy is supported by the Australian Research Council through CAASTRO†. The Centre for All-sky Astrophysics is an Australian Research Council Centre of Excellence, funded by grant CE11E0090.

References

Blandford, R. 2009, *AAS*, 213, 21301
Burton, M.G. *et al.* 2005, *PASA*, 22, 199.
Cui, X. 2010, *Highlights of Astronomy*, 15, 639
Green, J. *et al.* 2012, astro-ph 1208.4012
Ilie, C., Freese, K., Valluri, M., Iliev, I. & Shapiro, P. 2012, *MNRAS*, 422, 2164
wMcGregor, P., Hart, J., Stevanovic, D., Bloxham, G., Jones, D., Van Harmelen, J., Griesbach, J., Dawson, M., Young, P. & Jarnyk, M. 2004, *SPIE*, 5492, 1033
McGregor, P. J., Conroy, P., Bloxham, G. & van Harmelen, J. 1999, *PASA*, 16, 273
Mould, J. 2011, *PASA*, 28, 266
Thompson, R. *et al.* 2007, *ApJ*, 657, 669
Zhao, G-B., Zhan, H., Wang, L., Fan, Z. & Zhang, X. 2011, *PASP*, 123, 725

Discussion

CHARLING TAO: There are problems with the persistence of Sofradir arrays.

JEREMY MOULD: There are strategies for dealing with the persistence. But, thank you, this will need to be investigated for the Plan B detectors.

HANS ZINNECKER: What about L and M band?

JEREMY MOULD: This is feasible. However, even in the Antarctic the thermal background comes roaring up and one becomes uncompetitive with space for broadband.

Astrophysics from Antarctica
Proceedings IAU Symposium No. 288, 2012
M. G. Burton, X. Cui & N. F. H. Tothill, eds.

© International Astronomical Union 2013
doi:10.1017/S174392131201695X

A European vision for a "Polar Large Telescope" project

Lyu Abe[1], Nicolas Epchtein[1], Wolfgang Ansorge[2],
Stefania Argentini[3], Ian Bryson[4], Marcel Carbillet[1], Gavin Dalton[5],
Christine David[6], Igor Esau[7], Christophe Genthon[8], Maud Langlois[9],
Thibault Le Bertre[10], Rachid Lemrani[11], Brice Le Roux[12],
Gianpietro Marchiori[13], Djamel Mékarnia[1], Joachim Montnacher[14],
Gil Moretto[15], Philippe Prugniel[9], Jean-Pierre Rivet[1], Eric Ruch[16],
Charling Tao[17,18], André Tilquin[17] and Isabelle Vauglin[9]

[1]Laboratoire J.-L. Lagrange, Université de Nice Sophia Antipolis, CNRS, Observatoire de la Côte d'Azur, Nice, France
email: Lyu.Abe@oca.eu

[2]RAMS-CON Management Consultants, Assling, Germany

[3]CNR-ISAC, Rome, Italy

[4]UKTC, Edinburgh, UK

[5]UKTC, RAL, UK

[6]Institut Polaire Français (IPEV), Brest, France

[7]Nansen Environmental and Remote Sensing Center (NERSC), Bergen, Norway

[8]LGGE, Grenoble, France

[9]CRAL Lyon, Université Lyon 1, France

[10]LERMA, Observatoire de Paris, France

[11]CNRS-IN2P3, CC, Lyon, France

[12]LAM, Marseille, France

[13]EIE Group, Mestre, Italy

[14]Fraunhofer Institute (IPA), Stuttgart, Germany

[15]IPNL, Lyon, France

[16]SAGEM-REOSC, France

[17]CNRS-IN2P3-CPPM, Marseille, France

[18]Tsinghua Center for Astrophysics, Tsinghua University, HaiDian district, BeiJing 100084, China

Abstract. The Polar Large Telescope (PLT) project is primarily aimed at undertaking large, wide band synoptic astronomical surveys in the infrared in order to provide critical data to the forthcoming generation of observational facilities such as ALMA, JWST, LSST and the E–ELT, and to complement the observations obtained with them. Sensitive thermal IR surveys beyond 2.3 μm cannot be carried out from any existing ground based observatory and the Antarctic Plateau is the only place on the ground where it can be envisaged, thanks to its unique atmospheric and environmental properties, such as the turbulence profile (image quality), the low opacity and the reduced thermal background emission of the sky. These unique conditions enable high angular resolution wide field surveys in the near thermal infrared (2.3–5 μm). This spectral range is particularly well suited to tackling key astrophysical questions such as: i) investigating the nature of the distant universe, the first generation of stars and the latest stages of stellar evolution, ii) understanding transient phenomena such as gamma ray-bursts and Type Ia supernovae, iii) increasing our knowledge of extra-solar planets. Further instruments may broaden the expected science outcomes of such a $2-4$ m class telescope especially for the characterization of galaxies at very large distance to provide new clues in the mysteries of dark matter and energy. Efforts will be made to merge this project with other comparable projects within an international consortium.

Keywords. telescopes, surveys

1. Introduction

Advances in understanding the Universe arise from improvements in our capacity to scrutinize the sky to greater depth, in previously unexplored wavebands, with better photometric and astrometric accuracy, or with improved spatial, spectral and time resolution. Thanks to the dramatic progress in computing science and technology and the rapidly increasing capacity of data archiving, future sky surveys will completely change our future perception of the Universe. The next-generation instruments, and the surveys carried out with them, will maintain this progress.

The pivotal astrophysical problems of the next decade, recently soundly discussed and synthesized in the prospect surveys carried out in Europe and the USA (Cosmic Vision, ASTRONET, US Decadal Survey, etc...) consist of the determination of the nature of dark energy and dark matter, the study of the evolution of galaxies and the structure of our Galaxy, the opening of the time domain to single out faint variable objects and transient events, the deep exploration of small bodies in the Solar System, and the identification of habitable extra-solar planets. All require wide-field frequently repeated deep imaging of the sky in a wide range of spectral bands. In the optical domain, the Large Synoptic Survey Telescope (LSST), a 6.5 m wide-field telescope equipped with a $\sim$10 square degrees camera, will repeatedly survey the sky with deep short exposures.

Thanks to the recent industrial production of larger infrared arrays (up to 16 Mpixels) and to the assessment of exceptional sites in Antarctica, one can now seriously contemplate to carry out similar surveys in the infrared and particularly in a spectral range that has been little scrutinized systematically from the ground, so far. It spans between 2.3 and 5μm.

The Antarctic Plateau offers considerable advantages for astronomy (*see* recent review by Burton 2010), in particular to undertake infrared surveys beyond 2.3 μm. The atmosphere is very cold, thus, the thermal emission of the sky background is about 20 times less than in the best temperate sites in the K band, and because most of the atmospheric turbulence lies in the first few tens of metres above the ground, one can easily overcome its deleterious effect on the image quality (spread of the PSF) by installing the telescope on top of a tower above this turbulent layer and possibly by using ground layer adaptive optics techniques. An optimised combination of both would provide quasi permanently exquisitely resolved images, down to 0.3 arcsec, *i.e.* twice as a good as with the LSST itself! These outstanding prospects were debated during the last 5 years and particularly within the earlier ARENA EC network (Epchtein *et al.*, 2010) lead us to consider that time is ripe to propose within the European Union (EU) a conceptual design study for a so-called Polar Large Telescope aimed at achieving, for the first time, a deep synoptic sky survey in the thermal infrared.

2. Project context and concept

The framework of the EU is indeed ideal to carry out this design study because Europe is playing a leading role in astronomy in large scale astronomical sky surveys particularly in the infrared (e.g. IRAS, 2MASS, DENIS, VISTA, CFHT, *see* Price 2009) and because

Europe is also at the forefront of polar research. Two member states, France and Italy, have implemented and are successfully operating all year round a multi-disciplinary scientific station, Concordia, sited in the heart of Antarctica at Dome C. Other sites are currently operating or are under development on the Antarctic Plateau, namely, the Amundsen-Scott station at the South Pole, Dome A (Kunlun) and Dome F (Dome Fuji) exploited by scientists from the USA, China and Japan, respectively. All of them have or plan to develop an astronomical observatory benefiting from the Antarctic environment. These sites offer exceptional astronomical conditions demonstrated by the thorough monitoring studies of the critical atmospheric parameters carried out during the last 10 years mostly by the US, Australian, Italian and French teams (e.g. Philips *et al.*, 1999, Walden *et al.*, 2005, Travouillon *et al.*, 2009, Aristidi *et al.*, 2009, Gredel 2010). Although they share common obvious properties (cold, dry, low sky emissivity), they are not equivalent and before installing any sort of costly instrument, a thorough comparison of their properties, based on modelling and long period monitoring of the critical parameters in each site, is mandatory. It is also timely and urgent that the Europeans carefully evaluate the feasibility and their ability to study and build such a facility in the coming decade and possibly give to it a label of *Large European Research Infrastructure*.

The consortium that was set up in 2010 gathers the various and often distant skills and expertise required to successfully carry out a Design Study from astrophysics to atmospheric physics and polar engineering and logistics. The important work already done by the UNSW group during the PILOT phase A study would serve us as a basis for the PLT. But instead of proposing a multi-purpose telescope ranging from visible to FIR, the PLT would primarily focus on near–IR windows unaccessible from other mid-latitude sites. The PLT must be elaborated and advertised as a complementary tool to future large ground-based telescopes and space-borne observatories.

The key features and objectives of the PLT are: 1) to cover very large sky areas, repeatedly and thus explore systematically the time domain. PLT will do that for the first time in the infrared. 2) To improve the ultimate detection sensitivity at all wavelengths. This is the case of PLT at $2\mu m$. 3) To explore in depth new spectral domains. PLT will explore systematically the $2.3–4\mu m$ range. 4) To improve the angular resolution using exquisitely good site and/or using adaptive optics techniques. PLT will provide the best images ($0.3''$) ever at $2\ \mu m$ thanks to the unique turbulence properties of Antarctic sites and the use of an optimized adaptive optics system. 5) To increase the size of statistical samples of all sorts of astrophysical objects and phenomena. PLT will achieve this goal as a consequence of the previous items.

3. Scientific objectives

The overall objective of the PLT Project is to prepare the conceptual foundations for the first, world wide leading, near infrared research infrastructure in the Antarctic. With the PLT, European astronomical would be able to explore for the first time the time domain in the 2–5 µm spectral range at unprecedented sensitivity, angular and time resolutions, and will cover large selected areas of the sky totalising several thousands of square degrees. With its unique surveying capability, the PLT would be an enthralling tool for the entire European astronomical community to remain at the forefront of the new generation sky surveys in the E–ELT era.

The science cases have been already discussed at length (see e.g. in Zinnecker *et al.*, 2008) and precisely established to support a phase–A study of the former Australian project PILOT (Lawrence *et al.*, 2009abc), but a list of highest priorities matching the characteristics of the PLT is currently being scrutinized for the soon to come *PLT Science*

Book. The PLT Design Study will also investigate the most efficient and most promising strategy in terms of scientific impacts and observing time efficiency for the PLT. This will identify the optimal integration time of each visit of the selected targets field, the most suitable number of visits per elapse of time, the best bandpass of the filters, the number of colours in which each target area should be observed, etc.

At the present time four major astrophysical fields are identified that will take considerable benefit from the PLT, i) the investigation of the distant universe, ii) the stellar populations, formation and evolution in our Galaxy and in the Local Group, iii) the characterization of exo-planets, and iv) the identification and characterization of small bodies in the Solar System. These fields are in common with other instruments that will operate at the same period, and in particular the LSST, but the PLT will explore a different spectral range at an even smaller angular resolution. The PLT and the LSST data could be managed using similar procedures and both instrument teams could be able to alert each other during the overlapping period of operation for an optimal follow–up of transient events.

(*a*) **Distant Universe.** PLT will be the most powerful and sensitive wide field imaging collector in the near thermal infrared. It is thus particularly well suited to the exploration of the distant universe, notably wherever the dust extinction strongly hampers the optical observations, such as the disks and bulges of galaxies where the largest space density of stars – and thus of SN candidates – are found. Moreover, it will explore the time dimension during a period of 10 years at different time resolutions from minutes to years. A coordination with the LSST, space surveyors at all wavelengths (such as GAIA, EUCLID, etc.), and the E–ELT for even more resolved images and spectroscopy, all planned to operate during the decade 2020–30 will be extremely efficient. The most promising domains, already mentioned in the PILOT proposal encompass:

- A near-infrared search for pair-instability supernovae (via a dedicated periodically repeated wide field survey) and gamma-ray burst afterglows (via alerts from high energy satellite, LSST detections), events which represent the final evolutionary stages of the first stars to form in the Universe.
- A deep survey in the near-infrared to study galaxy structure, formation and evolution via the detection of a large sample of high redshift galaxies.
- A near-infrared search for Type Ia supernovae to obtain light curves that are largely unaffected by dust extinction and reddening, allowing tighter constraints to be placed on the expansion of the Universe; and a study of a sample of moderate-redshift galaxy clusters aimed at understanding galaxy cluster growth, structure and evolution.

(*b*) **Stellar formation and evolution, galactic ecology.**
- Thermal infrared imaging is ideally suited to probe the stellar content of our Galaxy and galaxies of the local groups, especially young stellar objects, stars in the late stage of evolution and very low mass stars. Repeated observations will provide hundred of thousands of light curves that will improve our knowledge of the mass loss process, enrichment of the interstellar medium in heavy element and the internal physical processes occurring in the AGB phase.
- An optical/near infrared survey of disk galaxies in the local group to study the processes of galaxy formation and evolution.
- An infrared survey of nearby satellite galaxies to trace their outer morphology, structure, age and metallicity.
- A deep repeated infrared survey of the Magellanic Clouds in order to understand the star formation, its history and the evolution processes in galaxies of different metallicities. The Magellanic Clouds will indeed receive special attention because

Table 1. *Main Site Characteristics*

	Location of the 3 main sites
Concordia	Dome C (alt.: 3,233m; lat.: 75S)
Kunlun	Dome A (alt.: 4,093m, lat : 80S)
Dome Fuji	Dome F (alt.: 3,810m, lat.: 77S)
	Operators
Concordia	IPEV (France); PNRA/ENEA (Italy)
Kunlun	Polar Research Institute of China
Dome Fuji	National Institute of Polar Research (Japan)
Median fraction of photometric time (Concordia)	~70%
Median value of free seeing (above turbulent layer, typically ~30 m) (Concordia)	0.3″
Median precipitable water vapour content (Concordia)	~300 μm (tbc)
External extreme temperature range (°C)	Summer: -20 to -40; Winter -40 to -90
Operation temperature range (°C)	$\geqslant -85°$
Fraction of sky observable from Dome C	40% (at 60° from zenith) (16 000 deg^2)
Declination range permanently observable from Dome C	$-90° < \delta < -30°$ (10 000 deg^2)

Table 2. *Main Telescope Characteristics*

FOV (unvignetted field)	$\geqslant 1°$
Effective clear aperture	2.5 m
Configuration	Ritchey-Chrétien, 2 Nasmyth foci, Alt-az field rotator
Etendue	5 m^{2}deg^2
Primary mirror material	Zerodur or SiC (tbd)
Surface quality (rms)	$\lambda/8$ at 2 μm
Coating of M1 and M2	Al or Au (lowest emissivity at 3 μm)
Spectral range of operation	2–5 μm (possibly extended to 10–30 μm)
Sky coverage	Goal: 5 000 deg^2 (at least 20 times a year)
Final f ratio	f/5 (primary f/1.2)
Diameter of 80% encircled energy spot at 2.4μm (optics)	0.2″ (goal)
Absolute pointing accuracy	5″ (goal: 1″)
Slew time (any direction in sky)	< 3 minutes
Tracking accuracy	< 0.2″ per period of 10 seconds
Total emissivity of the telescope at 2.4 μm	<5%
Height of tower (above ground level)	<30 m
Expected life time	$\geqslant 10$ yrs

they offer the opportunity to study in a relatively limited area a considerable number of stars of various metallicity. Moreover, they are continuously observable in optimal condition from Antarctica.

(*c*) **Exoplanet science, very low mass stars and brown dwarfs.**
  • The search for free-floating planetary mass objects.
  • the follow-up of gravitational microlensing candidate detections based on alerts from dedicated survey telescopes.
  • the collection of high precision photometric infrared light curves for secondary transits of previously discovered exoplanets.

(*d*) **Small bodies of the Solar System.** The PLT, with its unique sensitivity and angular resolution in the thermal infrared, will be particularly efficient in the detection of nearby small bodies and will provide critical additional data to the LSST.

4. The largest optical/IR telescope in polar environment

The following elements reproduce part of the proposal for a PLT Design Study (PLT DS) that was submitted to the European Commission in late 2010 (Epchtein & the PLT consortium). Although unsuccessful (ranked 6th in the final stage), it provides the basis and key elements that should be studied for the telescope, and upon completion,

Table 3. *Near Infrared Camera and system performances*

Array Type	HgCdTe HAWAII 4RG 4k×4k (Teledyne)
Pixel size	10 μm or 15 μm (tbd)
Scale	0.16/0.22″/pxl (tbd)
Field of View	⩾ 40′ × 40′
Final PSF FWHM (with tip-tilt and/or GLAO)	0.3″ across full FOV
Array operating temperature	40 K (cryocooler)
Cold stop	Yes T<100 K
Filter slots number	3 (minimum) to 8 (goal)
Filter set	K_d, L_s, L' (+ possibly: K_s, M', narrow bands, Grisms)
Read out time	5 sec
Integration time per frame (typical)	20 to 100 sec
Focal plane configuration	16 buttable Hawaii RG4
Single visit depth (5 /100 s) mAB	goal: Kdark ∼22 / L short ∼18 [AB mag]
Coadded depth mAB after one year	goal: Kdark ∼25.5/ L short ∼21 [AB mag]

Table 4. *Observing strategy and Data Management*

Coverage per period of 24 h	500 deg²
Number of visits/yr of selected targets	⩾20
Cumulative coverage/yr	100 000 deg² (i.e. 20° × 5,000°) (goal)
Raw pixel data per 24 h periods	0.5 TB to 1 TB
Yearly (200 days) archive amount	100 TB to 200 TB
Computational requirements	Necessary to process up to 1Tb per day locally (tbd)
	Necessary to archive, give access and process up to 1 Pb (remote centre) (tbd)
Bandwidth of communications Telescope to base	∼0.5 Gib/sec
Base to remote archive	Through high capacity physical media (hard disk, mag. tape)

would constitute the ground for future developments of large IR telescopes in Antarctica. The final PLT DS document can be retrieved from https://sites.google.com/site/pltantarctica/. The main characteristics of the PLT are summarized in Tables 1 through 4, and the sections below detail some of the listed features.

The PLT with its 2.5 m aperture, would be the largest optical/IR telescope ever studied operating in the extreme polar environment. Studies of the best technical solution to face the peculiar polar constraints (icing, wind, low ambient temperature and large short range temperature variations) will focus on:

- The telescope mechanical structure.
- The optical material (primary, secondary and tertiary).
- The enclosure and its effect on turbulence ("dome seeing" study).
- The high stiffness, low vibration and oscillation tower.
- The electronics devices, cables, optical fiber connection.
- The thermalization of the telescope in order to smooth the possible large fluctuation of external temperatures (especially if it is set up at several tens of metres).

The PLT Design Study will produce a cost efficient estimation of the requested level of automatization, robotization, intelligent remote control of the telescope, i.e., able to manage permanently the observing strategy. The risk and cost of using light mirrors made for instance of SiC for the primary mirror will be also technically assessed and estimated. A reduced weight of the primary mirror would have a considerable impact on the rest of the structure and consequently reduce drastically the logistics means (and induced costs and time delay) to deploy.

All these studies will have an important impact on industrial progress in the use of material and structures operating under extreme climatic conditions and will improve the know-how for other polar instruments.

4.1. *Wide field of view infrared camera*

The PLT IR camera is the largest ever proposed for the 2–5 μm range in terms of pixel number (250 Mpixels). Its FPA basically consists of an assembly of 16 buttable 4k×4k arrays covering a field of view of ∼40 square arcminutes, with a pixel size of 0.15 arcsec. Teledyne Technology Inc. in the USA produces suitable state-of-the art infared arrays for astronomy (Hawaii–4RG). Although arrays of this size are not yet commercially available, they are likely to be on the market by 2013, thus timely in the PLT schedule. This camera is particularly efficient and challenging, and the Design Study will have to identify possible show stoppers in its definition and its capacity to support the Antarctic environment conditions and consequently its high level of reliability and automatization.

4.2. *Adaptive optics techniques in polar atmospheric conditions*

One considerable advantage of the Antarctic sites is the unique behaviour of the atmospheric turbulent ground layer of the atmosphere. It is much thinner than above any other ground sites with a median measured value of its thickness of ∼30 m (Aristidi *et al.*, 2009), to be compared to the hundreds of metres usually found at the best astronomical sites (Mauna Kea, Paranal). If one can permanently benefit from seeing conditions close to 0.3 arcsec – basically twice as good as anywhere else on the Earth – this would provide an enormous advantage for larger instruments to be later implemented at Dome C or other comparable Antarctic sites. It is thus among the major impacts of the PLTDS to determine, essentially through modelling, whether one can fully benefit from the so called Ground Layer Adaptive Optics (GLAO) techniques across a large field of view of ∼40 square minutes to alleviate the effect of turbulence, and thus leading to reduce the height of the tower.

A major outcome of the PLTDS is to produce a full assessment model of GLAO able to reach this goal: **an image quality pattern of the order of 0.3 arcsec quasi permanently. This would be in itself a major advance in astronomical observations from the ground**.

These (mostly) theoretical studies, based on the results of the site assessment studies currently on going on site (turbulence profiles and seeing measurements) will also determine the optimal height of the tower supporting the telescope and the parameters of the GLAO system to be implemented (deformable mirror, number of actuators, frequencies). This study should ultimately lead to R&D studies and the implementation of a prototype of the GLAO subsystem that could be, for instance, assessed on an already existing telescope at Concordia such as IRAIT.

4.3. *A massive data pipeline and processing centre in an extremely remote environment*

The IR camera will yield one to several Tbytes of data per period of 24 hours and the estimated final data product amount after 10 years of operations is in the Pbyte range. Although there is no major technological challenge to achieve this task in the near future (the LSST will produce 10 times more data), the challenge is that the processing must be done essentially on-site because, due of the very small existing and foreseeable bandpass of the telecommunications (Inmarsat, Iridium), one cannot transfer electronically such a huge amount of data from Antarctica to the rest of the world. There will be a massive data centre in Europe (for instance at CC–IN2P3 in Lyon), but most of the raw data processing will have to be carried out on site.

The present design study will investigate the most efficient way to transfer the raw data from the telescope/camera to the dedicated control room (very likely to be set up in one of the existing Concordia buildings), to run, manage and maintain the data centre with only one person available during the winterover periods, to single out rapidly transient events

(SNe, GRBs, transits...) and to alert the community of these events with the shortest elapse of time. On the other hand, the PLT controler should be able to receive alerts from other survey telescopes and make appropriate observation decision, accordingly. Finally, the most appropriate and safe physical media on which the ~100 Tbytes data could be transferred bi-annually from Antarctica to the data centre in Europe will have to be investigated.

5. Provisional conclusion

Although time is not particularly favourable to make prospects of costly instruments in Antarctica, we, as well as several other groups around the world, are convinced that there is a bright future to the exploitation of a polar astronomical station equipped with performing instruments, dedicated to just a few well identified science cases and objectives. We believe that surveying the sky at an extreme depth, in new spectral bands, at high angular resolution, is one of the top priorities for a middle-size polar instrument. The PLT could provide invaluable data for several high priority science cases and could be in many cases be a cheaper and more flexible alternative to space missions. To optimize and secure its development, the instrument should be studied, built and operated by an international team, after a careful selection of the most appropriate Antarctic site. Therefore, in the meantime, we strongly recommend to proceed with coordinated campaigns of site qualification at the currently running stations, and possibly elsewhere, to foster the links between the groups involved in these studies all over the world, to specify the most exciting science cases and to carry out appropriate design studies and *in-situ* demonstrators (e.g. to test GLAO devices). Installing a facility in a harsh environment is not a question of competition, but clearly of world-wide cooperation.

We hope that this Symposium will be an opportunity to give a new impetus to this prospect in the future.

References

Aristidi, E., Fossat, E. Agabi, A. *et al.*, 2009, *A&A* 499, 955

Burton, M. G., 2010, *Astron. Astrophys. Rev.*, 18, 417

Dalton, G., 2010, Proc. SPIE Vol. 7735, 77351J

Epchtein, N. (coordinator) and the ARENA consortium, 2010, *A Vision for European Astronomy and Astrophysics at the Antarctic station Concordia, DomeC*, the ARENA roadmap, Novaterra (http://arena.oca.eu)

Gredel, R., 2010, EDP EAS Publication Series Vol. 40, 11

Lawrence, J. *et al.*, 2009, *PASA*, 46, 379(a), 397(b), 415(c)

Philips, A., Burton, M. G., Ashley, M. C. B., *et al.*, 1999, *ApJ* 527, 1009

Price, S. D., 2009, *Space Sc. Rev* 142, 233

Travouillon, T., Jolissaint, L., Ashley, M. C. B. *et al.*, 2009, *PASP*, 121, 668

Walden, V. P., *et al.*, 2005, *PASP*, 117, 300

Zinnecker, H., Epchtein, N., & Rauer R.(eds.), 2008, Procs. second ARENA Conference on *"The Astronomical Science Cases at Dome C"*, EDP EAS Ser. Vol. 33

Astrophysics from Antarctica
Proceedings IAU Symposium No. 288, 2012
M. G. Burton, X. Cui & N. F. H. Tothill, eds.

© International Astronomical Union 2013
doi:10.1017/S1743921312016961

Dome Fuji in Antarctica as a Site for Infrared and Terahertz Astronomy

Masumichi Seta[1], Naomasa Nakai[1], Shun Ishii[1], Makoto Nagai[1], Yusuke Miyamoto[1], Takashi Ichikawa[2], Naruhisa Takato[3] and Hideaki Motoyama[4]

[1]Institute of Physics, University of Tsukuba,
305-8571, 1-1-1 Ten-nodai, Tsukuba, Ibaraki, Japan
email: seta@physics.px.tsukuba.ac.jp

[2]Astronomical Institute, Tohoku University,
980-8578, Aramaki, Aoba, Sendai, Miyagi, Japan

[3]Subaru Telescope, National Astronomical Observatory of Japan,
96720, 650 North A'ohoku Place, Hilo, HI, USA

[4]National Institute of Polar Research,
190-8518, 10-3 Midoricho, Tachikawa, Tokyo, Japan

Abstract. Dome Fuji on the Antarctic high plateau may be a good site for terahertz astronomy because of its high altitude of 3,810 m and low average temperature of $-54°$C. We have demonstrated that the opacity at 220 GHz from Dome Fuji in summer is very good and stable; $\tau = 0.045 \pm 0.007$. We have developed a transportable 30 cm telescope to map the Milky Way in the CO (J=4–3) and the [CI] ($^3P_1$–$^3P_0$) lines at Dome Fuji from 2014. It has a 9′ beam. Physical conditions such as density and temperature of molecular clouds could be derived from a direct comparison of CO (J=4–3) and [CI] ($^3P_1$–$^3P_0$) with CO (J=1–0) taken by the Columbia–CfA survey. We are also developing a 1.2 m sub-millimeter telescope. It will be equipped with a dual superconducting device (SIS) receiver for 500/800 GHz. The 1.2 m telescope produces a 2.2′ beam at 492 GHz and could map a molecular cloud entirely. It could also observe nearby galaxies in the CO (J=4–3), CO (J=7–6), [CI] ($^3P_1$–$^3P_0$), [CI] ($^3P_2$–$^3P_1$) and in continuum emission between 460–810 GHz.

Keywords. Site testing, telescopes, sub-millimeter

1. Introduction

Recently several detectors for sub-millimeter (Sub-mm) and terahertz (THz) astronomy have become available such as the Superconductor-Insulator-Superconductor (SIS), Hot Electron Bolometer (HEB), Microwave Kinetic Inductance Detector (MKID), in addition to the Schottky barrier diode mixer. However THz astronomy is challenging because of strong absorption by the Earth's atmosphere. Sub-mm and THz astronomy are only conducted at dry and high attitude site such as Mauna Kea in Hawaii and the Atacama Desert in Chile. However, the atmospheric transparency is neither low nor stable enough for sub-mm and THz astronomy there. Dome Fuji on the Antarctic high plateau is one of several promising sites for THz and infrared astronomy because of its low average temperature of $-54°$C and high altitude of 3,810 m. In this paper we introduce radio astronomy plan at Dome Fuji. The plan for the infrared telescope at Dome Fuji is described by Okita (2012) in these Proceedings.

2. Dome Fuji Station

The Japanese Dome Fuji station is located at a latitude of 77°19′S and longitude of 39°42′E at an altitude of 3,810 m on the Antarctic plateau (Watanabe *et al.*, 1999). It is about 1,000 km away from the Japanese Syowa station. The annual average temperature is −54.4°C and the lowest temperature recorded is −79.7°C (Yamanouchi *et al.*, 2003). The maximum temperature in summer is −21.1°C. It is known that the average fraction of the sky obscured by clouds has been 30% from 1995–1997 (Yamanouchi *et al.*, 2003). The mean wind speed is 5.8 m s^{-1} and it rarely exceeds 10 m s^{-1}. These conditions are favourable for astronomy because a high percentage of practical observational time is available in Dome Fuji.

Now the winter over station is under snow and a new Dome Fuji station is planned. Astronomy is one of key sciences for the new Dome Fuji station. The new station may be placed about 60 km away from the present station, where older ice and drier atmospheric conditions are expected. A high deck type of building is planned for the new station to avoid snow accumulation around the buildings. The capacity for winter is designed to be ten people for the first stage, with the possibility of future expansion. Electric power of 100 kVA may be available. Inmarsat Satellite communication is possible from Dome Fuji. The construction hopefully will start in 2016 and be completed by 2018. Several summer traverses are planed during the construction periods, so astronomical data will be obtained soon.

3. Opacity at Dome Fuji

We measured atmospheric opacity at Dome Fuji from December 2006 – January 2007 (Ishii *et al.*, 2010) and January 2010 using a tipping radiometer. We show the radiometer at Dome Fuji in Figure 1.

The instrument measured the brightness temperature of the sky at 220 GHz with a beam of 63′. The scanning mirror was programmed to scan the sky every two minutes with the zenith angle between $Z = 0$ (sec $Z = 1.0$) and $Z = 70.5°$ (sec $Z = 3.0$), and the zenith opacity was derived by reducing the data for one scanning. The 220 GHz opacity is an important parameter needed to evaluate the site because it has been measured

Figure 1. Radiometer at Dome Fuji in 2010.

at various sites in the world. We show the results in Figure 2. The opacity was low and stable; $\tau = 0.045 \pm 0.007$ for about one month in summer from December 2006 to January 2007. The low and stable value of 0.054 was again obtained for about one week in the summer of January 2010. This low value of the opacity is comparable with that of the best season in the Atacama Desert in Chile.

Computer simulations show the winter opacity at Dome Fuji to be much better than that of the Atacama Desert (Ishii *et al.*, 2010). The sub-mm transparency is much improved in winter. It should noticed that new THz windows above 1 THz open at Dome Fuji. Low precipitable water vapor (PWV) of 0.025 mm was recorded in Dome A for winter (Yang *et al.*, 2010), and may have similar conditions to Dome Fuji. So the opacity at Dome Fuji must be very good in the winter season. We have plans to measure the opacity in winter in 2014.

4. Route for Dome Fuji

Dome Fuji station is about 1,000 km away from Syowa station. Icebreaker Shirase and snow sledges transport the most of astronomical instruments. Shirase operates only once a year. It leaves Japan in the middle of October and arrives at the coast near Syowa station in the middle of December, after picking up observers and fresh food at Fremantle in Australia. Shirase accommodates 56 boxes of 12-ft $(3.6\,\text{m} \times 2.4\,\text{m} \times 2.4\,\text{m})$ containers. It is desirable that all instruments are designed to be packed in the 12-ft container. A deck crane is used to transport the container from Shirase to the iced sea near Syowa station. The maximum weight of cargo on the 12-ft container is limited to be less than 15 t. However, astronomical instruments for Dome Fuji must be designed to be much lighter than 15 t. The traverse to Dome Fuji starts from the S16 point near to Syowa. Syowa station is on a small island and all cargo for Dome Fuji is transported to S16 either by a CH-101 helicopter or by snow sledges. The maximum weight is limited to be less than 2 t and 5 t for the helicopter and the sledge, respectively.

It takes about 3 weeks to go from S16 to Dome Fuji. We measured the shock on the cargo during transportation from Dome Fuji to S16. The maximum shock recorded every 20 minutes is shown in Figure 3. Strong shocks were recorded at the time of lifting the instrument to/from sledge at Dome Fuji and S16. However strong shocks were also recorded from January 18 to February 2. This time corresponds to the route between middle point and the Mizuho mid-way station, where rough snow surfaces occur. New snow sledges are under development so as to reduce the strong vibration during the

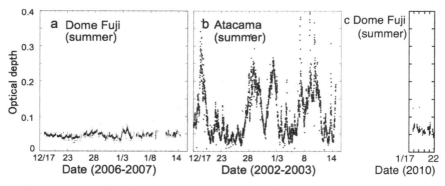

Figure 2. Zenith optical depth at 220 GHz measured at Dome Fuji in December 2006 – January 2007: (a) January 2010; (c) comparison with the Atacama Desert in the same season; (b) data for the Atacama consisting of the best records obtained from 1995 to 2004.

transportation on the rough snow surface. It is also possible to use a small airplane to fly to Dome Fuji directly but funds are limited for its use.

5. The 30 cm Radio Telescope

We have developed a prototype telescope for operation at Dome Fuji. It is a transportable sub–mm wave telescope (Ishii *et al.*, 2012). It has 30 cm diameter offset-Cassegrain antenna that produces a 9′ beam. It is equipped with a heterodyne SIS receiver operating at 460–492 GHz with a quasi-optical single sideband filter. It uses a 1-GHz bandwidth spectrometer that covers a velocity width of 600 km s^{-1} with a velocity resolution of 0.04 km s^{-1}.

Large scale galactic CO (J=1–0) surveys revealed the distribution of molecular cloud in the 1980s (e.g., Dame *et al.*, 1987, Scoville *et al.*, 1987). However the physical conditions of the molecular clouds remains unclear. Multiple line observations are required to derive conditions such as density and temperature. The Univ. of Tokyo–NRO CO (J=2–1) survey shows that distributions of CO (J=2–1) and CO (J=1–0) are similar in the Milky Way although it also shows peculiar ratios for selected regions (e.g., Sakamoto *et al.*, 1987). The CO (J=4–3) and CO (J=1–0) ratio may be useful to constrain the physical conditions of the molecular clouds. It is also important to have a clear picture of molecular cloud formation. The fine-structure line of neutral atomic carbon [CI] is an important tracer for phase transitions of interstellar matter from atomic to molecular. The 30 cm telescope is scientifically targeting survey observations in the CO (J=4–3) and [CI] ($^3P_1$–$^3P_0$) lines. The beam size of 9′ is designed to be comparable to that of Colombia–CfA CO (J=1–0) survey telescope so that a direct comparison of the data will be possible.

Transportability is an important requirement so the telescope can be assembled by only a few people. The maximum weight of each component is designed to be less than 60 kg and the total weight is about 700 kg. The total power consumption is also restricted to be less than 2.5 kW, so it can be operated by small electric generators.

We have already built the telescope and tested it in Switzerland, and in northern Chile. We succeeded in mapping the Orion molecular cloud in CO (J=4–3) following first light in both CO (J=4–3) and [CI] ($^3P_1$–$^3P_0$) toward Orion KL. The SSB system noise temperature including atmospheric loss was 3,000 K in Chile. Now we are upgrading the

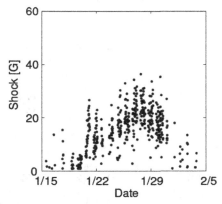

Figure 3. Maximum shock every 20 minutes during transportation from Dome Fuji to the S16 point near the Syowa station.

SIS receiver to install a sideband separating SIS mixer and the system noise temperature is expected to be less than 500 K in SSB operation.

6. Future Radio Telescopes at Dome Fuji

We are also developing a 1.2 m telescope. It is a modification of the 1.2 m telescope used on Mt. Fuji (Sekimoto *et al.*, 2000). The parabolic antenna produces a 2.2′ beam at 492 GHz. It is equipped with a 500/800 GHz SIS receiver. It is possible to detect both [CI] ($^3P_1$–$^3P_0$) at 492 GHz and [CI] ($^3P_2$–$^3P_1$) at 809 GHz simultaneously. It can also be tuned for CO (J=4–3) and CO (J=7–6). A continuum detector is also being installed.

We will be able to reveal the structure of entire molecular clouds in both the CO and [CI] lines with the moderate beam size. A sub–mm continuum camera will also be installed on the 1.2 m telescope. We will also observe several nearby galaxies, including the Large Magellanic Cloud and the Small Magellanic Cloud. In addition, several large telescopes such as a 10 m class THz telescope are also planned.

References

Dame,T. M., Ungerechts, H., Cohen, R., de Geus, E., Grenier, I., May, J., Murphy, D., Nyman, L., & Thaddeus, P. 1995, *ApJ*, 322, 706

Ishii, S., Seta, M., Nakai, N. Nagai, M., Miyagawa, N., Yamauchi,A., Motoyama, H., & Taguchi, M. 2010, *Polar Science*, 3, 213

Ishii, S., Seta, M., Nakai, N., Miyamoto, Y., Nagai, M., Maezawa, H., Nagasaki, T., Miyagawa, N. Motoyama, H., Sekimoto, Y., & Bronfman, L. 2012, *23rd Int. Symp. Space. THz Tech.*

Okita, H. 2012, *IAU Symposium 288*

Sakamoto, S., Hasegawa T., Handa, T., Hayashi, M., & Oka, T. 1997, *ApJ*, 486, 276

Scoville, N. Z, Yun M. S., Clements, D. P., Sanders D. B., & Waller, W. H. 2003, *ApJ Suppl.*, 63, 821

Sekimoto, Y., Yamamoto, S., Oka, T., Ikeda, M., Maezawa, H., Ito, T., Saito, G., Iwata, M., Kamegai, K., Sakai, T., Tatematsu, K., Arikawa, Y., Aso, Y., Noguchi, T., Maezawa, K., Shi, S., Saito, S., Ozeki, H., Fujiwara, H., Inatani, J., Ohishi, M., Noda, K., & Togashi, Y. 2000, *Review of Scientific Instruments*, 71, 7

Watanabe, O., Kamiyama, K., Motoyama, H., Fujii, F., Shoji, H., & Satou, K 1999, *East Antarctica, Annals of Glaciology*, 29, 176

Yamanouchi, T. Hirasawa, N., Hayashi, M., Takahashi S., & Kaneto, S. 2004, *Mem. Natl. Inst. Polar Res. Spec. Issue*, 94

Yang, H., Kulesa, C. A., Walker, C. K., Tothill, N. F. H., Yang, J., Ashley, M. C. B., Cui, X., Feng, L., Lawrence, J. S., Luong-van, D. M, McCaughrean, M. J., Storey, J. W. V. Wang, L., Zhou, X., & Zhu, Z. 2000, *PASP*, 122, 490

Astrophysics from Antarctica
Proceedings IAU Symposium No. 288, 2012
M. G. Burton, X. Cui & N. F. H. Tothill, eds.

© International Astronomical Union 2013
doi:10.1017/S1743921312016973

Opportunities for Terahertz Facilities on the High Plateau

Craig A. Kulesa[1], Michael C.B. Ashley[2], Yael Augarten[2],
Colin S. Bonner[2], Michael G. Burton[2], Luke Bycroft[2],
Jon Lawrence[3], David H. Lesser[1], John Loomis[4],
Daniel M. Luong-Van[2], Christopher L. Martin[5], Campbell McLaren[2],
Shawntel Stapleton[4], John W.V. Storey[2], Brandon J. Swift[1],
Nicholas F.H. Tothill[6], Christopher K. Walker[1] and Abram G. Young[1]

[1]Steward Observatory, University of Arizona, 933 N. Cherry Ave., Tucson, AZ 85721 USA
email: ckulesa@email.arizona.edu

[2]School of Physics, University of New South Wales, Sydney, NSW 2052, Australia

[3]Australian Astronomical Observatory, PO Box 915, North Ryde, NSW 1670, Australia

[4]Raytheon Polar Services Company, US Antarctic Program

[5]Physics and Astronomy Department, Oberlin College, 110 N. Professor St., Oberlin, OH 44074, USA

[6]University of Western Sydney, Locked Bag 1797, Penrith, NSW 2751, Australia

Abstract. While the summit of the Antarctic Plateau has long been expected to harbor the best ground-based sites for terahertz (THz) frequency astronomical investigations, it is only recently that direct observations of exceptional THz atmospheric transmission and stability have been obtained. These observations, in combination with recent technological advancements in astronomical instrumentation and autonomous field platforms, make the recognition and realization of terahertz observatories on the high plateau feasible and timely. Here, we will explore the context of terahertz astronomy in the era of *Herschel*, and the crucial role that observatories on the Antarctic Plateau can play. We explore the important scientific questions to which observations from this unique environment may be most productively applied. We examine the importance and complementarity of Antarctic THz astronomy in the light of contemporary facilities such as ALMA, CCAT, SOFIA and (U)LDB ballooning. Finally, building from the roots of THz facilities in Antarctica to present efforts, we broadly highlight future facilities that will exploit the unique advantages of the Polar Plateau and provide a meaningful, lasting astrophysical legacy.

Keywords. infrared: general, submillimeter, galaxies: ISM, ISM: clouds, stars: formation, molecular processes, instrumentation: spectrographs, instrumentation: interferometers, site testing

1. Introduction

Terahertz (THz) radiation, from 0.5 – 5 THz (600 – 60 micrometers wavelength), is one of the last regions of the electromagnetic spectrum which remains largely unexplored. This is partly due to the opacity of the Earth's atmosphere at these frequencies and partly due to the difficulty constructing terahertz detectors, spectrometers, and telescopes. Nestled between traditional radio astronomy at cm- and mm-waves using heterodyne receivers, and the bolometer and photoconductor detectors of infrared astronomy, the terahertz regime is a technological hybrid; a confluence that often necessarily inherits the most difficult aspects of both worlds. Thus, terahertz telescopes are often built in the style of radio telescopes but with optical figure requirements akin to infrared telescopes.

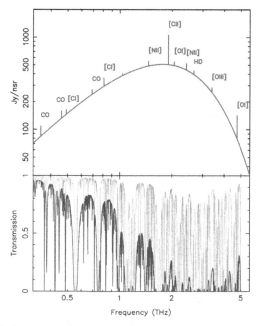

Figure 1. Schematic of the spectral characteristics observed toward an interstellar cloud, featuring a ∼30K continuum, an assortment of selected atomic and molecular spectral lines, underlined by exceptional atmospheric transmission observed from the ground (75 micrometers of precipitable water vapor at Dome A and Ridge A, Antarctica) (Yang *et al.* 2010) and airborne platforms such as SOFIA (8 micrometers of precipitable water vapor).

The sensitivity of terahertz telescopes hinges delicately on the water vapor content of the atmosphere above the telescope, as it does for the infrared. Heterodyne receivers are used for high resolution spectroscopy as at radio wavelengths, but the special challenges of fabricating submillimeter-scale quasioptical and waveguide structures, in addition to THz local oscillator sources, are both extreme and unique. Even the units in which terahertz astronomers speak reflect this unusual conglomeration of overlapping instrumentation – while heterodyne spectroscopists speak in units of frequency (GHz or THz), bolometer imaging cameras typically adopt wavelength units (micrometers), and the devotees of Fourier Transform Spectrometers and laboratory astrochemistry inevitably speak in wavenumbers (cm^{-1}). New researchers in terahertz astronomy, particularly from radio or optical regimes, often find the lingual partitioning of the field practically schizophrenic.

Regardless of language, the most diagnostic and luminous spectral signatures of many common elemental species lie at terahertz frequencies. These spectral lines are signposts of star and planet formation, the evolution of matter in galaxies, the rich astrochemistry of interstellar clouds, even the prebiotic building blocks of life. Furthermore, the reprocessing of visible and ultraviolet light by dust grains in interstellar clouds makes the continuum emission of star forming regions, circumstellar (pre-planetary) disks, and entire galaxies peak at terahertz frequencies. This continuum emission is often comparable to, if not significantly larger than, starlight directly generated at visible wavelengths. The spectroscopic signatures of pivotal atoms and molecules, coupled with bright continuum emission, leads to a rich spectrum of emission and absorption lines which is uniquely diagnostic of a wide variety of astrophysical phenomena (Figure 1).

Table 1 reflects a sample of important atomic and molecular species at terahertz frequencies, along with their observability from excellent mid-latitude sites, the summit

Table 1. Observability of Important THz Lines

Species	Freq (THz)	Midlatitude Ground[a]	Antarctic Ground[a]	Airborne[a]
C	0.492, 0.809	Y	Y	Y
CH	0.532, 0.536	N	Y	Y
H_2O	0.557, 1.113	N	N	N
HCl	0.635	Y	Y	Y
D_2H^+	0.691	Y	Y	Y
CO	1.037-1.497	N	Y	Y
CH^+	0.835	M	M	Y
OH^+	0.909	M	Y	Y
NH_2	0.953	M	Y	Y
NH	0.974	N	M	Y
NH^+	1.013	N	Y	Y
H_2O^+	1.115	N	N	N
HF	1.232	N	N	Y
H_2D^+	1.370	N	Y	Y
N^+	1.461	N	Y	Y
OH	1.835, 1.838	N	M	Y
H_2O_2	1.846	N	N	N
C^+	1.901	N	M	Y
O	2.060, 4.746	N	M/N	Y
HD	2.675	N	N	Y
O^{++}	3.394	N	M	Y

Notes:
[a] Excellent ground-based facilities such as those on the Chilean Atacama desert host a median value of 500 μm of precipitable water vapor (PWV); the best Antarctic sites such as Ridge A and Dome A denote best quartile winter conditions of 75 micrometers precipitable water vapor (PWV). The airborne facility (SOFIA) assumes 8 micrometers at an elevation of 13 kilometers. A zenith angle of 30 degrees is assumed throughout. A "Y" label implies 25% transmission or greater, "M" implies marginal transmission of 5-25%, whereas "N" implies typical transmission of less than 5%. Atmospheric transmission computed from the *am* model (Paine & Blundell 2004).

of the Antarctic Plateau, and airborne observatories. They encompass the fine structure transitions of elemental ions and atoms, particularly those of carbon, nitrogen and oxygen; the ground state transitions of pivotal light diatomic molecules, particularly hydrides; and the low-frequency vibrational modes of heavy molecules.

2. The Role of the Antarctic Plateau for Terahertz Exploration

The capabilities of ground-based, airborne and space instrumentation are leading to a renaissance in terahertz astronomy. In particular, the SPIRE-FTS, PACS and HIFI instruments aboard the *Herschel* Space Observatory have expanded the reach of terahertz spectroscopy by orders of magnitude. However, the lifetime of Herschel is relatively short; breaking new scientific ground will require renewed effort in terahertz spectroscopic instrumentation. In the post-Herschel world, this field will be dominated by three major players: the Atacama Large Millimeter Array (ALMA), the Stratospheric Observatory for Infrared Astronomy (SOFIA) and the 25-meter CCAT antenna on Cerro Chajnantor. The combination of ALMA and CCAT will provide excellent coverage of the traditional submillimeter-wave bands below 1 THz (300 μm) at high angular resolution and sensitivity, whereas SOFIA will provide access to the >1 THz universe at sub-arcminute resolution for up to 1000 hours per year (Figure 5). To maximize scientific impact, terahertz facilities on the Antarctic Plateau should complement these major facilities in a meaningful way. The exceptional atmosphere above the Ridge A and Dome A sites provides unique and important opportunities that represent expansion of capability and uniquely answer crucial astronomical questions by direct observation. Here, we present two examples that illustrate two unique classes of opportunities. The first application focuses on wide-field, large scale imaging spectroscopy of the Galaxy; the second on targeted terahertz interferometric studies of star forming regions.

2.1. *Large-Scale Spectroscopic Mapping of the Galaxy – the HEAT telescope at Ridge A*

The evolution of galaxies is determined to a large extent by the life cycles of interstellar clouds, as shown in Figure 2. The interstellar medium in the Galaxy is a patchy, clumpy medium that encompasses extremes of temperature and a wide range of densities (Cox 2005).

Interstellar clouds play a central role in cosmic evolution; they are simultaneously the sites of formation of all stars and planets, and the reservoirs of material that has been processed through previous generations of stars. Although they are among the coldest, least energetic objects in astronomy, they depend upon interactions with high energy photons and cosmic rays for much of their internal heating and chemical activity. They are largely comprised of molecular hydrogen and atomic helium, neither of which have accessible emission line spectra in the prevailing physical conditions in cold interstellar clouds. Thus, it is necessary to probe the nature of the cold interstellar gas via rarer trace elements. Carbon, for example, is found in ionized form (C^+) in neutral clouds, eventually becoming atomic (C), then molecular as carbon monoxide (CO) in dark molecular clouds. Just as there is a "carbon cycle" on the Earth, there is an equivalent life cycle of matter in the Galaxy that is traced well via carbon species. Critically, these three forms of elemental carbon are readily observable at terahertz frequencies (Table 1).

Although we are now beginning to understand star formation, the formation, evolution and destruction of molecular clouds remains shrouded in uncertainty. For example, the formation of interstellar clouds is a prerequisite for star formation, yet the process has not yet been identified observationally! Theories of cloud formation and destruction must be guided and constrained by observations of all atomic and molecular gas components and therefore should be comprised of observations spanning C^+, C and CO, over the full range of environments in which molecular clouds are constructed and destroyed.

After a handful of heterodyne measurements from the Kuiper Airborne Observatory (Boreiko & Betz 1991, 1995, 1997), the first landmark steps toward the dissection of Galactic interstellar material via the heterodyne measurement of the 1.9 THz line of C^+ were taken by the Herschel open time key program "GOT C+" (Figure 3). Analysis of diffuse clouds in a sample characterized by C^+ and HI emission but without CO demonstrates the presence of a significant diffuse warm, dark H_2 component (Langer *et al.* 2010; Velusamy *et al.* 2010; Pineda *et al.* 2010) that had been suggested by previous work (Grenier *et al.* 2005; Wolfire *et al.* 2010). The diversity of clouds that may be disentangled along even a single line of sight is nothing short of spectacular. While "GOT C+" now demonstrates what can be accomplished by combining C^+ emission with CO and HI maps, it is limited to 900 individual lines of sight through the Galaxy, summing to less than 0.1 square degree of sky. However fully sampled two-dimensional maps in C^+ and C must await dedicated survey facilities at THz frequencies.

To fill this need, the 60 centimeter aperture High Elevation Antarctic Terahertz (HEAT) telescope was installed near the summit of the Antarctic Polar Plateau, Ridge A, at 4040 meter elevation. Coupled to the UNSW-built PLATeau Observatory for Ridge A (PLATO-R), its survey operations began in January 2012 with Galaxy-wide mapping of the $^3P_2 - ^3P_1$ line of atomic carbon at 809 GHz. It will be joined by the lower lying 492 GHz line of atomic carbon and the 1900 GHz line of ionized carbon in 2013. These

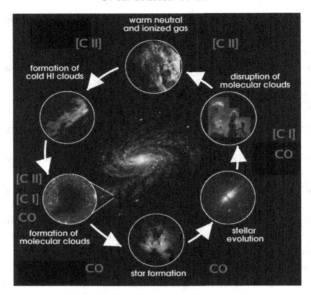

Figure 2. Sketch of the various stages in the life cycle of interstellar clouds.

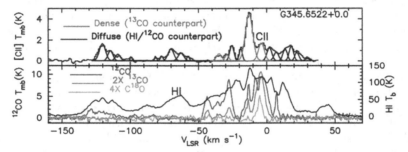

Figure 3. Spectra of C^+ obtained with Herschel/HIFI for the GOT C^+ program at $l = 345.65°$ and $b = 0°$, along with the CO data from the Mopra telescope and HI surveys. Many different types of interstellar clouds can be seen in this line of sight. The Gaussian decomposition for C^+ is also shown in the upper panel. Caption and figure from Langer *et al.* (2010).

surveys will connect the molecular cloud component with the atomic diffuse interstellar medium, illuminating the origin of molecular clouds and the origin of turbulence in the cold ISM. The first data from HEAT have already uncovered diffuse atomic carbon emission in the Galaxy and directly measured the dark molecular gas (H_2 clouds without CO emission). Initial estimates show this component to be almost 40% of the molecular mass assessed by CO emission alone (Figure 4). Much more of the cold, neutral ISM is in the form of clumpy photoionization regions (PDRs) than previously suspected.

HEAT has also shown the Ridge A site to be an exceptional location from which to perform terahertz astronomy from the ground. Based on the correlation of HEAT tipping measurements at 809 GHz (370 μm) and satellite-based soundings at infrared and microwave wavelengths from January through October 2012, the number of productive "submillimeter" days in which the mean daily atmospheric opacity τ_{810} at 810 GHz (370 μm) is $\leqslant 1.5$ has been 258, or almost 86% of the time. Even more impressive, the number of "terahertz" days in which the daily mean atmospheric opacity at 1.5 THz (200 μm) is $\leqslant 1.5$, equivalently $\tau_{810} \leqslant 0.5$, has been 85, or 28% of the time. In comparison to data returned from the APEX radiometer on the Chajnantor plain, there are 2.45× more

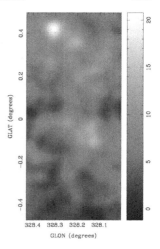

Figure 4. The international PLATO-R and HEAT observatory at Ridge A, Antarctica (left) and a first-light integrated intensity map of 0.6 square degrees of the Galactic Plane (right), observed in the 809 GHz fine structure line of atomic carbon. The widespread emission and filamentary structure highlights molecular gas not previously traced by CO.

Table 2. Exceptional THz transmission above Ridge A, Jan–Oct 2012

Metric	Ridge A	Chajnantor	Advantage
fraction($\tau_{810} \leqslant 1.5$)	87%	38%	2.3×
days $<\tau_{810}> \leqslant 1.5$	258	105	2.45×
fraction($\tau_{1500} \leqslant 1.5$)	32%	4%	8×
days $<\tau_{1500}> \leqslant 1.5$	85	5	17×

"fully submillimeter" days and 17× more "fully terahertz" days (Table 2) at Ridge A. Observing at 1.5 THz from Ridge A is not so different than observing at 810 GHz from South Pole, as was done at the AST/RO telescope for over a decade (Stark *et al.* 2001).

The atmospheric stability, critical to the construction of high-fidelity Galactic Plane maps, is a characteristic that is even more advantageous for Antarctic terahertz astronomy. Peterson *et al.* (2003) demonstrated that the submillimeter-wave sky noise is dominated by fluctuations in water vapor content, and is over 30 times lower at South Pole than at Mauna Kea or Chajnantor owing to the reduced water vapor content of the Antarctic atmosphere. Initial data at Ridge A in 2012, in comparison to the operation of the HEAT telescope at South Pole in 2011, suggests that the sky noise is less than half that of South Pole on average, and therefore scales somewhat more sharply than linear with total water vapor content.

2.2. *Beyond ALMA: Terahertz Interferometry*

The same exceptional site characteristics that make the summit of the Antarctic Plateau so well suited for large-scale mapping also render it ideal for terahertz interferometry. Indeed, the phase noise between arms of a terahertz interferometer is dominated by the time variation of water vapor along a line of sight, which is dramatically lower over the Antarctic Plateau than at the best developed mid-latitude sites. The parameter space filled by other major facilities shows opportunity for a high resolution facility at frequencies $\geqslant 1$ THz (Figure 5). Such a facility could essentially only be performed from a site such as Dome A or Ridge A, based on the excellent likelihood of stable

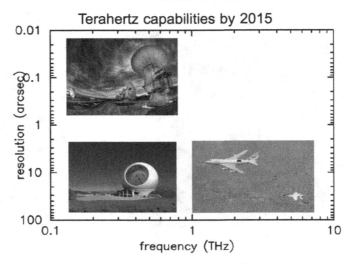

Figure 5. Major new facilities to be commissioned in this decade include ALMA, SOFIA and CCAT. A significant opportunity lies in terahertz interferometry, which will not be practical from airborne platforms nor productive from mid-latitude ground-based sites.

terahertz conditions for long periods of time. Furthermore, airborne and balloon-borne experiments are not conducive to the long baselines needed to achieve sub-arcsecond angular resolution at these frequencies. Not only would such a facility provide unprecedented angular resolution in the far-infrared, it would also act as an important low-cost pathfinder for future space-borne interferometers such as SPIRIT or TPF/I.

What might be the principal science drivers of an initial two or three element interferometer, providing high resolution terahertz astronomy?

The spectroscopic measurement of molecules in the dense cores of dark clouds, where high column densities ($N_H > 10^{22}$ cm^{-2}) and high volume densities ($n_H > 10^4$ cm^{-3}) are prevalent, are frustrated by the freeze-out of gas phase volatiles onto dust grains. In particular, CO, the common tracer for molecular hydrogen, is observed to form ices under such circumstances (Hotzel *et al.* 2002). Thus, infrared dark clouds (IRDCs) and actively collapsing protostellar cores are difficult to measure in the gas phase, since gaseous abundances are often reduced by gas-grain interactions. However, primordial hydrogenic ions like H_3^+ are especially important as they do not aggregate onto dust grains and therefore remain in the gas phase. While H_3^+ does not have a radio spectrum, its deuterated forms such as H_2D^+ are sensitively probed at terahertz frequencies. Thus, they constitute crucial spectroscopic probes of the velocity field in dark, starless, and collapsing cloud cores (Bergin & Tafalla 2007) (Figure 6). Light hydrides with ground state lines at terahertz frequencies are among the best probes of residual gas in planet-forming circumstellar disks.

3. Conclusions and Prospects

As preceded by the development of charge-coupled devices (CCDs) for visible light in the 1980's and large format photodiode arrays in the infrared during the 1990's, terahertz instrumentation and science is poised to make significant advances in capability in the coming decade. While it is likely that the unexpected discoveries will leave the greatest legacy, the impact of anticipated discoveries is no less exciting and encompasses the broad

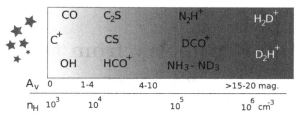

Figure 6. Schematic summary of the major gas phase probes of starless cores as function of depth and density into a cloud. In deep protostellar environments, only primordial molecules with terahertz spectra remain diagnostic. Adapted from Bergin & Tafalla (2007).

range of astrophysical studies, from Solar System astronomy to the formation of stars and galaxies at high redshift.

Furthermore, the opportunities provided by the Antarctic Plateau are important not only to the development of the terahertz waveband, but also as a technological testbed. Meaningful improvements to natal terahertz technologies are required in the context of development of large format focal plane arrays, wide instantaneous spectral coverage, and detector sensitivity. However, the number of terahertz observatories where new concepts in instrumentation can be tested and deployed for science is very small and needs to be developed further in the post-Herschel era. Ground based sites such as the summit of the Antarctic plateau (Ridge A and Dome A) therefore offer great promise in the terahertz atmospheric windows and represent an excellent proving-ground for promising technologies, which can then be productively applied to airborne, balloon-borne, and space-based platforms.

References

Bergin, E. A. & Tafalla, M. 2007, *Annual Review of Astronomy & Astrophysics*, 45, 339
Boreiko, R. T. & Betz, A. L. 1991, *Astrophysical Journal Letters*, 380, L27
Boreiko, R. T. & Betz, A. L. 1995, *Astrophysical Journal*, 454, 307
Boreiko, R. T. & Betz, A. L. 1997, *Astrophysical Journal Supplement*, 111, 409
Cox, D. P. 2005, *Annual Review of Astronomy & Astrophysics*, 43, 337
Grenier, I. A., Casandjian, J.-M., & Terrier, R. 2005, *Science*, 307, 1292
Hotzel, S., Harju, J., Juvela, M., Mattila, K., & Haikala, L. K. 2002, *Astronomy & Astrophysics*, 391, 275
Langer, W. D., Velusamy, T., Pineda, J. L., *et al.* 2010, *Astronomy & Astrophysics*, 521, L17
Paine, S. & Blundell, R. 2004, *Fifteenth International Symposium on Space Terahertz Technology*, 418
Peterson, J. B., Radford, S. J. E., Ade, P. A. R., *et al.* 2003, *Publ. Astronomical Society of the Pacific*, 115, 383
Pineda, J. L., Velusamy, T., Langer, W. D., *et al.* 2010, *Astronomy & Astrophysics*, 521, L19
Stark, A. A., Bally, J., Balm, S. P., *et al.* 2001, Publ. Astronomical Society of the Pacific, 113, 567
Velusamy, T., Langer, W. D., Pineda, J. L., *et al.* 2010, *Astronomy & Astrophysics*, 521, L18
Wolfire, M. G., Hollenbach, D., & McKee, C. F. 2010, *Astrophysical Journal*, 716, 1191
Yang, H., Kulesa, C. A., Walker, C. K., *et al.* 2010, *Publ. Astronomical Society of the Pacific*, 122, 490

Astrophysics from Antarctica
Proceedings IAU Symposium No. 288, 2012
M. G. Burton, X. Cui & N. F. H. Tothill, eds.

© International Astronomical Union 2013
doi:10.1017/S1743921312016985

Optical Interferometry from the Antarctic

Peter Tuthill

Sydney Institute for Astronomy,
School of Physics
University of Sydney
N.S.W. 2006 Australia
email: `p.tuthill@physics.usyd.edu.au`

Abstract. The unique atmospheric conditions which pertain in the high Antarctic plateau offer dramatic gains for many areas of Astrophysics. Optical Interferometry is among the most technologically demanding branches of modern instrumentation, and furthermore, is one which is most strongly limited by the stability of the atmosphere at the observatory site. The long-term potential for spectacular gains by implementing an interferometer on the high Antarctic plateau are presented.

Keywords. long baseline optical interferometry

1. Introduction

Many of the most topical astrophysical research frontiers entail observation of matter in the close environment of stars or highly luminous cores. Studies of exo-planets, star formation, stellar winds and active galactic nuclei are all limited by the extreme dynamic ranges, high angular resolutions and high measurement precision needed to discriminate the faint signals against the glare of a luminous central core. These considerations drive optical designs towards an interferometer, however at mid-latitude sites there is a heavy penalty from the seeing in the turbulent atmosphere.

Quantifying just how strong this penalty will be depends in detail on the specifics of the experiment, however interferometer performance will in general be strongly sensitive to all three fundamental seeing parameters: r_0 (spatial), t_0 (temporal) and θ_0 (angular) coherence properties of the incoming wavefront. One mental construct to crudely quantify a site's seeing metric is to consider a hyper-volume consisting of the product of these three numbers – the larger this volume the better – but even such a multiplicative scaling probably under-represents the importance of good seeing. As most adaptive opticians know, when the seeing is sufficiently bad, it is better to just do something else with the telescope: one gets far less than half the science when fighting two arcsecond seeing compared to one arcsecond seeing.

From the Antarctic plateau, all three of these atmospheric properties attain their most favourable values on the surface of the Earth, as described in the opening session of this conference. For many crucial observational schemes such as nulling interferometry or precision astrometric interferometry, the dramatically improved conditions will result in orders of magnitude increase in sensitivity on top of gains already available from the extreme low temperature and water vapour.

The potential for Antarctic interferometry, enabled by such conditions, is fundamentally different from that at a conventional site. An Antarctic plateau interferometer could rival the immensely more expensive space missions to what many regard as the most important keystone project in modern astronomy: the characterization of exo-planets and detection of objects down to one earth-mass. Furthermore, a wealth of other stellar astrophysics can be addressed by such a device. Observation of disks around young stellar objects should reveal substructure such as spiral density waves and gap clearing due to planet growth. Mass loss phenomena in evolved stars, and substructure such as disks and jets within dusty compact microquasars, will be within reach of imaging observations. Studies of fundamental stellar properties – sizes, effective temperatures, distances and (using binary stars) masses – will mean that almost every branch of stellar astronomy will benefit.

2. Optical interferometry prospects: a big picture view

The field of long-baseline optical interferometry presently finds itself at something of a crossroads. Interest in the long-dormant field of high resolution imaging was rekindled worldwide by the dramatic success of the Narrabri Stellar Intensity Interferometer of Hanbury Brown and Twiss in the late 1960's to early 1970's (Hanbury Brown *et al.* 1967a; Hanbury Brown *et al.* 1967b; Hanbury Brown *et al.* 1974). There followed a flurry of prototypes and pathfinders in the 1980's and what can be justifiably described as a first major generation of dedicated science interferometers, mostly constructed in the decade leading up to the turn of the century. Another decade on, and we now find that the majority of these projects (more than half a dozen) have run their course and now been mothballed (e.g. KI, MIRA) or decommissioned (e.g. PTI, IOTA, GI2T). Notably, the survivors have been the ones with the more ambitious basic architecture (larger apertures; longer baselines; more telescope stations) which now boast high and still rising science output. In particular, the VLTI dominates the Southern, and the CHARA array the Northern skies at milli-arcsecond resolution; both instruments boast a rich arsenal of science instruments capable of a variety of observational modes.

However, the emergence of what might be described as a "second generation" of interferometers which build upon the successes of the first is not at all apparent. The only major new optical interferometric facility nearing science readiness is the LBT Interferometer whose novel architecture may well deliver unique science reach, but which seems more accurately described as the first coherent-light focus of an ELT rather than a long-baseline interferometer. The hope for significant new interferometer projects in the present decade rests with MROI and a major upgrade to NPOI; both with an uncertain (at the time of writing) funding outlook. Important progress is of course being made with new instrument subsystems such as the ambitious GRAVITY beam combiner at VLTI. Few proposals for new-generation interferometers of any real scope and ambition have been advanced; none seems to have attracted any significant following.

Although such a bleak picture might seem discouraging, the unquestioned potential for the basic physics of interferometry to deliver unique and critical knowledge remains. In particular, space missions (SIM, TPF, Darwin) to detect and image exoplanets around the stars in the solar neighbourhood and to characterize their basic properties have been pushed well into the future by funding constraints. However, the drive to accomplish the science remains strong and will only become more pressing as new discoveries continue to flesh out the statistical prevalence of exoplanets.

As has been pointed out before, the Antarctic provides a near space-analogue in many key respects and is available for the prototyping and commissioning of testbeds at a fraction of the price and complexity of a spacecraft. Maintaining ongoing efforts to advance the field and inform the design of tomorrow's audacious space missions targeting exoplanetary characterization is of critical importance. Wherever such spacecraft require the very ultimate in angular resolution, then the basic physics of detection will always drive designs towards the principles of interferometry. This will only lend weight beyond the direct science benefits outlined here to maintain a vibrant research community pushing the boundaries of these techniques.

3. Architectures for optical interferometry

Early experiments and interferometer prototypes typically recorded quite simple visibility (or v^2) amplitude information from only two telescopes over a single spectral bandpass. Such data can be recovered with a correspondingly simple beam combiner: at its most basic a beamsplitter to combine the beams and a pair of photodetectors to record the starlight fringes. However as the field has flourished, the capabilities of the instruments and the complexity of the data forms recovered has proliferated. There are now a large number of variations of interferometer design which target specific science niches. Each can make quite different demands on the basic instrumentation and infrastructure, and when a concept design is transferred to the high Antarctic plateau, each might realize gains or challenges from quite distinct aspects of the working environment.

Major interferometers at mid-latitude sites such as CHARA and VLTI have migrated towards a model similar to classical research observatories in which the large infrastructure such as telescopes and delay lines is kept as generally applicable as possible, and a variety of more specialized instruments are available to accept the starlight at the final step and deliver scientific results. Such a "switchyard" approach which allows a variety of science to be produced is unlikely to be profitable for Antarctic deployment due to its extra complexity and risk. The first major hurdle in planning an Antarctic Interferometer is therefore deciding upon the science goal and consequently the architecture do deliver it. Selecting the best match comes down to: (1) delivery of unique, world-leading science; (2) the instrument which exploits the unique conditions to greatest advantage over competitors; and (3) is most robust and reliable to operate with minimal support in the challenging environment. Some initial discusison of the trade-offs involved is given below as the merits of several plausible architectures are debated, although it is well beyond the scope of the present work to arrive at any quantitative conclusions. The interested reader is strongly encouraged to seek out the Arena Roadmap (2010) which goes into many key issues in far more detail than is possible here, and in particular the results of Working Group 3 led by Vincent Coude du Foresto.

3.1. *A narrow-angle astrometric interferometer*

If we assume that the physical layout of telescopes in an array can be very well characterized, then the trajectory of the interference fringe envelope generated by each baseline as a function of time can be used to extract very precise astrometric data recording the position of the stellar target. Repeated measurements covering many stars can generate an accurate all-sky wide-angle astrometric catalog (e.g. NPOI see Hutter *et al.* 1998), although the precision with which the effective baseline can be measured does limit the final accuracy.

One way to dramatically boost the precision while keeping the baseline metrology requirements relatively modest is to implement a *narrow angle astrometric mode* in which the interferometer switches between observations of relatively close ($\lesssim$ few arcsecond) binary stars. Such an observing mode was pioneered with the PHASES instrument at the Palomar Testbed Interferometer (Muterspaugh *et al.* 2010), while several instruments worldwide such as GRAVITY and PRIMA at the VLTI and MUSCA at SUSI all employ a set of similar basic principles.

There are several specific advantages such an interferometer design would gain from Antarctic deployment. In particular, the fundamental limitation to the astrometric measurement process is an extremely strong function of the isoplanatic patch size (Lloyd *et al.* 2003). Of all the unique properties of Antarctic plateau seeing, perhaps this is the one which is most exceptional due to the concentration of the atmospheric turbulence in the near-field at the boundary layer. Furthermore, telescope apertures and baselines can be comparatively modest.

Mitigating against this design are several practical factors. Astrometric modes have proven to be relatively difficult and complex to implement at mid-latitude interferometers, and none has yet approached the fundamental performance limitation set by the atmosphere. Furthermore, any such instrument will face severe near-term competition from the Gaia mission which will achieve $\lesssim 20\,\mu$arcsec precision over a relatively deep whole-sky survey.

3.2. *A nulling interferometer*

An interferometric analog of a Coronagraph, Nulling interferometer designs hold out the promise of rejecting the overwhelming glare from a bright central star yet allowing the study of faint circumstellar material or structures in the immediate vicinity. The original idea of placing a destructive interference fringe to cancel the star dates back to Bracewell (1978). The first and most ambitious modern separate telescope nulling instrument was operated as a part of the Keck interferometer (Serabyn *et al.* 2012) which employed a 4-beam nulling configuration delivering significantly deeper nulls than a simpler 2-beam configuration could achieve.

Obtaining reasonable null depths motivates a requirement for relatively good wavefront quality. In the absence of a high-order adaptive optics system (both complex and expensive), this wavefront quality is one factor that pushes nulling designs towards longer wavelengths in the mid-infrared. Fortunately, much of the most profitable astrophysics is also best studied in this waveband. Varying forms of dusty disks from exo-zodiacal clouds to debris disks all offer considerable advantage to longer wavelength study. Furthermore, the early stages of brown dwarf and planetary assembly are those in which the object is both hottest and brightest, with contrast ratios becoming dramatically less challenging as one moves from the visible to the mid-infrared.

Nulling interferometry stands to make very significant gains when implemented on the Antarctic plateau. Long-wavelength operation benefits from good atmospheric transparency (low water vapor), and more importantly, from the thermal environment. Typical interferometer optical trains may have more than 20 reflections bringing the starlight to the detector. Reducing the self-induced thermal radiation from the mirror train itself by exploiting the cold ambient conditions will already deliver a profound impact on the sensitivity of the device, before one factors in extra gains due to the excellent seeing. A scientifically competitive nuller would, however, require some reasonable scale of infrastructure with moderately large apertures ($\gtrsim 0.5$ m) and designs may require considerable engineering in order to exploit the good seeing above the boundary layer.

3.3. *An imaging/closure phase interferometer*

Among the commonest functionalities for a modern optical interferometer is to recover complex visibility data products. Such devices are the most direct descendents from detection schemes employed in the earliest devices. Unfortunately extreme measurement precision, well beyond anything yet demonstrated, would be required for such instruments to reveal the presence of high contrast companions by exploiting visibility amplitude data alone. Much emphasis and effort has instead been focused on observables extracted from the Fourier *Phase*. Although the phase itself is corrupted by the atmospheric seeing, observables such as *differential phase* and particularly the *closure phase* hold the promise of pushing detection thresholds into the planetary regime for systems which are young and bright.

Indeed, within the context of aperture masking interferometry, the use of closure phase methods to recover high contrast companions well into the planetary mass regime has become well established in recent years (Tuthill *et al.* 2006; Lloyd *et al.* 2006) and indeed has delivered the first direct detection of a planetary candidate at the epoch of formation (Huélamo *et al.* 2011).

Practical implementation of such methods is not strictly prescriptive of the interferometer architecture nor even the observing wavelength, although reasonable extrapolations of successful current experiments would most likely come up with a straw-man near-infrared device with several ($4 \sim 6$) telescopes of moderate ($\gtrsim 0.5$ m) aperture. Gains over mid-latitude sites would mainly arise from superior seeing, provided the boundary layer problems can be solved, although the potential to leverage still further thermal gains towards somewhat longer wavelengths in L–band (3.6μm) is also attractive.

One potentially interesting variation to this design may be to explore the idea of a *closure-phase nulling interferometer*, as proposed by Chelli *et al.* (2009). This requires baselines sufficiently long to reach the first null of the stellar visibility function, which for nearby bright stars is of order ~ 300 m or so.

3.4. *Reformulating the optical interferometer*

Optical interferometer designs from the 20th century relied heavily on precision control with precisely engineered components implemented in (as much as possible) a thermally and mechanically stable environment. Following the same recipe may be costly and difficult in an Antarctic plateau context, particularly so when the requirement to be elevated some ~ 20 m above the ice to avoid the boundary layer seeing.

It is therefore interesting to ask whether all this concrete and precision-milled steel is really necessary to perform the basic functionality of an interferometer? One telling point to note is that for most functioning instruments, despite the money spent to deliver high degrees of inherent mechanical stability, it remains a part of normal procedure to search over some relatively wide interval for the fringes – often many millimetres. Small changes in optical alignment, together with errors introduced at moving parts and thermal/mechanical drifts, conspire to create pathlength imbalances.

I therefore propose a shift in the underlying philosophy of interferometer design. Rather than attempting to *control* optical surfaces and components to the micron level, instead we merely perform metrology adequate to *characterize* these paths in real time. Such information can be employed to drive a short-throw delay line to compensate for pathlength excursions which should be relatively slow (if care is taken to minimize mechanical vibration in the instrument). Tabulated below is a straw-man design which reformulates an interferometer along these principles.

Classical Interferometer Subsystem	Reformulation for Antarctica
Light Collecting Typically afocal telescopes or siderostats mounted on very stable concrete piers or (for designs with movable telescopes) kinematic mounts at fixed telescope base stations.	**Light Collecting** The basic optics can be similar, although mounted on towers to rise above the inversion layer. Located to sub-mm accuracy with some form of metrology grid (several technical solutions exist).
Beam Transport Convey light collected over an array spanning hundreds of metres to a common point. Typically evacuated steel pipes with an internal embedded chain of relay mirrors.	**Beam Transport** Use pre-matched lengths of optical fiber deployed to each telescope. For a continually moving telescope design (see next point) then dragging a fiber cable may be a viable option.
Delay Lines Starlight must be pathlength matched to micron precision with a varying delay according to the projected geometry of each interferometer baseline onto the sky. This is done with motorized carriages running on precision steel rails under laser metrology control. Ideally, the entire delay line is implemented in a vacuum.	**Delay Lines** Possibly delay lines running in a tunnel with a refrozen flat ice floor. More radically, implement continually moving telescopes on an ice plane thereby removing the need for any separate delay line (and performing the delay correction effectively in vacuum).
Beam Combination Beam combiners and all the optical processing that goes with them is usually achieved with one (or more) optical table of lenses, mirrors and beamsplitters.	**Beam Combination** Accomplish all optical processing functionality in integrated optics chips. These are ideal for challenging environments: hermetically sealed, alignment-free, and highly robust.

4. Conclusions

A discussion paper on prospects and designs for a future Antarctic interferometer has been presented. Several basic architectures and specific science cases seem promising enough to merit further exploration; in particular a mid-infrared nulling interferometer and a near-infrared imaging/closure phase instrument. From the perspective of fundamental physical principles, interferometry presents the natural way to obtain information on the finest possible angular scales and so reveal the physics of processes such as stellar and planetary assembly which occur on such a remote stage that milli-arcsecond resolution is required. The high Antarctic plateau offers conditions dramatically superior to any other known terrestrial site for this endeavor, delivering order-of-magnitude gains to instruments operating at fundamental noise limits set by the turbulent atmosphere. The author hopes, therefore, that the present dearth of major new activity in interferometry might be likened to a saturated solution of science potential with ever increasing concentration. All it takes is one crystal – perhaps an Antarctic snowflake – to suddenly cause a flurry of new activity to precipitate.

References

ARENA Roadmap 2010. Available online at arena.oca.edu

Bracewell, R. N. 1978, *Nature*, 274, 780

Chelli, A., Duvert, G., Malbet, F., & Kern, P. 2009, *A&A*, 498, 321

Hanbury Brown, R., Davis, J., Allen, L. R., & Rome, J. M. 1967, *MNRAS*, 137, 393

Hanbury Brown, R., Davis, J., & Allen, L. R. 1967, *MNRAS*, 137, 375

Hanbury Brown, R., Davis, J., Lake, R. J. W., & Thompson, R. J. 1974, *MNRAS*, 167, 475

Huélamo, N., Lacour, S., Tuthill, P., *et al.* 2011, *A&A*, 528, L7

Hutter, D. J., Elias, N. M., & Hummel, C. A. 1998, *Proc. SPIE*, 3350, 452

Lloyd, J. P., Martinache, F., Ireland, M. J., *et al.* 2006, *ApJL*, 650, L131

Lloyd, J. P., Lane, B. F., Swain, M. R., *et al.* 2003, *Proc. SPIE*, 5170, 193

Muterspaugh, M. W., Lane, B. F., Kulkarni, S. R., *et al.* 2010, *AJ*, 140, 1579

Serabyn, E., Mennesson, B., Colavita, M. M., Koresko, C., & Kuchner, M. J. 2012, *ApJ*, 748, 55

Tuthill, P., Lloyd, J., Ireland, M., *et al.* 2006, *Proc. SPIE*, 6272,

Astrophysics from Antarctica
Proceedings IAU Symposium No. 288, 2012
M. G. Burton, X. Cui & N. F. H. Tothill, eds.

© International Astronomical Union 2013
doi:10.1017/S1743921312016997

Preliminary design of the Kunlun Dark Universe Survey Telescope (KDUST)

Xiangyan Yuan[1,2,3], Xiangqun Cui[1,2,3], Ding-qiang Su[4,5,1], Yongtian Zhu[1,2,3], Lifan Wang[3,6], Bozhong Gu[1,2], Xuefei Gong[1,2,3], Xinnan Li[1,2]

[1] National Astronomical Observatories / Nanjing Institute of Astronomical Optics & Technology, Chinese Academy of Sciences, Nanjing 210042, China

[2] Key Laboratory of Astronomical Optics & Technology, Nanjing Institute of Astronomical Optics & Technology, Chinese Academy of Sciences, Nanjing 210042, China

[3] Chinese Center for Antarctic Astronomy

[4] Dept. of Astronomy, Nanjing University, 22 Hankou Road, Nanjing 210093, P. R. China

[5] Key Laboratory of Modern Astronomy and Astrophysics (Nanjing University), Ministry of Education, P. R. China

[6] Purple Mountain Observatory, Chinese Academy of Sciences

[7] email: xyyuan@niaot.ac.cn

Abstract. From theoretical analysis and site testing work for 4 years on Dome A, Antarctica, we can reasonably predict that it is a very good astronomical site, as good as or even better than Dome C and suitable for observations ranging from optical to infrared & sub-mm wavelengths. After the Chinese Small Telescope ARray (CSTAR), which was composed of four small fixed telescopes with diameter of 145 mm and the three Antarctic Survey Telescopes (AST3) with 500 mm entrance diameter, the Kunlun Dark Universe Survey Telescope (KDUST) with diameter of 2.5 m is proposed. KDUST will adopt an innovative optical system which can deliver very good image quality over a 2 square degree flat field of view. Some other features are: a fixed focus suitable for different instruments, active optics for miscollimation correction, a lens-prisms that can be used as an atmospheric dispersion corrector or as a very low-dispersion spectrometer when moved in / out of the main optical path without changing the performance of the system, and a compact structure to make easier transportation to Dome A. KDUST will be mounted on a tower with height 15 m in order to make a full use of the superb free atmospheric seeing.

Keywords. Dome A, astronomical sites, telescopes, KDUST, seeing

1. Introduction

Four years of site led by the Chinese Center for Antarctic Astronomy (CCAA) shows that the highest Antarctic inland plateau Dome A (latitude 80°22′02″S, longitude 77°21′11″E, elevation 4,093 m) can be reasonably predicted as an excellent site for ground-based astronomical observation, which would be as good as Dome C where the mean seeing is about 0.27″ above 30 m from the ground (Lawrence *et al.* 2004) or even better (Yang *et al.* 2010; Bonner *et al.* 2010). China has been actively promoting Antarctic astronomy since 2005. The first-generation Chinese Antarctic optical telescope is known as Chinese Small Telescope Array (CSTAR; Liu & Yuan 2009) composed of four identical telescopes with diameter 145 mm and four different filters (G, R, I and open band). CSTAR was deployed on Antarctic Dome A in January 2008 and mainly used for variable sources detection and site testing. The Antarctic Survey Telescopes AST3 are the second-generation Chinese telescopes on Dome A. AST3 is composed of three large field of view optical telescopes matched with G, R or I filters (Yuan & Su 2012), with entrance pupil diameter 500 mm, primary mirror diameter 680 mm, and field of view 2.92° × 2.92° corresponding to a 10K × 10K CCD. The main scientific goals include a survey of Ia supernovae to study the dark energy of the universe, micro-lensing to search for exoplanets and to find new variable sources. The first AST3 telescope for I band was installed on Dome A in January 2012. The other two AST3 telescopes will be installed there in 2014. Then will come the third-generation Chinese Antarctic telescopes, including a 2.5 m optical/infrared telescope and a 5 m sub-mm telescope. The 2.5 m Kunlun Dark Universe Survey Telescope, abbreviated to KDUST, will make a full use of the superb free atmospheric seeing and deliver high resolution image over a large field of view. The sensitivity of KDUST in the optical band and NIR will be better than 8 m telescopes on temperate sites, and even comparable with a ~ 30 m telescope on a temperate site in the K–band. The main science goals of KDUST are: weak lensing to study dark matter, Type Ia supernovae and the dark energy of the universe, the Galaxy and exoplanets.

2. Preliminary design of KDUST

Technical challenges. Dome A has excellent astronomical observing conditions such as continuous observing time for around 4 month per year, low turbulence boundary layers and very good free atmospheric seeing, very low sky background in the thermal IR etc. But the very low temperature (ranges from −30° to −83°C) and low air pressure, high relative humidity, harsh transportation conditions and very limited working time on site per year bring many challenges in both logistics and infrastructure (Burton *et al.* 2010) for developing a 2.5 m high resolution telescope for Dome A. Some technical challenges are listed here, for example, study on large FOV and compact optical system with high resolution image quality which can be matched with the excellent seeing conditions on Dome A; fast *f*-ratio mirror manufacturing technology; deicing and removing snow from a large optical surface without degrading mirror seeing; precise tracking, pointing & focusing under very low temperatures; thermal and humidity control technology (Saunders *et al.* 2008); assembly & alignment of a large telescope on site in a limited time; antivibration transportation for precise devices; remote control and long-time unattended operation; huge data acquisition, storage and pipeline; clean and automatic energy supply of about 20 kW for KDUST in the polar environment.

Main characteristics of KDUST. After study and comparison of several types of optical system which can all deliver very good images over large fields of view, such as the R–C system with correctors (SDSS etc.), Paul-Baker systems (LSST etc.), TMA systems (SNAP etc.) and a LAMOST-type system, we have chosen to use the Chinese 2.16 m type coudé system. The optical layout and image spot diagram are shown in Fig. 1. This system differs from the traditional coudé system, with an aspheric tertiary mirror which can achieve high resolution image quality over a large field of view, fixed focus and flat focal plane. The main characteristics of KDUST are:

Diameter: 2.5 m

Field of view :~ 1.5°

F-ratio: ~ 9.5

Plate scale: ~ $0.1''/10\mu$m

Working wavelength range: 400–2200 nm

Image quality: 80% in less than $0.3''$

Supporting tower: 15 m

Low-dispersion spectrum (~ $10''$ length)

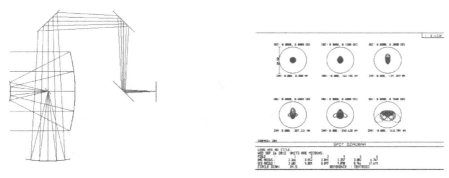

Figure 1. Optical layout (left) and spot diagram (right) for KDUST (the circle shows 0.3″).

The 2.5 m primary mirror has a fast *f*-ratio of about 1.05, thus it is quite challenging to produce this mirror. We are considering using a temperature optimized lightweighted Zerodur blank and stress lap to polish this mirror. Active optics will be necessary to correct the mis-collimation error since the tolerance for both secondary and tertiary mirror is tight. A pair of lens-prisms can be designed near the exit pupil, which is near the edge of the primary to compensate for atmospheric dispersion. This lens-prism can be moved in and out of the main optical path without changing the performance of the system. It can also be used in a contrary way at sites like Dome A or space where the sky brightness is very low, thus it can produce a very low dispersion spectrum of about 10 arcseconds length in lieu of multicolour photometry for celestial objects over the whole FOV. The flat mirror near the exit pupil can be used as a tip-tilt mirror to remove most of the turbulence above the telescope and tower wind shake. The model of the telescope with its compact structure and 15 m high tower and open dome are shown in Fig. 2.

Figure 2. Model of KDUST with compact structure, open dome and 15 m tower.

The structure of the KDUST Project. Five workpackages are planned for the KDUST project, including site & facility, telescope unit, instruments, data & operations and project management. A flow diagram outlining this structure is shown in Fig. 3. Each of the workpackages involves several work groups. For example, the telescope unit includes

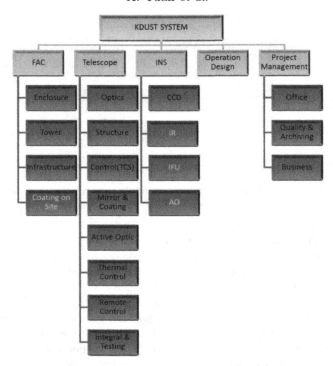

Figure 3. A structure diagram for the KDUST project.

optics, structure, telescope control system, mirror & coating, thermal control, remote control, integration and testing, undertaken by eight groups. The instrument workpackage includes the first generation instruments for a large array of imaging CCDs, and consideration of second generation instruments such as IR detectors, IFUs etc.

At this stage the KDUST project is at the phase of submitting proposals for the Twelfth Five-Year National Science Project. Related key technologies are also being studied and tested at the same time.

This work is supported in part by National Natural Science Foundation of China (NSFC 11190011 and 11190013).

References

Bonner, C. S. *et al.* 2010, *PASP*, 122, 1122
Burton, M. G. *et al.* 2010, *EAS Publications Series*, 40, 125
Lawrence, J. S., Ashley M. C. B., Tokovinin A. & Travouillon Tony 2004, *Nature*, 431, 278
Liu, G., Yuan X. 2009, *Acta Astronomica Sinica*, 50, 224
Saunders, W. *et al.* 2009, *Proc. SPIE*, 7012, 70124F
Yang, H. *et al.* 2010, *PASP*, 122, 490
Yuan, X., Su D. 2012, *MNRAS*, 424, 23

Astrophysics from Antarctica
Proceedings IAU Symposium No. 288, 2012
M. G. Burton, X. Cui & N. F. H. Tothill, eds.

© International Astronomical Union 2013
doi:10.1017/S1743921312017000

The SCAR Astronomy & Astrophysics from Antarctica Scientific Research Programme

**John W. V. Storey[1], Lyu Abe[2], Michael Andersen[3],
Philip Anderson[4], Michael Burton[1], Xiangqun Cui[5],
Takashi Ichikawa[6], Albrecht Karle[7], James Lloyd[8], Silvia Masi[9],
Eric Steinbring[10], Tony Travouillon[11], Peter Tuthill[12]
and HongYang Zhou[13]**

[1] School of Physics, University of New South Wales, Australia
email: j.storey@unsw.edu.au

[2] CNRS Laboratoire Lagrange, Universite de Nice Sophia Antipolis, Nice, France

[3] Dark Cosmology Centre, University of Copenhagen, Copenhagen, Denmark

[4] The ScottishAssociation for Marine Science, Oban, United Kingdom

[5] Nanjing Institute for Astronomical Optics and Technology, Nanjing, China

[6] Department of Physics, Tohoku University, Aoba, Sendai, Japan

[7] Department of Physics, University of Wisconsin, Madison, USA

[8] Department of Astronomy, Cornell University, Ithaca, USA

[9] Physics Department, University of Rome "La Sapienza", Rome, Italy

[10] Herzberg Institute of Astrophysics, Victoria, Canada

[11] California Institute of Technology, Pasadena, USA

[12] School of Physics, Sydney University, Sydney, Australia

[13] University of Science and Technology of China, Hefei, China

Abstract. SCAR, the Scientific Committee on Antarctic Research, is, like the IAU, a committee of ICSU, the International Council for Science. For over 30 years, SCAR has provided scientific advice to the Antarctic Treaty System and made numerous recommendations on a variety of matters. In 2010, Astronomy and Astrophysics from Antarctica was recognized as one of SCAR's five Scientific Research Programs. Broadly stated, the objectives of Astronomy & Astrophysics from Antarctica are to coordinate astronomical activities in Antarctica in a way that ensures the best possible outcomes from international investment in Antarctic astronomy, and maximizes the opportunities for productive interaction with other disciplines. There are four Working Groups, dealing with site testing, Arctic astronomy, science goals, and major new facilities. Membership of the Working Groups is open to any professional working in astronomy or a related field.

Keywords. Telescopes, Atmospheric Effects, Instrumentation, Site Testing, Antarctica.

---

1. Introduction

The *Astronomy & Astrophysics from Antarctica* (AAA) Scientific Research Program kicked-off in 2011. The objectives of AAA are to coordinate astronomical activities in Antarctica to ensure the best possible outcomes from international investment in Antarctic astronomy, and to maximize the opportunities for productive interaction with other disciplines. To achieve this, AAA is working to:

- Coordinate site-testing experiments to ensure that results obtained from different sites are directly comparable and well understood,
- Build a database of site-testing data that is accessible to all researchers,

- Increase the level of coordination and cooperation between astronomers, atmospheric physicists, space physicists and meteorologists,
- Extend existing Antarctic site-testing and feasibility studies to potential Arctic sites; for example, in Greenland and Canada,
- Define and prioritise current scientific goals,
- Create a roadmap for development of major astronomical facilities in Antarctica,
- Stimulate international cooperation on major new astronomical facilities in Antarctica.

In order to carry out these tasks, AAA has created four Working Groups. These Working Groups are open to all interested researchers, and are:

- A. Site testing, validation and data archiving (chair: Tony Travouillon),
- B. Arctic site testing (chair: Eric Steinbring),
- C. Science goals (chair: Michael Burton),
- D. Major new facilities (chair: Peter Tuthill).

In the following sections we describe the progress made so far.

2. Working Group A: Site Testing, Validation and Data Archiving

2.1. *Scope and Aims*

A large number of studies have been conducted, particularly over the past 15 years, into the site conditions at the various Antarctic stations. To a significant extent, however, these studies have been carried out using different instruments and indeed different techniques at the different sites. This makes inter-comparison of the sites difficult, and direct assessment of the relative merits of different sites is often unreliable. Atmospheric models and satellite data have also been used to generate predictions of site conditions and to further our understanding. Again, the different approaches used by different groups can sometimes make comparisons problematic.

The first task for Working Group A is therefore to create a database of all published papers that contain site testing data. This database will then link to the original data, where publicly available, or provide contact details for the authors where the data are not in a readily accessible form.

Secondly, Working Group A will identify gaps in our knowledge of the different sites, and recommend programs to acquire and make available the missing information.

2.2. *Current status*

Work is currently underway to identify and catalogue all published papers on Antarctic site testing. Each paper is to be tagged according to a non-exclusive set of 33 different descriptors:

1. Methodology
(*a*) In situ/ground-based observations
(*b*) Satellite observations
(*c*) Modelling
(*d*) Instrument design

2. Location
(*a*) Arctic
(*b*) Antarctic
(*c*) South Pole
(*d*) Dome A
(*e*) Dome C
(*f*) Dome F

(*g*) Ridge A
(*h*) McMurdo/balloon
(*i*) Greenland
(*j*) Northern Canada
(*k*) Other

3. Wavelength

(*a*) γ-ray/X-ray,
(*b*) UV
(*c*) Optical
(*d*) Infrared
(*e*) THz
(*f*) Millimetre/Sub-mm
(*g*) Particle
(*h*) Other

4. Parameter

(*a*) Meteorological data
(*b*) Atmospheric transparency
(*c*) Sky brightness and stability
(*d*) Seeing and integrated turbulence
(*e*) Free atmosphere turbulence profiles
(*f*) Boundary layer turbulence
(*g*) Precipitable water vapour
(*h*) Aerosols/ice crystals
(*i*) Ozone
(*j*) Other

It is hoped to have the bulk of the catalogue completed by June 2013.

3. Working Group B: Arctic Site Testing

3.1. *Scope and Aims*

This section defines the implementation plan of the the Arctic Site Testing Working Group (ASTWG). The ASTWG has connections to the known site-testing activities in the Arctic, and seeks to be inclusive to all. Unlike the other three Working Groups, ASTWG does not involve any specific activity in Antarctica, although there are several areas of overlapping interest. The intention here is to outline efforts already underway in the Arctic, to indicate similarities and differences with those undertaken in the Antarctic, and thereby facilitate improved communication and knowledge sharing. This can further assure the compatibility of datasets, and potentially guide cross-linked investigations or spur new activities of possible benefit for both Poles.

3.2. *Background*

The two major landmasses nearest the North Pole are Greenland (Denmark) and Ellesmere Island (Canada). Neither of these provide elevations as high as attainable on the Antarctic plateau, but they are comparably cold and dry. A geographical parallel is provided by the Greenland icecap, reaching 3,200m at Summit (72N). At high northern latitudes, the polar vortex functions in the same way as in the Antarctic, although in detail there are differences. Away from the icecaps, a strong and stable surface-based thermal inversion peaked at about 1,000m develops and remains during much of winter. Several locations rise above this. One is Barbeau Peak (82N, 2,600m) within Quttinirpaaq National Park on Ellesmere Island, but mountains nearer the coast top 1,400m to 1,900m.

Meteorological records date back decades for several major bases and research stations in the Arctic. The most northern is the military outpost at Alert (82N), on Ellesmere Island. It is provided with broadband communications via microwave repeater stations linking to Eureka (80N), where geosynchronous satellites become visible. Eureka is accessible year-round by air, and by ship in summer. This is a manned weather station, providing among many other meteorological and atmospheric observations, twice-daily balloon aerosondes. Nearby is the Polar Environment Atmospheric Research Laboratory (PEARL) on a 600m high ridge which is linked by a road. Another scientific base to the south is located at Resolute Bay (76N), which along with Eureka provides aircraft support for field logistics. On Greenland, the United States military base at Thule (76N) provides access to the Summit station (72N) via support from the National Science Foundation.

Serious interest in astronomical site testing in the High Arctic was initiated with the first conclusive evidence of excellent free-atmospheric seeing above the Antarctic glacial plateau (Lawrence *et al.* 2004). The analogy to the Greenland icecap initiated observations at Summit (Andersen & Rasmussen 2006, Andersen, Pedersen & Sorensen 2010). An independent investigation into the feasibility of observations from high and very remote mountain peaks on northwestern Ellesmere Island was also undertaken (Steinbring *et al.* 2008, Wallace *et al.* 2008), anticipating the avoidance of boundary conditions associated with the surface of an ice plateau (Hickson *et al.* 2010, Steinbring *et al.* 2010, Steinbring *et al.* 2012). A strong advantage for a high mountain site over the icecap has not yet been proved, and in the meantime a compromise afforded by the year-round logistics provided by the mid-elevation site at PEARL is also being investigated.

3.3. *Astronomical Site-Testing Observations in the Arctic*

Compared to the Antarctic, astronomical site testing in the Arctic is only in the early stages, with the first work commencing in 2006. Several university and government laboratory groups are collaborating to deploy various instruments, either of their own design or closely following standard ones. Observations are taking a path typical of mid-latitude sites. Initial meteorological and basic sky-quality measurements are now leading to more detailed studies, first with observations obtained during brief campaigns. Characterisation of the scientific potential of the sites starts by deploying small semi-autonomous telescopes, desirably followed by fully robotic versions.

3.3.1. *Instrumentation Deployment*

Several investigations of which the ASTWG is familiar have been undertaken or will be underway in the next two to three years:

• Basic meteorological and sky-clarity observations. Small autonomous weather stations and cameras were constructed by National Research Council of Canada Herzberg Institute of Astrophysics (NRC/HIA) and deployed on three mountains on northwestern Ellesmere Island. These data can be compared with manned sea-level stations, e.g. those of Environment Canada – counterpart to the US National Oceanographic and Atmospheric Organization (NOAO) – that have records dating back over 50 years.

• Sky-monitoring cameras. Data from existing atmospheric-science instrumentation at PEARL are being analysed. Of key interest are all-sky camera images taken by the University of New Brunswick beginning in 2007. Some sky surface-brightness measurements have also been taken in the optical and near-infrared using instruments from the University of Toronto.

• Sub-mm radiometer. A commercial 225 GHz tipping radiometer provided by Academia Sinica Institute of Astronomy and Astrophysics (ASIAA) in Taiwan was operated at PEARL in the winter of 2010/11, and has since winter 2011/12 been operating at Summit in Greenland (see Matsushita *et al.* paper in these Proceedings).

• Specialised optical turbulence measuring instruments. A Polaris-tracking instrument was installed on a 39m high mast at Summit by the University of Copenhagen. At Eureka, two Sound Detection and Ranging (SODAR) units surplus from TMT site testing have been operated nearby for two winters. At the same time the Arctic Turbulence Profiler (ATP), a lunar scintillometer built by University of British Columbia, and similar to an instrument at Cerro Tololo Inter-American Observatory, has been undergoing field-readiness testing on the PEARL roof. The intention is to re-deploy it at a remote Ellesmere mountain site. Balloon-borne thermosondes are also being considered for Eureka.

• Small telescopes. A Differential Image Monitoring Measurement (DIMM), a clone of the unit at the Canada-France-Hawaii Telescope, was built by University of Toronto and utilised on a portable 0.35m telescope. That, combined with Multi-Aperture Seeing Sensor (DIMM/MASS) has also been deployed in 2012, with further observations planned. This might be improved to include a variant of Single-Star Scintillation Detection and Ranging (SCIDAR).

3.3.2. *Initial Results to be Followed Up*

Preliminary Arctic results from these studies, either presented in the literature or known to the ASTWG, are encouraging:

• Remote Ellesmere Island mountain sites are accessible by helicopter in summer and above the thermal inversion in winter, providing reliable logistics and good weather (Steinbring *et al.* 2010). Skies are clear 65% of the time or better, as expected from satellite measurements. All-sky camera measurements indicate that photometric conditions occur for a significant fraction of that time, typical of the best mid-latitude sites such as Mauna Kea, Hawaii (Steinbring *et al.* 2012).

• Skies are dark, as expected in the optical based on latitude, and with indications that this follows into the near infrared consistent with atmospheric conditions. The atmosphere is extremely dry, with mean precipitable water vapour below 2mm at sea level from Eureka in winter, allowing for good 225 GHz transparency at PEARL, possibly opening useful sub-mm windows (Sivanandam *et al.* 2012, Asada *et al.* 2012).

• The boundary layer appears to be thin at PEARL, according to ATP measurements. For photometric conditions, the median boundary-layer seeing in winter below 10m is better than 0.55″, improving to 0.45″ at 20m. The first wintertime data taken with DIMM are consistent with those results; seeing of 0.6″ to 1.2″ under clear skies from the 6m high PEARL roof.

• Some Polaris-tracking seeing measurements taken at Summit in summer show seeing as good as 0.5″ at 39m, suggesting conditions comparable to the Antarctic icecap (Andersen, Pedersen & Sorensen 2010).

3.4. *Future Steps*

Clearly, more site testing is needed to fully realise the potential of Arctic sites. Within Canada this features as a priority among small projects within the 2010 Long Range Plan for Canadian Astronomy, which recommends that the conditions at the PEARL site be characterised, as well as at least one of the remote mountain sites. If justified by the site conditions, this would lead to the development and construction of a 1 to 4m class optical/NIR telescope. That recommendation does not provide funding, although incentive could come with published results. Costs for Arctic logistics are relatively inexpensive, considering the remoteness of the locations. For example, transport from northern

Canadian cities to Eureka is only approximately \$20 per kilogram by air, and considerably less via the yearly sea voyage. The AST seeks to encourage the follow-up of initial results, along with the other items highlighted below:

- Carrying on with current instrumentation. Summit station can easily support field-readied autonomous instruments such as the ASIAA radiometer. Winter campaigns at Eureka take advantage of year-round logistics and existing atmospheric research programs underway. Coordination with groups there and elsewhere is sought, e.g., other site-testing groups or those in atmospheric or space physics research such as aurora observations and orbital-debris tracking.
- Improvement in site-testing apparatus. Advancements in field-readying instrumentation for polar environments are always on-going, with some special aspects for the High Arctic, especially survivability under conditions of rime-icing and in high winds. Installation of a stand-alone 6m tall site-testing tower near the PEARL facility is planned. This can reduce the deleterious effects of heat and turbulence associated with the PEARL building itself, and so provide more reliable seeing measurements.
- Further study of the high mountain sites on Ellesmere Island. That could include re-deployment of the ATP, although this would require a reliable source of power. Both wind-power and fuel-cell generators were investigated but not fully developed during the initial feasibility study. One possibility would be a PLATO-like system, similar to the compact unit built for Ridge A.
- Initiation of scientific productivity. A 0.5m wide-field optical imaging telescope, primarily for planet-hunting via transits - is being readied for PEARL as a collaboration between the Dunlap Institute, NRC/HIA, and other Canadian government departments. It takes advantage of the already excellent infrastructure in place at Eureka, and is not strongly dependent on the local seeing conditions near PEARL. It is to act as a pathfinder to future, larger instrumentation yet to be designed. Engineering studies for a 2m class telescope comparable to those for the Antarctic Domes are to follow. The deployment of a sub-mm antenna for Very-Long Baseline Interferometry (VLBI) from Summit is being investigated.

SCAR can facilitate these advancements in a number of ways, all of which fall largely within the activities recommended under the 'Coordination' section of Working Group C (see §4.7). Some key interactions of particular value for the Arctic are as follows:

- Provide a forum for increasing awareness among the Antarctic and broader scientific community of Arctic site-testing activities, the currently known conditions, and efforts to characterise those for scientific utility.
- Encourage interaction among users of the different polar research facilities, potentially leading to improved use of sites otherwise not considered. For example, the reversed seasons of the poles may provide opportunities for more rapid deployment or field-readying of instruments. Of particular note are the year-round logistics available at Eureka.
- Providing opportunities for the international astronomical site-testing community to meet and discuss methods and results. This might be in the form of sponsoring conferences and workshops, as well as encouraging or developing other web-based formats. This can help ensure complementarity of data, sharing of engineering or scientific resources, and possibly reveal new synergies between programs at various sites.

4. Working Group C: Science Goals

4.1. *Scope and Aims*

The Science Goals Working Group (SG) is one of four groups reporting to the Scientific Committee on Antarctic Research (SCAR) Astronomy & Astrophysics from Antarctica (AAA) Scientific Research Programme. This document defines the implementation plan for the SG. The main objective is to define the kinds of astrophysical observations and experiments that are best conducted from Antarctica, to outline some of the science investigations such observations would facilitate, and to suggest ways by which these may be achieved.

4.2. *Background*

Extensive site testing activities undertaken by the international science community in Antarctica over the past two decades have established that the continent provides a unique environment that is favourable for many kinds of astronomical research programs. This is on account of the cold, dry and stable air found above the high Antarctic plateau, as well as the pure ice below. The opportunities for astronomy are found across both the photon and particle spectrum. The summits of the Antarctic plateau provide the best seeing conditions, the darkest skies and the most transparent atmosphere of any earth-based observing site. The astronomy that has been conducted from Antarctica so far includes optical, infrared, terahertz and sub-millimetre astronomy, measurements of the cosmic microwave background radiation anisotropies, solar astronomy, as well as high-energy astrophysics involving the measurement of cosmic rays, gamma rays and neutrinos. Antarctica is also a rich source for meteorites.

Many sources have been drawn upon for the material presented here, in particular the reviews of Burton (2004), Storey (2005), Storey (2009) and Burton (2010), as well as the report from the "Astrophysics from the South Pole" workshop held in Washington DC in April 2011. Several books have also been devoted entirely to astronomy in Antarctica, the proceedings of conferences held on the subject. These provide sources for further reading on the science ideas presented here, as well as on many others. The books include the proceedings of the American Institute of Physics conference on Astronomy in Antarctica in Newark, USA in 1989 (Mullan, Pomerantz & Stanev 1990), the Astronomical Society of the Pacific symposium on Astronomy in Antarctica in Chicago, USA in 1997 (Novak & Landsberg 1998), the Concordia station workshop in Capri, Italy in 2003 (Fossat & Candidi 2003) and the three European ARENA conferences, held in Roscoff, France in 2006 (Epchtein & Candidi 2007), Potsdam, Germany in 2007 (Zinnecker, Epchtein & Rauer 2008) and Frascati, Italy in 2009 (Spinoglio & Epchtein 2010).

4.3. *Site Conditions for Astronomical Observations in Antarctica*

As a result of the site testing campaigns conducted in Antarctica (e.g. see the Working Group A implementation plan) we know that the following facets of the environment on the Antarctic plateau provide exceptional conditions for many types of astronomical observation:

- The darkest infrared sky background of any ground-based location due to being the coldest place on the Earth's surface, the flux dropping by factors between 20 and 100 in the thermal infrared (from 2.2–30μm) from good temperate sites.
- The most transparent skies of any ground-based site, from 3μm to 3mm, due to the lowest precipitable water vapour columns, 3–5 times lower than at the best temperate sites.

- Clear skies, with photometric conditions occurring an exceptional 75–90% of the time at some high plateau locations.
- The high stability of the sky background, in particular when above the thin, turbulent surface boundary layer, important for accurate sky subtraction across infrared to millimetre wavelengths.
- Exceptional seeing conditions above a narrow surface boundary layer, typically twice as good as from excellent temperate sites and sometimes falling below 0.1″ in the visual. In addition, there are larger isoplanatic angles and longer coherence times than found at good temperate latitude sites (typically 2–3 times as good).
- Long interrupted periods of darkness during winter months.
- Reduced scintillation noise, 3–4 times lower than at temperate sites.
- Aerosol levels up to 50 times lower than at temperate sites, together with minimal pollution.
- The circumpolar vortex, allowing high-altitude balloons to circulate the continent in summer with near-constant elevation.
- Vast quantities of pure, clear ice approximately 3 km deep, suitable for as both a target for high energy particles and a conduit for observing the resultant Cherenkov radiation from the particle interactions in the ice.
- No aircraft contrails in winter.
- Antarctica is the quietest continent, seismically.

4.4. *Astronomical Techniques Advantaged by Antarctica*

The site conditions outlined above provide advantages for many kinds of astronomical observations. Here we list techniques that would benefit from them, and the nature of the gain achievable over good temperate-latitude sites.

- Optical: improved image quality.
- Infrared: low sky & telescope backgrounds, together with improved sky stability, in addition to better image quality.
- Terahertz: windows opened for regular observation.
- Sub-millimetre: improved sky transmission and improved sky stability.
- Millimetre: improved sky stability.
- Time series astronomy: high duty cycle measurements, with stable conditions and long, uninterrupted periods of darkness.
- Precision photometry: low scintillation noise.
- Infrared interferometry and adaptive optics: improved values of coherence time, isoplanatic angle and Fried parameter, together with the temperature stability for delay lines.
- Cosmic rays: low energy threshold due to the proximity to geomagnetic pole.
- Neutrinos: use of vast quantities of pure ice as an absorber as well as for the deployment of detectors to observe the radiation tracks produced when high energy neutrinos interact with nuclei in the ice.
- Balloons: long duration flights in a constant environment, due to the Antarctic circumpolar vortex.
- Solar: excellent daytime seeing, low coronal sky brightness and high duty cycle for measurements during summer months.
- Paleo-astronomy: fossil signatures of externally induced atmospheric disturbances embedded within ice columns.

We highlight two specific examples further here:

- In the field of infrared astronomy Antarctica can facilitate big science with small telescopes. A telescope on the plateau, with an aperture four times smaller in diameter to a good temperate-latitude telescope, will have similar sensitivity for many applications.

• In the sub-millimetre and THz bands, Antarctica facilitates big dishes and big interferometers. The vast expanse, low winds and stable conditions of the Antarctic plateau places less restrictions on the scope that possible facilities may have than when placed at high-altitude temperate sites.

4.5. *Astronomical Science Programs Conducted in Antarctica*

Astronomy programs have been conducted in Antarctica since the IGY, 50 years ago. We briefly summarise some of these programs here. This list provides a practical guide to the scope of activities it is possible to accomplish in this field in Antarctica.

• Meteorites: most productive place to search for meteorites on account of their relative ease of identification (Nagata *et al.* 1975), including discovery of meteorites from Mars (McKay *et al.* 1996).

• Helioseismology, through extended observations of solar oscillations (e.g. Grec, Fossat & Pomerantz 1980).

• Cosmic microwave background anisotropies: determination of the flat Universe (BOOMERanG; de Bernardis *et al.* 2000), first detection of CMBR polarization (DASI; Kovac *et al.* 2002), high angular scale anisotropy (ACBAR; Kuo *et al.* 2004), Sunyaev-Zeldovich effect in galaxy clusters (SPT; Staniszewski *et al.* 2009).

• Sub-millimetre & terahertz astronomy: first detection of [CI] in the Magellanic Clouds (AST/RO; Stark *et al.* 1997), mapping of [CI] and high-J CO across the warm gas of the Central Molecular Zone (Martin *et al.* 2004), first ground-based spectrum of the THz [NII] line (Oberst *et al.* 2006), measurement of large-scale toroidal magnetic fields through dichroic polarization (Viper; Novak *et al.* 2003), determination of cosmic infrared background contributions (BLAST; Devlin *et al.* 2009), mapping of the cold dust cores across the Vela Molecular Ridge (Netterfield *et al.* 2009).

• Infrared astronomy: mapping of PAHs emission across star forming complexes (SPIREX; NGC6334 – Burton *et al.* 2000, Carina – Brooks *et al.* 2000), IR–excesses from disk around young stars (Chamaeleon I – Kenyon & Gomez 2001, 30 Doradus – Maercker & Burton 2005, NGC3536 – Maercker, Burton & Wright 2006).

• High-energy astrophysics.
  ○ For cosmic rays, the spiral nature of the solar magnetic field (McCracken 1962).
  ○ For gamma rays, search for point sources using the SPASE and SPASE–2 air shower arrays (Smith *et al.* 1989, Dickinson *et al.* 2000).
  ○ The detection of high energy neutrinos with AMANDA (Andrés *et al.* 2001), AMANDA–II (Ackermann *et al.* 2005) and IceCube (Ahrens *et al.* 2004 and see references under Abbasi *et al.* 2011b, Abbasi *et al.* 2011c); in the TeV to EeV energy range IceCube is now able to detect $\sim 50,000$ atmospheric neutrinos, 1 billon air showers and 100 billion atmospheric muons per year. IceCube performs searches for cosmic neutrinos at a wide energy range as well as neutrino physics, cosmic rays physics and anisotropy searches.
  ○ Search for the Askaryan effect (ultra-high energy neutrinos) with RICE (Kravchenko *et al.* 2008) and, from balloons, with ANITA (Gorham *et al.* 2010) and with the Askaryan Radio Array (Hanson *et al.* 2012).
  ○ In addition, an ice core timeline has been found to show changes in atmospheric composition possibly attributable to ionization events caused by gamma rays from historical supernovae (Motizuki *et al.* 2010).

4.6. *Science Opportunities for Antarctic Astronomy*

Several extensive science cases have been published encompassing observatory-style science that could be advantageously carried out with Antarctic telescopes, in particular across the optical to sub-millimetre wavelength domain (e.g. Burton *et al.* 1994, Burton *et al.* 2005, Lawrence *et al.* 2009a, 2009b, 2009c, Epchtein 2010). There have also been many descriptions given for specific science investigations (e.g. as found in the three ARENA conference proceedings). Identifying any particular science case as the feature for a science program is, of course, fraught with difficulty, as ideas constantly evolve as the nature of scientific enquiry does. However, we may indicate some of the programs that have received wide interest:

4.6.1. *Optical and Infrared Astronomy*

First Light in the Universe

• A near infrared search for pair-instability supernovae at redshift 6 and beyond. Here the wide-field, high sensitivity capabilities of an Antarctic telescope would provide a powerful synergy for the James Webb Space Telescope (JWST). It would probe a different region of the pair-instability supernovae parameter space, i.e. allowing detection of these theoretically predicted objects if they are either faint and common (JWST), or bright and rare (Antarctica).

• Identifying gamma-ray bursters at high redshifts. Due to their high intrinsic luminosity, gamma-ray bursts are a powerful probe of the Universe at a range of cosmological distances. By following-up in the near infrared every high energy satellite gamma-ray burst alert in the observable sky, an Antarctic telescope would be expected to find a number of objects in the redshift range z=6–10, if they exist. Wide field infrared surveys would also probe the rate of occurrence of high redshift gamma-ray burst orphan afterglows to lower number densities than possible with any other facility.

• The 1.25 to 2.5μm range is traversed by the Lyman break at redshifts 12 to 26, allowing identification of sources in the early phase of the epoch of reionization. Light signatures of the development of supermassive black holes and the formation of globular clusters are to likely to be found during this crucial period. Deep surveys of significant regions of sky will be required. If JWST is restored and flies, objects discovered in these surveys will have their spectra analysed by JWST. Otherwise the ELTs will tackle this task. Either way, the large fields of infrared Antarctic telescopes will be of tremendous value.

• The 2 to 4μm range corresponds to the rest-frame optical range for galaxies at $2 < z < 5$. An Antarctic telescope would be well suited to extend the determination of the physical parameters, largely used today in the local universe, to the high-redshift universe. The H–Lyα line is one of the main tracers of star formation at all redshifts. It lies in the K (2μm) band for $z = 2$ to 3, the peak for star bursts in the evolution of galaxies. In the L band (3–4μm) band the $z = 4$ to 5 range can be explored. These two domains are crucial to understanding of the star formation history of the universe. The low sky thermal emission and the wide field of view would also allow a wide-area 2.4μm K_{dark} survey to find more distant galaxies at lower limiting masses than possible with other facilities. Additionally, the combination of deep K_{dark} survey data with longer wavelength Spitzer "warm mission" observations, would allow the identification of significant numbers of Balmer break galaxies in the redshift range $z = 5$–7. These evolved galaxies represent the very earliest stellar populations to form in the Universe.

The Equation of State of the Universe and the Dark Universe (Dark Energy and Dark Matter)

• Dusty supernovae: with a wide field-of-view combined with a high infrared sensitivity and wide wavelength coverage an Antarctic telescope should be able to detect and obtain infrared light-curves for larger numbers of Type Ia supernovae at higher redshift than possible with any other current facility. Infrared supernova data, with reduced effects of dust extinction and reddening, also show less dispersion in SN1a peak magnitudes, and allow tighter constraints to be placed on the parameters of the expansion of the Universe.

• Weak lensing imaging at $2.4\mu m$ in Antarctica offers a unique capability to image distant galaxies with high sensitivity in this band, over wide fields with an angular resolution comparable to their size due to the excellent seeing. Obvious applications are the more precise determination of independent masses for the oldest clusters found by the South Pole Telescope – a crucial diagnostic of structure formation – or weak lensing studies at larger depths, or on smaller scales, than will be possible with, e.g., LSST or PANSTARRS, to distinguish between cosmology variants. Similarly this capability also benefits measurements of baryonic acoustic oscillations, whose constant size provides a standard cosmological ruler, to determine the angular diameter distance relationship with redshift.

• Optical weak lensing: the image quality over wide fields allows lensing studies to a depth not otherwise achievable from the ground in the *riz* bands. Uniquely, Antarctic telescopes can take full advantage of orthogonal transfer arrays, allowing fast-guiding over arbitrarily large fields. The aim would again be to push lensing studies to redshifts not accessible to LSST or PANSTARRS.

• The cosmic web. The high-redshift intergalactic medium could be studied over large fields of view through observations of resonant UV lines, in particular from $Ly\alpha$ at 121.6nm at redshifts of $z \sim 2$–3, in contrast to the pencil-beam sight lines currently provided through use of quasar absorption lines. Such observations would provide direct probes of large scale structure and the dark matter distribution in the early Universe. They are facilitated by continuous access to a dark and stable sky, with low extinction and scattering, to enable accurate spectral cancellation of the sky background emission, but does not need high spatial resolution. Such conditions are found, for instance, at the South Pole.

Stellar Populations

• Stellar populations: the combination of high sensitivity, wide field of view, and good spatial resolution would enable an Antarctic optical/infrared telescope to conduct a range of studies of stellar populations in nearby galaxies. These range from detailed studies of the Magellanic clouds in new wavelength bands, and to studies of the high surface brightness inner parts of disc galaxies out to the edge of the Local Group. Multiple epochs of observation could be conducted for the Magellanic Clouds, so allowing light curves to be determined.

• Asteroseismology: the good image quality over wide fields, the low scintillation noise, and the long-time duration observations possible, allows for precision milli-magnitude optical photometry of entire star clusters simultaneously. This means that asteroseismology can be undertaken en masse for stellar populations of uniform age and metallicity in order to probe stellar activity.

The Galaxy and Galactic Ecology

• Mid-infrared spectrophotometric surveys searching for signatures of embedded protostars, crystalline silicates, and circumstellar disks around young stellar objects and brown dwarfs.

• The site conditions on the summits of the Antarctic plateau makes possible sensitive measurements in the mid-infrared of two of the lowest energy lines of molecular hydrogen, emitted at 12 and $17\mu m$, which could not be undertaken from temperate sites. This makes it possible to directly image the warm environment of molecular clouds, on the arcsecond scale, over wide regions of the Galactic plane. These lines arise from the surface layers of molecular clouds, warmed by the interstellar radiation field or in clouds heated by turbulence. They comprise much of the molecular environment of the Galaxy, outside the cold, dense cores where star formation is initiated. The turbulent injection of energy and its dissipation is suspected of being the primary agent responsible for the initiation and regulation of star formation through molecular clouds, and so is a central facet affecting the Galactic ecology of the star-gas cycle. The turbulence distribution and strength could be directly probed through measurement of these H_2 lines over wide angular scales.

Exo-planets

• Exoplanets by the transit technique. The observation of known transits, selecting the brightest targets, is the only way to characterise the planetary atmospheres, through their composition (molecules, clouds, hazes) and temperature profiles, by recording the stellar spectrum at low resolution during a transit. From the difference between two spectra, one obtained with the planet in front of its host star (i.e. the primary transit) and the other of the planet + star, one can derive the transmission spectrum of the planet. Observations in the 2 to $5\mu m$ range, where the emission is dominated by common molecules such as H_2O, CH_4 and CO_2, are recommended. With a measurement of the secondary transit (planet behind its star) as well, an emission spectrum of the planetary day-side can also be obtained.

• Exoplanets by the micro-lensing technique. This technique is powerful method for providing statistical results on exoplanets in the parameter space of [mass, distance to star] that are not accessible by other methods, making possible the detection of Earth-mass planets from one to a few AUs from their host star. For the best chance of detecting the perturbation of the light curve due to a planet, continuous monitoring is necessary during the few days around the light maximum. Photometry undertaken in the $2\mu m$ K band, rather than in the visible, significantly reduces the extinction toward the Galactic Bulge, which, in addition, is rich in old star populations having their maximum flux in this band. The two Magellanic Clouds are also favourable targets for such a program.

• The inner parts of debris disks and exo-zodiacal dust clouds could be examined in a range of stellar systems using an infrared nulling interferometer. Such physical features are of considerable importance in characterising the nature of planetary systems, and more-over such warm zodiacal dust around nearby stars may hinder future space missions aimed at the direct detection and study of Earth-like planets.

• Direct detection of exo-planets. The highest spatial resolution can be obtained through infrared interferometry, also providing spatial filtering in high contrast imaging, necessary for the direct detection and characterisation of planets orbiting around other stars. Optimally, such an instrument would be placed in space, but the extreme cost of such a mission likely precludes it in the near future. Infrared interferometric observations are suited from the Antarctic plateau due to the combination of slow, low-altitude turbulence, low water vapour content and low temperature, as well as the completely temperature stable sub-ice environment for housing the delay lines.

4.6.2. *Terahertz and Sub-millimetre Astronomy*

The Formation of Molecular Clouds

Molecular clouds can be formed from large structures of atomic gas that envelope spiral arms of galaxies. Observations made in other galaxies do not show us how this occurs as they lack the resolution needed to resolve the processes at work; the structures must be studied within our own Galaxy. They must include the diagnostic emission features that trace the three main types of clouds: atomic clouds (where H is dominant), clouds with only hydrogen in molecular form (known as "dark" clouds for they have little or no CO emission), and fully molecular clouds (i.e. with CO and other molecules). The problem is a challenging one, for there are no direct emission signatures from the newly-formed hydrogen molecules that mark where molecular clouds manifest themselves – hydrogen molecules will all be in their ground state unless the gas is heated to temperatures > 100 K. The process of molecular cloud formation thus needs to be inferred from trace species, by measurement of C^+, N^+, C, H and CO lines, arising from ionized, atomic and molecular gas. Thus THz observations of the [CII] 158μm and [NII] 205μm lines, combined with sub-mm and mm observation of [CI] (371, 610μm) and CO (e.g. 652μm, 870μm, 1.3mm, 2.6mm) and existing HI (21cm) data, are needed.

The Origins of Stellar Mass

Stars arise from molecular cores, and their mass must be related to the physical environment existing in these natal cores and the size of their gas reservoirs. The pre-stellar and proto-stellar core mass functions are therefore important quantities to measure in order to determine the physical basis of the initial mass function (IMF). Core mass and luminosity, together with density and temperature profiles, can be determined through measurement of a core's spectral energy distribution. For the coldest cores this will peak in the THz and sub-mm bands; for instance a 15 K dust core will emit most strongly at $\sim$ 1.5 THz (200μm).

The Galactic Star Formation Rate: calibrating the "Schmidt" law

Star formation in external galaxies has commonly been estimated using the "Schmidt" law, which relates the star formation rate to the gas surface density, with a threshold surface density below which star formation is suppressed. This law has been used with great success despite its simplicity as a power law relationship, being based on just empirical data, for it is not possible to resolve the individual star forming regions in other galaxies. However, it does not explain why the rate of star formation has such a correlation. THz observations, together with sub-mm and mm supporting data, will make it possible to directly examine the relationship, and so allow this question to be addressed. This is possible because they permit calibration of the law by measuring the two relevant quantities, the star formation rate and the gas column density. The N^+ line intensity is directly proportional to the flux of ionizing photons, which in turn arises mostly from short-lived massive stars, so providing a measure of the star formation rate. To find the gas surface density requires measurement of the atomic and molecular components of the gas over the same region, and is given by combining data from four species (C^+, C, CO & H). Taken together, these provide a determination of the amount of material existing along a sight line.

The ISM of the Magellanic Clouds

The LMC and SMC are the nearest galaxies to us, and provide laboratories where galactic-scale processes can be readily studied. As low-metallicity systems they are also used as templates to compare with star formation in our Galaxy and so for conjecture how star formation may occur in the early universe. The sub-mm and mm molecular lines have proved to be hard to measure in the LMC/SMC, with only the CO J=1–0 line readily mappable in reasonable integration times. The strength of the two bright THz lines, [NII] and [CII], will therefore be particularly valuable for understanding the molecular environment and the star formation occurring within the Magellanic Clouds. The N^+ line intensity can be used to provide a measure of the star formation rate across each Galaxy, and the C^+ line is the dominant cooling line of their ISMs.

Templates for Understanding High Redshift Extra-galactic Emission

As principal cooling lines in the interstellar medium, C^+ and N^+ emission will provide primary diagnostic tools for studying the most distant galaxies with ALMA, where the lines will be red-shifted into the sub-millimetre bands. However such galaxies will be spatially unresolved, with only global properties of their emission measurable. Observations in our own Galaxy which encompass a wide range of physical conditions are therefore essential in providing a template for use in interpreting the emission from galaxies in the early universe.

4.6.3. *Cosmic Microwave Background Radiation (CMBR)*

The exceedingly good stability of the sky in the millimetre-bands, together with the possibility of extended observations of a section of the celestial sphere at near constant zenith angle, have allowed several frontier investigations into the CMBR, as outlined above. While this has been largely led by investigations conducted at the South Pole, high plateau sites away from the Pole may offer an advantage in not having exactly the same zenith angle, allowing for cross-linked scans and drift removal techniques to further improve the background subtraction and removal of instrumental artefacts.

There are two areas of CMBR research with clear potential for undertaking new science from Antarctica: small-scale temperature anisotropies and the B–mode polarization power spectrum.

Small-scale Temperature Anisotropies

Clusters of distant galaxies can be readily detected through the Sunyaev-Zel'dovich (SZ) effect, measuring across the peak of the CMBR spectrum (which occurs around 1mm wavelength) at high angular resolution, as has been demonstrated through the success of the 10m South Pole Telescope (SPT). The brightness of the SZ–effect is independent of redshift, so that high-redshift clusters are equally easy to find as those nearby, providing the measurements have sufficient spatial resolution. The magnitude of the SZ-effect depends on the thermal energy of the cluster, and so of its mass, a parameter that cosmological models can predict. Combined with subsequent red-shift determinations, the mass distribution of large-scale structures can then be determined. This provides completely independent constraints on the nature of the Dark Energy that is presumed to be driving the accelerating Universe than the direct measurements of cosmological expansion inferred from observations of distant supernovae.

B–mode Polarization

Polarization measurements of the CMBR provide a means to understand dark matter, dark energy, neutrino mass and the epoch of Inflation. Density perturbations at the epoch of Recombination ($z \sim 1000$) result in "E–mode" polarization of the CMBR, as was first measured by the DASI experiment at the South Pole (Kovac *et al.* 2002). Their signal is about 1% the level of the CMBR itself. Additional signals in the CMBR can be generated by gravitational waves. Unlike the density fluctuations, these need not obey parity, and therefore can generate what are known as "B–mode" polarization, a further $10^2 - 10^3$ times weaker again then the E–mode signal. There are two likely contributors to any B–mode polarization. The first are gravity waves generated during the epoch of Inflation, occurring at, or near, the GUT (Grand Unified Theory) energy scale. They might generate signals on angular scales of 1 degree. A second, stronger means of producing B–modes is through gravitational lensing of the CMBR signal on its passage to us, which can convert some of the E–mode polarization into a B–mode signal. This is expected to peak on angular scales of a few arcminutes. The lensing signal depends on the dark energy equation of state, the spatial curvature of the Universe, and the sum of the neutrino masses, so any measurement of it provides constraints on these parameters.

4.6.4. *High Energy Astrophysics: Neutrinos and Cosmic Rays*

An intriguing questions in physics is the origin and evolution of the cosmic accelerators that produce the highest energy particles known: ultra-high energy cosmic rays

(UHECR) with energies up to $\sim 10^{20}$ eV for a single particle. These can "only" travel for ~ 100 Mpc from their origin before interacting with a microwave background photon, in the process creating an UHE ("cosmogenic") neutrino. Cosmic rays at lower energies will be deflected by intergalactic magnetic fields and high energy gamma rays will get absorbed by interaction with lower energy photons. The neutrino can travel virtually unimpeded throughout the Universe, and so provides a messenger from the creation of those high energy particles in nature, if they can be detected. Detection can occur as a result of their interactions with nuclei in large volumes of a dense dielectric, and the subsequent particle cascades that are produced. The Antarctic icecap provides the most suitable such medium, as its clarity also allows the resultant Cherenkov radiation pulse to be detected and traced back along the direction of the incoming neutrino, and so towards its source. This has been exploited by the construction of the IceCube neutrino telescope, sensitive to neutrinos in the 10^{11-18} eV energy range detectable via their optical Cherenkov light. At higher energy radio frequency emission dominates, and this has driven the construction of the RICE and ANITA experiments seeking to measure the radio pulse caused by the resulting Askaryan effect. More recently the construction of the first phase of the Askaryan Radio Array has begun at the South Pole.

Astrophysical Sources of Neutrinos

A galactic or extra-galactic source of neutrinos would be manifested with IceCube as a statistically significant clustering of high energy neutrinos on the sky, a small excess on the background created by atmospheric muons and neutrinos (Abbasi *et al.* 2011a). At energies above one PeV there is essentially no background from atmospheric neutrinos. At lower energies the excess would be defined as a clustering in direction and or time. No such excess has yet been detected, though IceCube was only fully-completed in 2011.

Gamma-ray bursts (GRBs) may also be sources of cosmic rays and therefore of high-energy neutrinos. Coincidence timing over the few tens of seconds when the gamma rays are seen (by space-based telescopes) may establish the connection, as well the source if the GRBs are also optically identified. A neutrino detection would provide evidence for proton acceleration in the progenitors, and so of the processes that fuel them.

IceCube is also highly sensitive to the neutrino burst from the core collapse in a supernova explosion in our galaxy. With 99% on-time operation it is prepared to record such an event with high precision.

Cosmic Rays

A surface detector array called IceTop is integral part of IceCube making it a versatile cosmic ray laboratory. The detailed measurements of cosmic rays will allow a very precise measurement of the energy spectrum as well as the composition of cosmic rays from PeV to EeV energies. The flux of about 100 billion muons per year can be used to measure for the first time the anisotropy of cosmic rays in the Southern sky to a precision of 10^{-4} on small and large scales (Abbasi *et al.* 2011d).

Extension to Lower Energy Neutrinos

IceCube may be extended to lower energy neutrinos, in the GeV range, using a "close-packed" array of photomultiplier strings in the centre of the array. 6 such strings have been included as part of IceCube itself ("Deep Core"), which lowers the energy threshold to ~ 10 GeV. A further ~ 20 strings would lower the threshold to ~ 1 GeV. At such energies neutrino measurements would be sensitive to indirect searches for dark matter and to precision measurements of atmospheric neutrino oscillations. The lower energy threshold also allows for a dramatically increased sensitivity in the indirect search for dark matter. Dark matter annihilation e.g. in the Sun allows IceCube to explore a large parameter space of dark matter models by looking for an excess of neutrinos from the sun at energies above 10 GeV. There are visions for much denser instrumentation which could allow the measurements of the proton decay.

Dark Matter

Dark matter might be detected directly through scintillation occurring in pure crystals where a dark matter interaction has occurred. The recoiling nuclei in a pure crystal of sodium iodide can emit photons which are detectable with photomultiplier tubes buried in the ice. An annual modulation in such a signal, the result of the Earths motion around the Sun, provides a means of extracting the dark matter signal from the background flux.

Cosmogenic Neutrinos

The interaction of the highest energy cosmic rays ($\sim 10^{20}$ eV) with a cosmic microwave background photon ($\sim 10^3$ eV) results in the production of a charged pion. This then decays to produce "cosmogenic" neutrinos that may continue, unimpeded, across the entire Universe. This is the Greisen-Zatsepin-Kuzmin (GZK) effect (Greisen 1966, Zatsepin & Kuzmin 1966). An image of the neutrinos from such a source would reveal a small halo surrounding the emitting source of the cosmic rays, whose size is determined by where they interact with CMBR photons. These neutrinos might be detected by the Askaryan effect, a shower of particles caused by the interaction of such a cosmogenic GZK neutrino with a nucleus and a subsequent charge anisotropy that develops in the particle cascade as the positrons annihilate. This results in a plug of charge, a few cm across, travelling faster than light in its medium, and producing a pulse of Cherenkov radiation, which then manifests itself observationally as a pulse of radio waves at a frequency of ~ 1 GHz.

The most promising location to employ the Askaryan Effect is within the Antarctic icecap. The ice is several km thick over the Antarctic plateau and relatively transparent to radio waves. By placing an array of radio antennae beneath the ice surface a neutrino detector can be constructed with an effective volume given by the surface area covered by antennae multiplied by the thickness of ice. Accurate timing of the radio pulses can then be used to reconstruct the direction of the incoming neutrinos to within $\sim 6°$ for an instrument about 10 km across. The strength of the radio signal gives information on the energy of the neutrinos, and provides a probe of the very highest energy interactions known in nature. IceCube is likely too small to reliably detect the small fluxes of such GZK neutrinos; a collecting volume 2–3 orders of magnitude larger is required. Such a detection experiment is proposed for the Askaryan Radio Array at the South Pole.

4.6.5. *Solar Astronomy*

High plateau sites provide unique conditions for solar astronomy. The periods of excellent daytime seeing (as observed at Dome C), combined with low coronal sky brightness (due to minimal aerosols and low water vapour), IR–accessibility and high duty cycle observations (the Sun being continuously above the horizon for several months, 1,700 hours with a Fried parameter $r_0 > 12$ cm allowing adaptive optics), offer interesting prospects for permanent high angular resolution solar studies (i.e. $<< 0.1''$) and for coronal access (cf. Damé *et al.* 2010). These conditions facilitate continuous observations at high resolution of the corona-chromosphere interface, notably for direct measurements of the magnetic field in the corona and chromosphere (prominences), as well as of waves (MHD or Alfvén) passing through them (and contributing to heating?). The magnetic investigation of the Sun, from the convection zone to the corona, could be undertaken from Antarctic solar telescopes, requiring instrumentation for 2D spectro-imaging, spectro-polarimetry, magneto-seismology and for chromospheric & coronal magnetometry.

A first mid-size facility is proposed, AFSIIC (Antarctica Facility for Solar Interferometric Imaging and Coronagraphy), using 3 off-axis telescopes of 50cm diameter each (1.4m equivalent telescope), and high resolution up to $0.06''$, allowing access these objectives with the proper flux, angular resolution and coronagraphic potential. Coronal sky conditions associated to high resolution (direct access to the free atmosphere seeing at temperature gradient inversion in the afternoon) provide unique conditions opening the possibility to test future coronal space missions like ESA ASPIICS/PROBA–3. Support infrastructure and logistics have been studied and evaluated to optimise observing conditions, noticeably a 30m tower (cf. Dournaux *et al.* 2010) to place the observatory over the very thin surface turbulent layer of Dome C, 70% of the time (i.e. almost 100% of the clear accessible time).

Another non-negligible advantage to be considered is complementarity with northern hemisphere observation sites that will suffer winter (non) observing conditions while excellent summer conditions prevail on Antarctica high plateau sites.

4.7. *Coordination*

Antarctic astronomy is not a special kind of astronomy; it is all astronomy that is facilitated by the special conditions of Antarctica. The astronomy that is conducted in Antarctica should only be that astronomy which can be done better there, whether it is by improved capability, or by its cost effectiveness compared to other locations, whether they are Earth-based or in space. For instance, transportation costs to Antarctica are of order $10 per kg of material, compared to $15,000/kg for a payload launched on a rocket to low Earth orbit (the typical cost of space hardware is also about $140,000/kg). Antarctica thus provides a much cheaper alternative to space, as well as the ability to recover from failures. Nevertheless, conducting astronomy in Antarctica does face special challenges, largely on account of the geo-political environment. Antarctica is not owned by anyone, it is international territory. It functions as an international science reserve, and activities conducted there are thus facilitated by international collaboration and co-operation, more so than from anywhere else. There is a special need for bodies like SCAR and the IAU (International Astronomical Union) to play a role in nurturing such activities and bringing together prospective partners to make the science happen.

SCAR can facilitate such interactions in a number of ways, which we highlight below:

• Providing a forum for the international community to meet, and in them showcasing the opportunities provided by Antarctica for astronomy.

• Bringing prospective partners together through awareness of the different, and complementary skills and capabilities they possess, so that these can all be used to tackle a particular project.

• Provide a forum to facilitate moving beyond national priorities, to setting objectives for international facilities, while taking account of disparate national capabilities and interests.

• Sponsoring conferences, workshops and meetings, to highlight the opportunities, and making the international science community aware of them. While SCAR cannot determine the new science that should be done in Antarctica, it can publicise the scientific techniques that Antarctica facilitates, so that individual scientists become aware of new opportunities to pursue ideas they have.

• In particular, by promoting such linkages with the IAU, SCAR may educate the astronomical science community about opportunities they may otherwise be unaware of. The IAU host a tri-annual General Assembly (similar to the bi-annual SCAR meeting), and SCAR could usefully have a presence at that Assembly. There is an IAU Working Group for Antarctic astronomy through which networks can be developed. In this vein, the IAU allocated this Symposium (IAU Symposium 288, "Astrophysics from Antarctica") for the General Assembly held in Beijing in China in August 2012. This was the first IAU Symposium to ever be held on the theme of astronomy in Antarctica.

• Through its websites, SCAR can provide information summarising the state of knowledge and practice in Antarctic astronomy. This would primarily be through giving direct linkages to other sites where that information is being actively generated.

5. Working Group D: Major New Facilities

5.1. *Scope and Aims*

This section is intended to capture the essence of discussions of this Working Group at the Taronga Zoo meeting (July 2011). Suggestions were wide-ranging and some do not lend themselves to easy categorisation. Organisation of this section attempts to group things according to four major themes.

5.2. *The International High Plateau Station*

Although national polar programs have many advantages in terms of leveraging resources from government and industry bodies, it was noted that the exclusive population of national bases also generates significant disadvantages and complications. They are often set up in an environment where political motives are competitive rather than collegial. When they become the vehicle for national prestige, the ethos of rapid and fluid international collaboration which is the lifeblood of the most successful modern scientific endeavours becomes difficult to support. This motivated the idea of some form of international station which would have some form of relatively open-door policy to national membership from a wide range of countries. This follows the trend of international astronomical observatories such as the Gemini partnership or the European Southern Observatory. Much of the same logic applies, with immediate and profound advantages gained by sharing infrastructure and creating a crucible for cross-fertilisation with scientists from many nations working together. This is surely a recipe for the construction and maintenance of healthy international collaborations.

Such an International High Plateau Station would have strong resonance with the spirit of the Antarctic Treaty itself. Perhaps most importantly of all, it would provide a vehicle for smaller countries (such as Indonesia or Denmark) with Antarctic aspirations to share the otherwise prohibitive costs of running an entire program alone. The democratisation of access to the Antarctic would surely yield long term benefits with a far wider national demographic to draw upon for new ideas in the exploration of the scientific potential of working there. It is not clear if such a station would be sited completely separately from existing infrastructure, or whether there could be a case for leveraging from existing supply and communications lines. Such an initiative already has a precedent in the ARENA program, albeit operating with a European rather than an International theatre. A further advantage of such an organisational structure is the security of commitment provided by international treaty level co-operative agreements: surely a lesson from space missions and major observatories we would do well to emulate.

5.3. *What are the major enabling technologies needed for the next steps?*

In no particular order of importance, enabling technologies required for new generation initiatives to take form on the ice were discussed.

• PLATO already forms an excellent model for providing much of the requirements for science from the High Plateau.

• Ground Layer Adaptive Optics (Inexpensive and likely to become far more common at mid-latitude observatories in coming years).

• Stable tower designs (and foundations that do not sink or differentially shift).

- Mitigating the effects of high relative humidity (but low absolute humidity).
- High bandwidth communications back to home institutions.
- Versatile devices and computing platforms which are robust and operate with low power demands (e.g. efficient Cryo-coolers).

5.4. *Are there simple strategies to adopt to help facilitate better collaborative links?*

- Again, PLATO sets an excellent precedent here in providing a "standard platform".
- Do we need a data standard/format/definitions for site testing data metrics?
- Push for portable science verification instruments which can be shared between sites.
- Is it possible to define standard hardware interfaces?

5.5. *Example science domains for major new initiatives*

For the present, this is an incomplete listing intended as a starting point for fleshing out of areas for growth of scientific capacity.

- High energy/particle astrophysics (already mature)
- CMB astronomy (already mature)
- Balloon-borne platforms (mature)
- Meteor sample programmes (mature)
- THz astronomy
  - Single dish
  - Interferometry
- IR astronomy
  - Wide-field imaging
  - Continuous cadence observing for variable/transitory phenomena
- Visible light adaptive optics (AO)
  - Hubble will likely be de-orbited someday and the demand for visible-light wide-field, high angular resolution imaging will suddenly be thrust upon ground based observatories. It is doubtful if any of the available flavours of AO at mid-latitude observatories will deliver sufficiently in this regard.
- Solar astronomy
  - Continuous cadence observing
- Optical interferometry
  - Several architectures possible: astrometric/nulling/imaging/closure phase.
  - Each has quite distinct science reach and distinct advantage from the Antarctic.

Acknowledgements

In addition to the authors of this paper there have been many contributors to the SCAR AAA SRP Working Group reports†. They include: WGA–Jon Lawrence, Geoff Sims; WGB–Ray Carlberg, Ming-Tang Chen, Paul Hickson; WGC–Hans Zinnecker, Luc Damé, Nicolas Epchtein, Yuko Motizuki, Jeremy Mould, Remko Stuik, Charling Tao; WGD–Michael Ashley, Nicolas Epchtein, Jian Ge, Gil Moretto, Carl Pennypacker, Jason Rhodes, Nick Suntzeff, Andre Tilquin, Lifan Wang, Don York and Wei Zheng.

References

Abbasi, R., Abdou, Y., Abu-Zayyad, T., *et al.* 2011a, *ApJ*, 740, 16
Abbasi, R., Abdou, Y., Abu-Zayyad, T., *et al.* 2011b, *Phys. Rev. D.*, 83, 092003

† See www.astronomy.scar.org

Abbasi, R., Abdou, Y., Abu-Zayyad, T., *et al.* 2011c, *ApJ*, 732, 18

Abbasi, R., Abdou, Y., Abu-Zayyad, T., *et al.* 2011d, *Phys. Rev. D.*, 83, 012001

Ackermann, M., Ahrens, J., Bai, X., *et al.* 2005, *Phys. Rev. D.*, 71, 077102

Ahrens, J., Bahcall, J. N., Bai, X., *et al.* 2004, *Astroparticle Physics*, 20, 507

Andersen, M. I., Pedersen, K., & Sørensen, A. N. 2010, *Highlights of Astronomy*, 15, 634

Andersen, M. I. & Rasmussen, P. K. 2006, *IAU Special Session*, 7

Andrés, E., Askebjer, P., Bai, X., *et al.* 2001, *Nature*, 410, 441

Asada, K., Martin-Cocher, P. L., Chen C.-P., Matsushita, S., Chen, M.-T., Inoue, M., Huang, Y.-D., Inoue, M., Ho, P. T. P.., Paine, S. N., & Steinbring, E., 2012, to appear in SPIE Conf. Series 8844, 'Ground-based and Airborne Telescopes IV', eds. L.M. Stepp, R. Gilmozzi, H.J. Hall

Brooks, K. J., Burton, M. G., Rathborne, J. M., Ashley, M. C. B., & Storey, J. W. V. 2000, *MNRAS*, 319, 95

Burton, M. G. 2004, *Organizations and Strategies in Astronomy*, Vol. 5, 310, 11

Burton, M. G., Ashley, M. C. B., Marks, R. D., *et al.* 2000, *ApJ*, 542, 359

Burton, M. G., Lawrence, J. S., Ashley, M. C. B., *et al.* 2005, *PASA*, 22, 199

Burton, M., Aitken, D. K., Allen, D. A., *et al.* 1994, *PASA*, 11, 127

Burton, M. G. 2010, *A&A Rev.*, 18, 417

Damé, L., Andretta, V., & Arena Solar Astrophysics Working Group Members 2010, *EAS Publications Series*, 40, 451

de Bernardis, P., Ade, P. A. R., Bock, J. J., *et al.* 2000, *Nature*, 404, 955

Devlin, M. J., Ade, P. A. R., Aretxaga, I., *et al.* 2009, *Nature*, 458, 737

Dickinson, J. E., Gill, J. R., Hart, S. P., *et al.* 2000, *Nuclear Instruments and Methods in Physics Research A*, 440, 114

Dournaux, J.-L., Amans, J.-P., Damé, L., & Le Moigne, J. 2010, *EAS Publications Series*, 40, 477

Epchtein, N. & Candidi, M. 2007, *EAS Publications Series*, 25

Epchtein, N. & Zinnecker, H. 2010, *Highlights of Astronomy*, 15, 622

Fossat, E. & Candidi, M. 2003, *Memorie della Societa Astronomica Italiana Supplementi*, 2, 3

Gorham, P. W., Allison, P., Baughman, B. M., *et al.* 2010, *Phys. Rev. D.*, 82, 022004

Grec, G., Fossat, E., & Pomerantz, M. 1980, *Nature*, 288, 541

Greisen, K. 1966, *Physical Review Letters*, 16, 748

Hanson, K., ARA Collaboration 2012, *Journal of Physics Conference Series*, 375, 052037

Hickson, P., Carlberg, R., Gagne, R., *et al.* 2010, *Proc. SPIE*, 7733,

Kenyon, S. J. & Gómez, M. 2001, *AJ*, 121, 2673

Kovac, J. M., Leitch, E. M., Pryke, C., *et al.* 2002, *Nature*, 420, 772

Kravchenko, I., Adams, J., Bean, A., , *et al.* 2008, *International Cosmic Ray Conference*, 3, 1229

Kuo, C. L., Ade, P. A. R., Bock, J. J., *et al.* 2004, *ApJ*, 600, 32

Lawrence, J. S., Ashley, M. C. B., Bailey, J., *et al.* 2009a, *PASA*, 26, 379

Lawrence, J. S., Ashley, M. C. B., Bunker, A., *et al.* 2009b, *PASA*, 26, 397

Lawrence, J. S., Ashley, M. C. B., Bailey, J., *et al.* 2009c, *PASA*, 26, 415

Lawrence, J. S., Ashley, M. C. B., Tokovinin, A., & Travouillon, T. 2004, *Nature*, 431, 278

Maercker, M. & Burton, M. G. 2005, *A&A*, 438, 663

Maercker, M., Burton, M. G., & Wright, C. M. 2006, *A&A*, 450, 253

Martin, C. L., Walsh, W. M., Xiao, K., *et al.* 2004, *ApJS*, 150, 239

McCracken, K. G. 1962, *JGR*, 67, 447

McKay, D. S., Gibson, E. K., Jr., Thomas-Keprta, K. L., *et al.* 1996, *Science*, 273, 924

Motizuki, Y., Naka, Y., & Takahashi, K. 2010, *Highlights of Astronomy*, 15, 630

Mullan, D. J., Pomerantz, M. A., & Stanev, T. 1990, *American Institute of Physics Conference Series*, 198,

Nagata, T. 1976, *Meteoritics*, 11, 181

Netterfield, C. B., Ade, P. A. R., Bock, J. J., *et al.* 2009, *ApJ*, 707, 1824

Novak, G., Chuss, D. T., Renbarger, T., *et al.* 2003, *ApJL*, 583, L83

Novak, G. & Landsberg, R. 1998, *Astrophysics From Antarctica*, 141,

Oberst, T. E., Parshley, S. C., Stacey, G. J., *et al.* 2006, *ApJL*, 652, L125

Sivanandam, S., Tekatch, A., Welch, D., Abraham, B., Graham, J., & Steinbring, E., 2012, to appear in SPIE Conf. Series 8844, 'Ground-based and Airborne Telescopes IV', eds. L.M. Stepp, R. Gilmozzi, H.J. Hall

Smith, N. J. T., Gaisser, T. K., Hillas, A. M., *et al.* 1989, Very High Energy Gamma Ray Astronomy, 55

Spinoglio, L. & Epchtein, N. 2010, *EAS Publications Series*, 40

Staniszewski, Z., Ade, P. A. R., Aird, K. A., *et al.* 2009, *ApJ*, 701, 32

Stark, A. A., Bolatto, A. D., Chamberlin, R. A., *et al.* 1997, *ApJL*, 480, L59

Steinbring, E., Carlberg, R., Croll, B., *et al.* 2010, *PASP*, 122, 1092

Steinbring, E., Leckie, B., Welle, P., *et al.* 2008, *Proc. SPIE*, 7012

Steinbring, E., Ward, W., & Drummond, J. R. 2012, *PASP*, 124, 185

Storey, J. W. V. 2005, *Antarctic Science*, 17, 555

Storey, J. W. V. 2009, *Association Asia Pacific Physical Societies Bulletin*, 19, 4

Wallace, B., Steinbring, E., Fahlman, G., *et al.* 2008, *Advanced Maui Optical and Space Surveillance Technologies Conference*,

Zatsepin, G. T. & Kuz'min, V. A. 1966, *Soviet Journal of Experimental and Theoretical Physics Letters*, 4, 78

Zinnecker, H., Epchtein, N., & Rauer, H. 2008, *EAS Publications Series*, 33

Astrophysics from Antarctica
Proceedings IAU Symposium No. 288, 2012
M. G. Burton, X. Cui & N. F. H. Tothill, eds.

© International Astronomical Union 2013
doi:10.1017/S1743921312017012

Preliminary daytime seeing monitoring at Dome A, Antarctica

Chong Pei[1,2,3,4] Zhengyangg Li[1,2,3] Hualin Chen[1,2,3] and Xiangyan Yuan[1,2]

[1] National Astronomical Observatories/Nanjing Institute of Astronomical Optics & Technology, Chinese Academy of Sciences, Nanjing 210042, China

[2] Key Laboratory of Astronomical Optics & Technology, Nanjing Institute of Astronomical Optics & Technology, Chinese Academy of Sciences, Nanjing 210042, China

[3] Graduate University of Chinese Academy of Sciences, Beijing 100049, China

[4] email: `cpei@niaot.ac.cn`

Abstract. Sites on Antarctic plateau have unique atmospheric properties that make them better than any mid-latitude sites as observatory locations. From site testing measurements over 4 years on Dome A carried out by the Chinese Center for Antarctic Astronomy, we can reasonably predict that Dome A is as good as or even better than Dome C, which has been proved to be the best astronomical site by now, and suitable for high angular resolution observations. Seeing monitoring is necessary for planning large scale ground-based optical astronomical telescopes. In 2012, the 28th Chinese Antarctic Scientific Expedition carried out preliminary daytime seeing monitoring using a Differential Image Motion Monitor (DIMM) placed at a height of 3.5m. The median seeing was found to be 0.8″. This will be the foundation of future research that obtains comprehensive and long-period monitoring of the site's optical parameters.

Keywords. Dome A, Seeing, Isoplanatic angle.

1. Theory

Fried (1966) introduced a parameter of atmospheric coherence length r_0 (Roddier (1981)), and he derived the relation of $\varepsilon_{FWHM} = 0.98\lambda/r_0$, generally expressed in units of arcseconds. A DIMM measures the variance of differential image motion which is related to wavelength, telescope diameter and Fried parameter r_0 as: $\sigma_d^2 = K\lambda^2 r_0^{-5/3} D^{-1/3}$ (Jean & Casiana (1995)), so $\varepsilon_0 = 0.98(\sigma_d^2)^{3/5}(D/\lambda)^{1/5}K^{-3/5}$. The isoplanatic angle can be estimated from the fluctuations of stellar flux received by an annular sub-aperture of this site testing instrument as $\theta_0^{-5/3} = A\sigma_I^2\cos z^{-8/3}$.

2. Instrument

A 35 cm Celestron commercial telescope tube with focal length of 3910 mm matched with a pupil mask and CCD camera was mounted on the first of three Antarctic Survey Telescopes (AST3-1) which has high pointing and tracking accuracy and could work at the very low temperature at Dome A. The special software with graphic user interface is designed for displaying real-time images, computing results and running status and long-distance control. The Dome A seeing monitoring system are formed by these hardware and software, shown in Figure 1. Three glasses with wedge angles of 50″ coated with ITO film to keep the frost and ice away from the surface in winter have been located on the mask attached to the telescope entrance pupil.

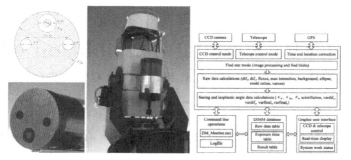

Figure 1. Left: the hardware of the Dome A seeing monitoring system. Right: organisation of the software of the Dome A seeing monitoring system.

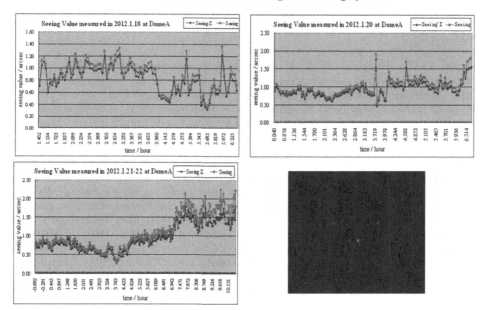

Figure 2. The results of seeing measurement and the image captured by the CCD with an exposure time of 5 millisecond.

3. Result

The successful operation of this seeing monitoring system enables us firstly to obtain the preliminary daytime seeing results for three days at Dome A by observing Canopus (-0.72 mag). The results are plotted in the following charts. In Figure 2, the upper left one is the result obtained during 01:22 to 06:20 on January 18, which shows seeing values centralised in the range of 0.87″ ∼ 0.97″ and accounts for 22%. The upper right one is the result obtained during 00:02 to 07:18 on January 20, which shows seeing values centralised in the range of 0.72″ ∼ 0.83″ and accounts for 27%. The lower left one is the result taken during 23:18 on January 21 to 10:59 on January 22, and shows seeing values centralised in the range of 0.7″ ∼ 0.8″ and accounts for 19.7%.

References

Roddier, F. 1981, *Progress in optics*, 19, 281

Jean, V. & Casiana, M. 1995, *PASP*, 107, 265

Astrophysics from Antarctica
Proceedings IAU Symposium No. 288, 2012
M. G. Burton, X. Cui & N. F. H. Tothill, eds.

© International Astronomical Union 2013
doi:10.1017/S1743921312017024

Testing the bimodal distribution of long gamma-ray bursts in the cosmological rest-frame

Christian Vásconez[1], Nicolás Vásquez[1,2] and Ericson López[1]

[1]Quito Astronomical Observatory, Escuela Politécnica Nacional,
POBox 17-01-165, Quito, Ecuador
email: christian.vasconez@epn.edu.ec

[2]Dept. of Physics, Escuela Politécnica Nacional,
Quito, Ecuador

Abstract. Among the several methods of classifying gamma-ray bursts (GRBs), the duration parameter has lead to the canonical classification of GRBs of long and shorts. However, the canonical classification of bursts has recently seen the emergence of a third type of GRB, which is present in a recent large burst sample from the Swift observatory. The high redshifts and the cosmological distances are directly confirmed for long bursts only, while for the short ones there is only indirect evidence for their cosmological origin. Cosmological objects should not only be redshifted in energy but also extended in time because of the expansion of the Universe. Meanwhile, an anticorrelation between the hardness and the duration is found for this subclass in contrast to the short and the long groups (Horvath *et al.* (2010)). Despite the differences among these three groups, it is not yet clear whether the third group represents a physically different phenomenon. In this scenario, we want to study the bimodal distribution of long bursts, focusing their temporal properties in the source location (burst frame). We have determined a temporal estimator in the cosmological rest-frame from a sample of 60 Swift's GRBs. If GRBs are at cosmological distances, then the burst profiles should be stretched in time due to cosmological time dilation by an amount proportional to the redshift, $1 + z$ (Chang (2001)). Complementary, we use the hardness ratio between the soft emission (15–50 keV) and hard X-ray emission (50–150 keV) in order to analyze the bimodal distribution of long bursts in the time-energy plane.

Keywords. gamma ray bursts, classification

1. Introduction

One of the most intriguing questions is to know if the canonical classification of bursts in the observer frame is hiding some intrinsic properties due to cosmological effects (Chang (2001)). Although there has been some progress in the burst classification into three categories (Horvath *et al.* (2010)), classification in the burst frame are still under discussion.The analysis of the emission time (t50) (Mitrofanov *et al.* (1999)) and isotropic energy reveals a subclass of long GRBs when temporal estimators are cosmologically corrected. A previous analysis revealed that different time estimators from standard T90 and T50, provided a better duration of the activity of the burst engine (t50). This temporal estimator, called emission time, correlates better with the isotropic energy and shows a bimodal distribution. A subclass of dim, low redshift and longer burst in a sample of 15 GRBs simultaneously detected by Swift and Suzaku missions was determined (Vasquez and Kawai, 2010), suggesting more than one progenitor for long GRBs. To investigate whether or not this bimodality is robust, we present a complementary analysis of an extended sample of long GRBs with known z.

2. Data and analysis

Swift bursts with known redshift are more than 200, among this data we chose a sample of long and bright GRBs, where brightness is defined as a burst with a BAT fluence at least of 30 ergs cm^{-2} during the period January 2006 to June 2012. There are 60 GRBs that held these characteristics but 8 where not analyzed because the lack of data or low signal to noise ratio. Then, 52 bursts were studied. We extracted spectra and light curves using Heasoft 6.11 and the analysis was done using XSPEC 12.7. The spectral model was a cut-off power low normalized to 50 keV. In order to look for spectral difference inside the sample we defined the hardness ratio (HR) as $\frac{(S_{15-50} - S_{50-150})}{(S_{15-50} + S_{50-150})}$, and we plot HR as a function of z in Figure 1.

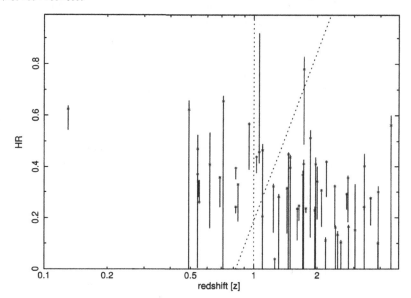

Figure 1. HR distribution as a function of z, suggested also a bimodal behavior. Hard burst associated with $z > 1$ seems to conform a different group from those GRBs with $z < 1$ that are X-ray rich.

3. Discussion

The evidence of a subclass of long GRBs is present in the redshift hardness ratio plane, which reinforce the possibility of a different engine or progenitor for GRBs which are X-ray rich with low redshift. The bimodality of long GRBs in the afterglow phase has been reported which is interpreted as possible difference of explosion mechanism between both types of long bursts (Dainotti *et al.* (2010)). In order to confirm the validity of this proposal, we are computing the emission times of the bursts in the studied sample to reproduced the bimodality in the temporal HR plane.

References

Chang H. Y. 2001, *ApJL*, 557, 85-88

Dainotti M. G., Willingale R., Capozziello S., Cardone V. F. & Ostrowski M. 2010, *ApJL*, 722, 215-219

Horváth I., Bagoly Z., Balázs L. G., de Ugarte Postigo A., Veres P. & Mészáros 2010, *ApJ*, 713, 552-557

Mitrofanov I., *et al.* 1999, *ApJ*, 522, 1069-1078

Page K. L., *et al.* 2007, *ApJ*, 663, 1125-1138

Vásquez N. & Kawai N. 2010, *ApJ*, 713, 552-557

Astrophysics from Antarctica
Proceedings IAU Symposium No. 288, 2012
M. G. Burton, X. Cui & N. F. H. Tothill, eds.

© International Astronomical Union 2013
doi:10.1017/S1743921312017036

Dome C site testing: long term statistics of integrated optical turbulence parameters at ground level

E. Aristidi, A. Agabi, E. Fossat, A. Ziad, L. Abe, E. Bondoux, G. Bouchez, Z. Challita, F. Jeanneaux, D. Mékarnia, D. Petermann and C. Pouzenc

Laboratoire Lagrange, Université de Nice, Parc Valrose, F-06108 Nice Cedex 2

Abstract. We present long term site testing statistics based on DIMM and GSM data obtained at Dome C, Antarctica. These data have been collected on the bright star Canopus since the end of 2003. We give values of the integrated turbulence parameters in the visible (wavelength 500 nm). The median value we obtained for the seeing are 1.2 arcsec, 2.0 arcsec and 0.8 arcsec at respective elevations of 8m, 3m and 20m above the ground. The isoplanatic angle median value is 4.0 arcsec and the median outer scale is 7.5m. We found that both the seeing and the isoplanatic angle exhibit a strong dependence with the season (the seeing is larger in winter while the isoplanatic angle is smaller).

Keywords. site testing, atmospheric effects, optical turbulence

1. Introduction

The AstroConcordia program has been, up to now, dedicated to the qualification of the site of Dome C, Antarctica for astronomical purposes. After almost 10 years of operation since the first results (Aristidi *et al.* 2003) we could measure long terms characteristics of the turbulent atmosphere. In this contribution we present the results of a decade of turbulence monitoring. We give statistics of the seeing, the isoplanatic angle, and the outer scale during the polar winter and summer. This paper is indeed an update of previously published work: Aristidi *et al.* (2009) and Ziad *et al.* (2008).

2. Results

All results presented here come from telescope-based instruments which observe a bright star in the visible and perform continuous measurements of the seeing, the isoplanatic angle and the outer scale. These instruments named DIMM, Thetameter and GSM are described in previous papers (Aristidi *et al.* 2005, 2009, Ziad *et al.* 2008). Several DIMM were placed at different elevations to sample the seeing inside the turbulent surface layer. One was at ground level (elevation h≃3m) as a part of GSM, one at h≃8m (this DIMM is taken as the standard one), and a DIMM was placed over the top of the calm building of Concordia at h≃20m in 2005 and 2012.

Statistics for the three parameters are given in Table 1. A big difference is observed between the summer and the winter for the seeing and the isoplanatic angle. The latter is sensitive to high altitude turbulent layers and decrease in winter as the polar vortex drives strong winds in the upper atmosphere.

The seeing is dominated by a strongly turbulent surface layer (SL) whose height is around 30 m. This layer is partly caused by a huge thermal gradient near the ground in

	Winter (Apr – Sept)	Summer (Dec – Jan)	Total
Seeing $h = 3m$ [arcsec]	2.4 [1.8 – 3.2]	1.0 [0.7 – 1.3]	2.0 [1.4 – 2.8]
Seeing $h = 8m$ [arcsec]	1.7 [1.0 – 2.4]	0.7 [0.4 – 0.8]	1.2 [0.6 – 1.9]
Seeing $h = 20m$ [arcsec]	0.8 [0.4 – 1.5]	-	0.8 [0.4 – 1.5]
Isop. angle [arcsec]	3.2 [2.3 – 4.4]	6.4 [4.5 – 8.5]	4.0 [2.7 – 5.8]
Outer scale [m]	-	-	7.5 [5.1 – 11.3]

Table 1. Statistics of the seeing at 3 elevations, the isoplanatic angle and the outer scale (Von-Kàrman model). Median value is given as well as the 1st and 3rd quartiles (between []).

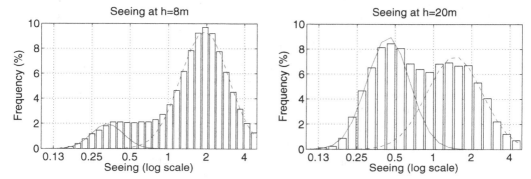

Figure 1. Left: Seeing histogram at an elevation of $h = 8m$ for the period April to September. Right: Idem for $h = 20m$ from available data in 2005 and 2012. Red (solid) and blue (dashed) lines are Gaussian fit of the two bumps. The area under them gives the fraction of time spent by the DIMM inside (dashed line) or outside (solid line) the SL. At $h = 20m$ the DIMM is in the free atmosphere 50% of the time (12% at $h = 8m$).

winter, and disappears partly in summer (Aristidi *et al.* 2009). The winter seeing is then highly dependent of the height of the telescope pupil. Fig. 1 presents the winter seeing histograms at $h = 8m$ and 20m. The two-bumps structure corresponds to situations where the DIMM is either inside or outside the SL, and the area under each bump gives the fraction of time spent in each situation. At $h = 20m$, the telescope is half of the time in the free atmosphere (FA) and this percentage will increase with h.

Regarding the outer scale, we here confirm the preliminary value of 7.5m published in Ziad *et al.* 2008). The measurements were made at ground level at $h = 3m$. No particular variability was observed between the summer and the winter.

3. Conclusion

In this paper we gave an update of the statistics of optical turbulence parameters and confirm the general trends published before: poor conditions at ground level in winter, but superb situation at an elevation of a few tens of meters above the snow where the FA can be attained during a significant amount of time. In the FA the median seeing is around 0.4″ (Aristidi *et al.* 2009).

References

Aristidi, E., Agabi, A., Vernin, J., *et al.* 2003, *A&A*, 406, L19
Aristidi, E., Agabi, A., Fossat, E., *et al.* 2005, *A&A*, 444, 651
Aristidi, E., Fossat, E., Agabi, A., *et al.* 2009, *A&A*, 499, 955
Ziad, A., Aristidi, E., Agabi, *et al.* 2008, *A&A*, 491, 917

Astrophysics from Antarctica
Proceedings IAU Symposium No. 288, 2012
M. G. Burton, X. Cui & N. F. H. Tothill, eds.

© International Astronomical Union 2013
doi:10.1017/S1743921312017048

Airglow and Aurorae from Dome A, Antarctica

Geoff Sims[1], Michael C. B. Ashley[1], Xiangqun Cui[2], Jon R. Everett[1], LongLong Feng[3,4], Xuefei Gong[2,4], Shane Hengst[1], Zhongwen Hu[2,4], Jon S. Lawrence[5,6], Daniel M. Luong-Van[1], Anna M. Moore[7], Reed Riddle[7], Zhaohui Shang[4,8], John W. V. Storey[1], Nick Tothill[9], Tony Travouillon[7], Lifan Wang[3,4,10], Huigen Yang[4,11], Ji Yang[3], Xu Zhou[4,12] and Zhenxi Zhu[3,4]

[1] School of Physics, University of New South Wales, Sydney NSW 2052, Australia
email: g.sims@unsw.edu.au

[2] Nanjing Institude of Astronomical Optics & Technology, Nanjing 210042, China

[3] Purple Mountain Observatory, Nanjing 210008, China

[4] Chinese Center for Antarctic Astronomy, China

[5] Department of Physics and Astronomy, Macquarie University, Sydney NSW 2109, Australia

[6] Australian Astronomical Observatory, Sydney NSW 1710, Australia

[7] Caltech Optical Observatories, Pasadena, CA, USA

[8] Tianjin Normal University, Tianjin 300074, China

[9] University of Western Sydney, Sydney NSW, Australia

[10] Department of Physics and Astronomy, Texas A&M University, College Station 77843, USA

[11] Polar Research Institute of China, Shanghai 200136, China

[12] National Astronomical Observatories, Chinese Academy of Science, Beijing 100012, China

Abstract. Despite the absence of artificial light pollution at Antarctic plateau sites such as Dome A, other factors such as airglow, aurorae and extended periods of twilight have the potential to adversely affect optical observations. We present a statistical analysis of the airglow and aurorae at Dome A using spectroscopic data from Nigel, an optical/near-IR spectrometer operating in the 300–850 nm range. The median auroral contribution to the B, V and R photometric bands is found to be 22.9, 23.4 and 23.0 mag arcsec^{-2} respectively. We are also able to quantify the amount of annual dark time available as a function of wavelength; on average twilight ends when the Sun reaches a zenith distance of 102.6°.

Keywords. Dome A, site testing, airglow, aurora, sky brightness

1. Airglow and Aurorae

Using the Nigel spectrometer (Sims *et al.* 2010), we investigated the sky brightness contribution from the dominant emission lines in the optical photometric bands: B (391.4 and 427.8 nm); V (557.7 nm); and R (630.0 and 636.4 nm). A sample histogram for the V band is shown in Figure 1.

2. Twilight

At high latitudes the total amount of useable dark time is critically dependent on the solar zenith distance at which it becomes dark. One interesting consequences of the low aerosol content of the Antarctic atmosphere is that it results in the sky darkening

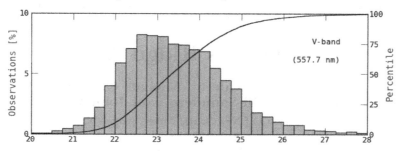

Figure 1. Probability histogram showing the contribution of the 557.7 nm line to sky brightness in the *V* band.

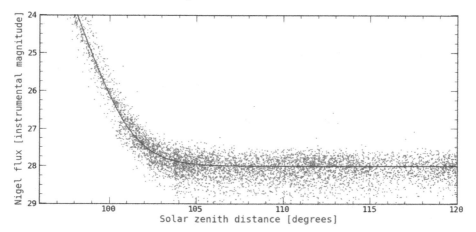

Figure 2. Determining the solar zenith distance at which twilight ends.

faster than at temperate latitudes. Figure 2 shows data from one wavelength interval, demonstrating that twilight effectively ends at a solar zenith distance of ∼ 103°.

3. Future Work

Auroral statistics for 2009 from Nigel have recently been published (Sims *et al.* 2012). Our data also complement photometric images from the Gattini camera (Moore *et al.* 2010) which uses *B*, *V* and *R* filters. This will allow the background sky brightness at Dome A to be disentangled from the various airglow and auroral emission lines (Moore *et al.*, in prep.).

4. Acknowledgements

The author acknowledges the Astronomical Society of Australia (ASA) for providing travel support to attend this meeting.

References

Moore, A. *et al.* 2010, *Proc. SPIE*, 7733, 77331S
Sims, G. *et al.* 2010, *Proc. SPIE*, 7733, 77334M
Sims, G. *et al.* 2012, *Publ. Astron. Soc. Pac.*, 124, 637

Astrophysics from Antarctica
Proceedings IAU Symposium No. 288, 2012
M. G. Burton, X. Cui & N. F. H. Tothill, eds.

© International Astronomical Union 2013
doi:10.1017/S174392131201705X

Where is Ridge A?

Geoff Sims[1], Craig Kulesa[2], Michael C. B. Ashley[1], Jon S. Lawrence[3,4], Will Saunders[4] and John W. V. Storey[1]

[1] School of Physics, University of New South Wales, Sydney NSW 2052, Australia
email: g.sims@unsw.edu.au

[2] University of Arizona, Tucson, USA

[3] Department of Physics and Astronomy, Macquarie University, Sydney NSW 2109, Australia

[4] Australian Astronomical Observatory, Sydney NSW 1710, Australia

Abstract. First identified in 2009 as the site with the lowest precipitable water vapour (PWV) and best terahertz transmission on Earth, "Ridge A" is located approximately 150 km south of Dome A, Antarctica. We use three years of data from the Microwave Humidity Sensor (MHS) on the NOAA-18 satellite and recent ground-based measurements from Ridge A to probe the PWV variations and stability over the high Antarctic plateau.

Keywords. Ridge A, site testing, PWV, terahertz

---

1. Introduction

The terahertz (THz) region of the electromagnetic spectrum remains one of the last truly unexplored spectral regimes, however the efficient absorption of THz radiation by atmospheric water vapour, oxygen and ozone limit observations to the highest and driest locations on Earth. In 2009, various existing and potential observing sites in Antarctica were compared based on a variety of criteria (Saunders *et al.* 2009), with Ridge A appearing to have the lowest PWV, and hence best THz transmission, of them all.

2. NOAA-18 satellite

The PWV extractions from MHS/NOAA-18 were calibrated and ground-truthed in 2010 using the Pre-HEAT instrument at Dome A (Yang *et al.* 2010). We then applied these same calibrations to data from the region surrounding Dome A. The 2008–2010 winter median PWV is plotted as a function of location in Figure 1.

3. HEAT at Ridge A

During early 2012, the High Elevation Antarctica Terahertz (HEAT) telescope at Ridge A observed a 24 hour period of exceptionally stable PWV ($< 10~\mu$m variance). Interestingly, contemporaneous PWV from the MHS/NOAA-18 showed an apparent high dependence of PWV on the viewing angle of the satellite. While this is nominally taken into account during the extraction process (Miao *et al.* 2001), additional adjustments were required to obtain a suitable correlation. A comparison is shown in Figure 2.

4. Future Work

The exceedingly low water vapour content of the Antarctic atmosphere warrants a more thorough and careful treatment of satellite data for the purpose of PWV extraction.

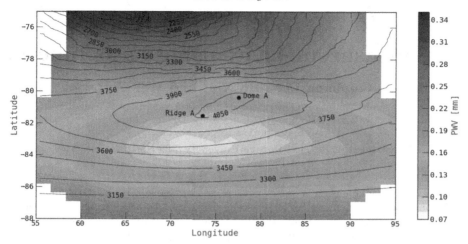

Figure 1. Winter median PWV (during 2008–2010) as a function of location using the MHS instrument on the NOAA-18 satellite. Elevation contours are shown in metres.

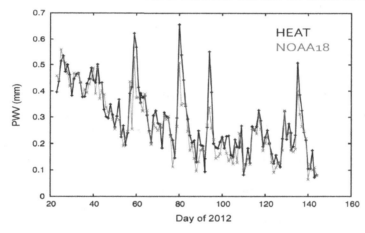

Figure 2. Comparison of PWV as measured by HEAT and the NOAA-18 satellite. The calibration of NOAA-18 satellite was obtained by a comparison with HEAT data during a time of stable PWV.

Ground truthing at one site can not necessarily be generalised to even nearby locations, as is the case for Dome A and Ridge A (separated by less than 150 km). We intend to use years of archival satellite data, calibrated with ground-based observations, to obtain reliable PWV maps for the Antarctic continent.

5. Acknowledgements

The author acknowledges the Astronomical Society of Australia (ASA) for providing travel support to attend this meeting.

References

Miao, J. *et al.* 2001, *J. Geophys. Res.*, 106, 10187
Saunders, W. *et al.* 2009, *Publ. Astron. Soc. Pac.*, 121, 976
Yang, H. *et al.* 2010, *Publ. Astron. Soc. Pac.*, 122, 490

Astrophysics from Antarctica
Proceedings IAU Symposium No. 288, 2012
M. G. Burton, X. Cui & N. F. H. Tothill, eds.

© International Astronomical Union 2013
doi:10.1017/S1743921312017061

Shape measurement by using basis functions

Guoliang Li[1,2], Bo Xin[2] and Wei Cui[2]

[1]Purple Mountain Observatory, Chinese Academy of Sciences,
2 West Beijing Road, Nanjing 210008, China.
email: guoliang@pmo.ac.cn

[2]Department of Physics, Purdue University, 525 Northwestern Ave.,
West Lafayette, Indiana 47907

Abstract. Gravitational lensing is one of most promising tools to probe dark energy and dark matter in our Universe. Lensing by larger-scale structures distorts the shape of background galaxies. For ground-based observations the shape is further distorted by atmospheric turbulence and optical distortions. Many algorithms were proposed to measure the shear signal but the systematic biases are still too large to be acceptable for the larger-sky surveys in the future. I will present our new algorithms for PSF reconstruction and shape measurements based on several sets of basis functions.

Keywords. gravitational lensing, shape, model fitting, basis function.

1. Fast calculating of model parameters

To measure the cosmic shear, the problem in hand is to determine the intrinsic (or pre-seeing) shape of a galaxy based on its observed shape. In practice, the galaxy image is noisy, which makes deconvolution challenging. As a first step, it is necessary to accurately quantify the observed shape of galaxies based on noisy images. In this article, we present a set of efficient algorithms specifically for this task. Usually, no less than six parameters are needed to model the light distribution of an object, including the centroid position (x, y), the ellipticity (g_1, g_2), the normalization I_0, and the profile parameter(s). High-dimensional parameter search is very time consuming, especially when the number of objects is large, and tends to be trapped at local minima. Instead of brute-force fitting, we propose to derive the centroid position, ellipticity, and normalization of a light distribution numerically and thus reduce the number of parameters that need to be determined through fitting. Similar effort has been previously undertaken. Miller *et al.* (2007) and Kitching *et al.* (2008) proposed a fast-fitting algorithm in Fourier space, which also takes into account the effects of point spread function (PSF). Here, we will only focus on how to reproduce the observed images of galaxies and stars respectively. Armed with these two kind of smooth images, deconvolution could be easily performed and the intrinsic shape of the galaxies could be estimated.

To test the efficiency of our algorithm, we simulate 500 galaxy images based on the Sersic model. The ellipticities are chosen randomly and Gaussian noise was added on the center position and the values of each pixel. Then the center and the ellipticity was estimated by using the above algorithms. Meanwhile we use the MINUIT minimization program (James & Roos 1975) as the brute-force method to fit these parameters. We show the differences in centroiding and the ellipticity in Fig.1. We can see that the calculated results are a little worse than that of model fitting but they are still good enough. And we should keep in mind that: 1) the process is accelerated by a factor of magnitude, 2) the shape roughly estimated parameters is not our final goal. They could be used as the

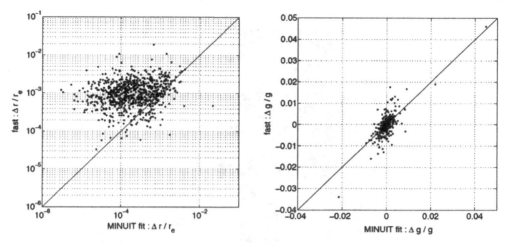

Figure 1. The differences in centroiding (L) and the ellipticity (R) which are derived by our fast calculating algorithm and by brute-force model fitting.

input parameters to create corresponding basis to improve the real image fitting further. More details can be found in Li *et al.* (2012)

2. PSF reconstruction by using gaussianlets and moffatlets

A point source will be observed as an extended image because the photons are affected by several complex physics processes. This means that the PSF can not be represented by a simple model function. It is well known that a set of complete and orthogonal basis functions could compose any shape of image. The only problem is how many basis functions we need. Therefore we construct two kinds of basis function to model the PSF and test their efficiency. For simplicity, we just consider the ellipticity of the PSF but ignore the higher-order moments.

Gaussianlets is a reduced version of shapelets (Massey & Refregier 2005) where we only keep the basis functions with $m = 0$, i.e., the basis functions are circular symmetric and contain no angular components. For a given image of star, we perform our fast calculating to derive the center and the ellipticity. Then the image is fitted by using a Gaussian function with same center and ellipticity. The resulting width is used as the only parameter to create the basis function set. All of the basis functions are reshaped into elliptical symmetry according to the measured ellipticity and linear minimizing is performed to obtain the coefficients for each basis function

Moffalets is another kind of basis function the 0th order of which is the Moffat function itself. Moffat model has been believed one of the optimal model for the PSF. Now with the high order corrections we could improve the model to an new level. Our gaussianlets and moffatlets has been tested in the star challenge of GREAT10 program and got high scores.

3. Galaxy reconstruction by using sersiclets

The Sersic model is successful to fit the light distribution of galaxies. But it is not always successful because of the complexity of real galaxies. Shapelets were first proposed to mimic the light distribution of any type of galaxies since they are complete and orthogonal basis functions. But shapelets are not optimal because they are not compact

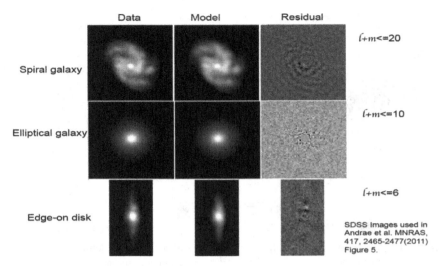

Figure 2. Image reconstruction by using elliptical sersiclets.

enough. From this point of view, Andrae *et al.* (2011) proposed another kind of basis function based on the Sersic model which are called Sersiclets. However either shapelets and Sersiclets need very high orders to reproduce an high ellipticity. Therefore we extend the standard sersiclets to elliptical sersiclets as same as we did for gassianlets and moffatlets and reproduced the three images which are used in their work. We show the fitting results as Fig. 2.

4. Summary

In this poster we present a set of algorithms to reconstruct the image of star and galaxy in a quick and accurate way. 1) For the issue of model fitting, we propose a new way to search the parameters separately. The center and ellipticity can be derived by calculations and our algorithm thus decrease the fitting dimension dramatically. 2) Gaussian and Moffat models are usually used to represent the PSF but they are not perfect because the PSF is too complex to be described by just one or two parameters. We proposed two kinds of basis function named gaussianlets and moffatlets to implement high-order correction, so that we could reproduce any kind of PSF accurately in principle. 3) We also improved the reconstruction of galaxy images by using elliptical sersiclets. They are able to reproduce the light distribution of any type of galaxy with higher efficiency than what shapelets can achieve.

References

Andrae, R., Melchior, P., & Jahnke, K. 2011, *MNRAS*, 417, 2465
Kitching, T. D., Miller, L., Heymans, C. E., van Waerbeke, L., & Heavens, A. F. 2008, *MNRAS*, 390, 149
James, F. & Roos, M. 1975, *Comput. Phys. Commun.* 10, 343
Li, G. L., Xin, B., & Cui, W. 2012, *arXiv:*1203.0751
Massey, R. & Refregier, A. 2005, *MNRAS*, 363, 197
Miller, L., Kitching, T. D., Heymans, C., Heavens, A. F., & van Waerbeke, L. 2007, *MNRAS*, 382, 315

Astrophysics from Antarctica
Proceedings IAU Symposium No. 288, 2012   © International Astronomical Union 2013
M. G. Burton, X. Cui & N. F. H. Tothill, eds.   doi:10.1017/S1743921312017073

A Multi-Aperture Scintillation Sensor for Dome A, Antarctica

Hualin Chen[1,2,3,4] Chong Pei[1,2,3] and Xiangyan Yuan[1,2]

[1] National Astronomical Observatories/Nanjing Institute of Astronomical Optics & Technology,
Chinese Academy of Sciences, Nanjing 210042, China

[2] Key Laboratory of Astronomical Optics & Technology, Nanjing Institute of Astronomical
Optics & Technology, Chinese Academy of Sciences, Nanjing 210042, China

[3] Graduate University of Chinese Academy of Sciences, Beijing 100049, China

[4] email: hualinchen@niaot.ac.cn

Abstract. Site-testing measurements by the Australian group has already shown that Dome
C on the Antarctic plateau is one of the best ground-based astronomical sites. Furthermore,
Dome A, the Antarctic Kunlun Station, as the highest point on Antarctic inland plateau, where
a Chinese Antarctic scientific expedition team first reached in 2005, is widely predicted to
be an even better astronomical site by the international astronomical community. Preliminary
site-testing carried out by the Center for Antarctic Astronomy (CAS) also confirms Dome A
as a potential astronomical site. Multi-aperture scintillation sensors (MASS) can measure the
seeing and isoplantic angle, the turbulence profile, etc., which are very important site-testing
parameters that we urgently need. The MASS site testing at Dome A is presented here, and
includes the method of processing data and the hardware for the extreme conditions of Dome
A, Antarctica.

Keywords. Antarctic Kunlun Station, Dome A, Scintillation, Seeing, Isoplanatic angle,
Turbulence profile.

---

1. Theory

Theory of wave propagation shows: Scintillation $s^2 \sim \int C_n^2 \cdot Q(h) \cdot dh$, Seeing $r_0 \sim$
$[\int C_n^2 \cdot dh]$, Isoplanatic angle $\theta_0 \sim [\int C_n^2 \cdot h^{5/3} \cdot dh]^{3/5}$, where $Q(h)$ is weight function
(WF). It depends on the receiving aperture geometry, light spectrum and the assumed
theory of turbulence.

2. Data processing

Scintillation indices calculation and their errors. A set of instantaneous scintillation
indices (SI), including normal, differential and differential exposure scintillation indices,
are calculated using raw photo counts from a given number of channels during the base
time 1 s. To obtain correct SIs from raw data, we must take into account non-linearity of
photo counters, subtract the contribution from photo noise, and extrapolate the indices to
zero-exposure time for removing the bias caused by finite exposure time, make correction
for background include dark count of PMTs and light from the sky.

Weighting functions. The weighting function (WF) depends on spatial turbulence spectrum and on the shape and size of receiving aperture. In addition, some de-correlation
of scintillation within the bandpass is also expected. So, for an altitude h, polychromatic
weighting functions table for both normal and differential SIs are pre-computed.

Profile restoration. A low-resolution turbulence profile (TP) restoration method is
adopted which is fitting the measured indices to a model with only a few layers, then

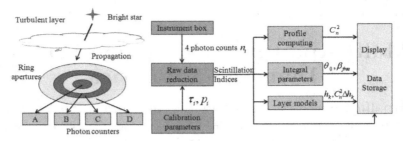

Figure 1. Left: principle of turbulence profile measurement with the MASS. Right: organization of the data processing in the MASS.

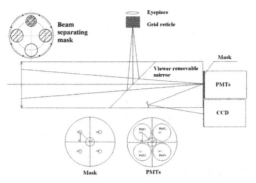

Figure 2. Simplified layout of the MASS instrument.

restoring these thick slabs of turbulence. Both fixed-layers with pre-defined altitudes and floating-layers with only three layers located at any altitudes restoration techniques are used. In fitting models to the data, we search for model parameters that minimize the merit function.

Seeing, isoplantic angle and atmospheric time constant. Because each SI is an integral over altitude of C_n^2 multiplied by some WF. Thus, any linear combination of SIs corresponds to the linear combinations of WFs. If the linear combinations of WFs approximates h^0 or $h^{3/5}$, then seeing or isoplanatic angle can be estimated from liner combinations of SIs directly, without profile restoration. The atmosphere time constant can be estimated from differential exposure scintillation index as: $\tau_{\mathrm{de}} = \mathrm{KT}[\sigma_{\mathrm{DE}}^2]^{3/5}$.

3. Hardware

A beam separating mask in the telescope entrance pupil separates the star light into six clusters. Two clusters image the star on the CCD, functioning as a differential image motion monitor (DIMM) measuring the whole atmosphere seeing. The other four clusters converge on PMTs through the telescope respectively. The PMTs output a pulse sequence corresponding to the photon sequence counting for the scintillation index, measuring the free atmosphere seeing, the isoplanatic angle and the turbulence profile. Low temperature protection is used for printed circuit boards, PMTs, CCDs and other important models. Thw data processing computer is in a nearby instrument house. The combination of DIMM and MASS is simple and robust, and should work at Dome A.

References

http://www.ctio.noao.edu/~atokovin/profiler/.

Astrophysics from Antarctica
Proceedings IAU Symposium No. 288, 2012
M. G. Burton, X. Cui & N. F. H. Tothill, eds.

© International Astronomical Union 2013
doi:10.1017/S1743921312017085

Design and field testing of the Fish-Eye lens for optical atmospheric observations

Ivan Syniavskyi[1], Yuriy Ivanov[1], Sergey Chernous[2] and Fred Sigernes[3]

[1] Main Astronomical Observatory NAS of Ukraine
27 Akademika Zabolotnoho str., Kiev, 03680, Ukraine
email: `syn@mao.kiev.ua`

[2] Polar Geophysical Institute of the Kola Science Centre RAS,
16 Fersman str. Apatity, Murmansk region, 184209, Russia
email: `chernouss@pgia.ru`

[3] The University Centre in Svalbard (UNIS),
Longyearbyen, N-9171, Norway
email: `fred.Sigernes@unis.no`

Abstract. The Fish-Eye lens MAO-08 is intended for observations of weak extended objects (aurora, twilight and dawn phenomena, stratospheric clouds, etc.) in narrow spectral bands with variable passband filters VARISPEC and in white light. It is valuable for astronomical observation for the target – background problem when we need to estimate the spectral transparence of the atmosphere. Besides having high power this lens can be used for meteor observations. It has been tested during the winter field conditions.

Keywords. Fish-Eye lens; Interference Filter; Auroral Observations, Hyaline Observations.

1. Introduction

The Fish-Eye lens is most often used for observations of atmospheric phenomena such as aurora and nightglow (Elvey & Stoffregen(1957), Lebedinsky (1961), Sandahl *et al.* (2008)). For observations of broad band, commercial fish-eye lenses can be used coupled to CCD detector. However, when it comes to obtaining narrow band images of auroral emissions lines with high temporal resolution, the task of researchers is much more complicated. The filters require to be collimated light in order to obtain optimum bandpass. A standard technique is to use an telecentric lens system (Mende *et al.* (1977), Sandahl *et al.* (2008)). The MAO-08 lens does not use this system and based on calculations as a whole.

2. Main Parameters

The ultra wide-angle lens MAO-08 can be used with interference filters, variable spectral bandwidth filters (VARISPEC), or without any filters (white light). Figure 1 shows the optical diagram of the lens. The first part converts the 180° field of view (FOV) to a narrow beam of 6°, where interference filters may be inserted. The back part focuses the collimated light onto the detector plane. The design is compact with unique optical characteristics. The basic technical data:

- FOV: 180 degrees;
- F/value: 0.82;
- Spectral range: 430 – 750 nm;

- Resolution: center – 100 l/mm; edge – 70 l/mm;
- Dimensions (with covers): length 235 mm and diameter 103 mm

3. Field testing

Lens tests in field conditions have confirmed high quality of images. (Fig. 2a) is image of the ray auroral arc, obtained at Barentsburg in white light. There are a plenty of constellations and stars in this figure and even the Milky Way is seen. (Fig. 2b) shows the weak auroral folds in the emission at 557.7 nm, obtained with narrow-band interference filter ($\Delta\lambda \sim 2$ nm).

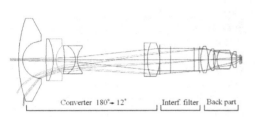

Figure 1. Layout of the MAO-08 lens.

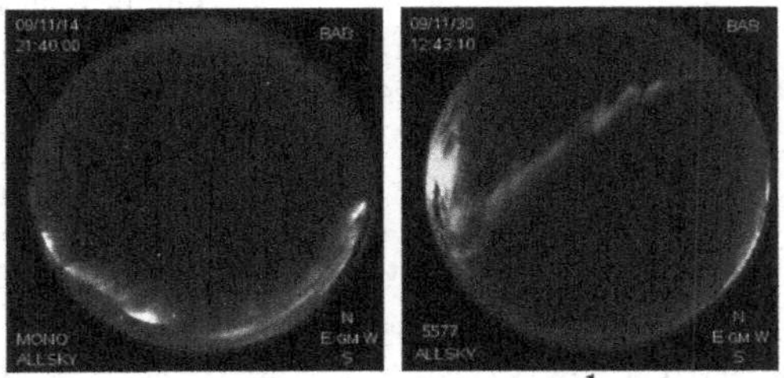

Figure 2. All-sky images of aurora: a) in white light, b) at $\lambda \sim$557.7 nm.

A first season of observations was spent during the period November 2011 – April 2012 in the Kjell Henriksen Observatory (KHO) in Svalbard located at the archipelago Svalbard 1000 km north of mainland Norway (78°N, 16°E), and Barensburg research Station, PGI (78.093°N, 14.208°E).

4. Conclusion

The MAO-08 lens has a large aperture, which exceeds this parameter of the other lenses for auroral research. It allows filters to be inserted into the lens. Pointed out is the possibility of obtaining simultaneously two images in two emissions of the same aurora by the single CCD camera.

5. Acknowlegements

This work was financially supported by The Research Council of Norway through the project named: Norwegian and Russian Upper Atmosphere Co-operation on Svalbard part 2 # 196173 / S30 (NORUSCA2).

References

Elvey, S. T. & Stoffregen, W. 1957, *Ann. Int. Geophys. Year*, 5, 117
Mende, S. B., Eather, R. H., & Aamodt, E. K. 1977, *App. Optics*, 16, 1691
Lebedinsky, A. I. 1961, *Ann. Int. Geophys. Year*, 11, 133
Sandahl, I., Sergienko, T., & Brandstorm, U. 2008, *JASTP*, 70, 2275

Astrophysics from Antarctica
Proceedings IAU Symposium No. 288, 2012
M. G. Burton, X. Cui & N. F. H. Tothill, eds.

© International Astronomical Union 2013
doi:10.1017/S1743921312017097

Solar Eclipses Observed from Antarctica

Jay M. Pasachoff

Williams College - Hopkins Observatory and California Institute of Technology
email: eclipse@williams.edu

Abstract. Aspects of the solar corona are still best observed during totality of solar eclipses, and other high-resolution observations of coronal active regions can be observed with radio telescopes by differentiation of occultation observations, as we did with the Jansky Very Large Array for the annular solar eclipse of 2012 May 20 in the US. Totality crossing Antarctica included the eclipse of 2003 November 23, and will next occur on 2021 December 4; annularity crossing Antarctica included the eclipse of 2008 February 7, and will next occur on 2014 April 29. Partial phases as high as 87% coverage were visible and were imaged in Antarctica on 2011 November 25, and in addition to partial phases of the total and annular eclipses listed above, partial phases were visible in Antarctica on 2001 July 2011, 2002 December 4, 2004 April 19, 2006 September 22, 2007 September 11, and 2009 January 26, and will be visible on 2015 September 13, 2016 September 1, 2017 February 26, 2018 February 15, and 2020 December 14. On behalf of the Working Group on Solar Eclipses of the IAU, the poster showed the solar eclipses visible from Antarctica and this article shows a subset (see www.eclipses.info for the full set). A variety of investigations of the Sun and of the response of the terrestrial atmosphere and ionosphere to the abrupt solar cutoff can be carried out at the future eclipses, making the Antarctic observations scientifically useful.

Keywords. eclipse, atmosphere, ionosphere

Scientific observations for the total eclipses ordinarily include electronic imaging of the corona to compare with simultaneous space observations of the Sun. Our links to maps and other items of coordination can be found at http://www.eclipses.info, the site of the IAU Program Group on Public Education at the Times of Eclipses and of the IAU Working Group on Eclipses. Special filters must be used to reduce the solar disk to a safe intensity during the partial phases; only during totality can one look safely without filters.

Since the paths of annularity or totality cross only a narrow band on Earth, observatories and ground stations in the path can get useful scientific observations. For example, the solar occultation that my group observed with the Jansky Very Large Array in 2012 should improve the spatial resolution of solar active regions, bringing radio observations to a finer resolution than ground-based optical observations or space optical or ultraviolet observations. Observations of temperature, ionospheric response, and other parameters during eclipses can be interpreted to show the response of the terrestrial atmosphere and ionosphere to an abrupt cutoff of insolation. For the total solar eclipses that are uniquely available from Antarctica (next on 2021 December 4), a wide range of imaging and spectroscopy should be obtained to help follow the Sun over the solar-activity cycle.

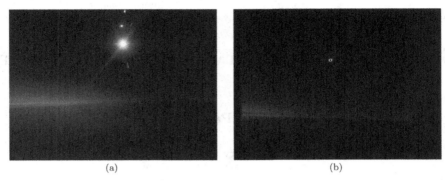

Figure 1. Wide-angle views out the windows of the Qantas flight to view the 2003 total solar eclipse over Antarctica, showing the umbral shadow approaching (left) and the solar corona (right) while the airplane was immersed in the shadow.

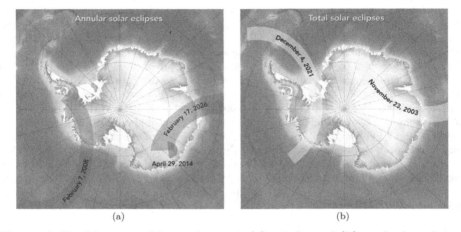

Figure 2. Notable past and future Antarctic **(a)** annular and **(b)** total solar eclipses.

Figure 3. Circumstances of the 2021 December 4 total solar eclipse shown in *Google Maps*. Courtesy of Xavier Jubier. http://xjubier.free.fr/en/site_pages/solar_eclipses/TSE_2021_GoogleMapFull.html

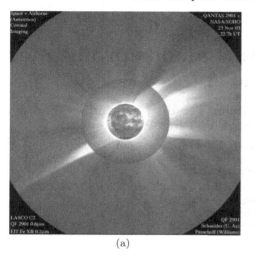

(a) (b)

Figure 4. (a) An image composite of outer corona with the second (C2) telescope of the Large Angle Spectroscopic Coronagraph on the European Space Agency/National Aeronautics and Space Administration's Solar and Heliospheric Observatory (SOHO/LASCO C2), mid and inner corona from QF 2901 (airborne over Antarctica) by JMP and Glenn Schneider (U. Arizona), and the chromosphere with SOHO's Extreme-ultraviolet Imaging Telescope (EIT), all nearly contemporaneous. **(b)** On the right is a wide-field mosaic image of the 2003 total solar eclipse made from several different exposures made from QF 2901, to compensate for the wide dynamic range of the solar corona, which falls off in intensity by a factor of about 1,000 within about one solar radius.

With a total eclipse and an annular eclipse each occurring somewhere on Earth about every 18 months but less often than every 300 years at a given location, we should be using all telescopes and other scientific instruments in eclipse paths to their fullest possible extent.

Acknowledgements

JMP's research on the solar eclipses of 2012 is supported by grant AGS-1047726 from the Solar Physics Program of the Division of Atmospheric and Geospace Sciences of the US National Science Foundation. For the 2003 eclipse, Jay Pasachoff was part of the EurAstro Association team on the Croydon Travel chartered 747 over Antarctica. I thank Michael Zeiler (`eclipse-maps.com`) for the maps. See `http://www.williams.edu/astronomy/eclipse`.

References

Pasachoff, Jay M. 2009, "Solar Eclipses as an Astrophysical Laboratory", *Nature*, June 11, **459**, 789-795, DOI 10.1038/nature07987.

Pasachoff, Jay M. 2009, "Scientific Observations at Total Solar Eclipses", *Research in Astronomy and Astrophysics*, **9**, 613-634. `http://www.raa-journal.org/raa/index.php/raa/article/view/182`.

Astrophysics from Antarctica
Proceedings IAU Symposium No. 288, 2012
M. G. Burton, X. Cui & N. F. H. Tothill, eds.

© International Astronomical Union 2013
doi:10.1017/S1743921312017103

SCIDAR: an optical turbulence profiler for Dome A

Li-Yong Liu[1,2], Yong-Qiang Yao[1], Jean Vernin[2], Merieme Chadid[2], Hong-Shuai Wang[1] and Yi-Ping Wang[1]

[1]National Astronomical Observatories, Chinese Academy of Sciences,
20A Datun Road, Chaoyang District, Beijing, China
email: liuly@nao.cas.cn

[2]Université de Nice-Sophia Antipolis, Observatoire de la Côte d'Azur, CNRS-UMR7293, Lab.
Lagrange,
Parc Valrose, 06108, Nice Cedex 2, France
email: jean.vernin@unice.fr

Abstract. This paper introduces a plan to detect turbulence profiles at Dome A with a Single Star Scidar (SSS), to enhance our understanding of the characteristics of the site. The development of a portable monitor for profiling vertical atmospheric optical turbulence and wind speed is presented. By analyzing the spatial auto and cross-correlation functions of very short exposure images of single star scintillation patterns, the SSS can provide the vertical profiles of turbulence intensity $C_n^2(h)$ and wind speed $V(h)$. A SSS prototype is already operational at Ali in Tibet which will be improved in order to become fully robotic and adapted to extreme weather conditions that prevail at Dome A in Antarctica.

Keywords. Scidar, Optical turbulence, Dome A

1. Introduction

On the Antarctic plateau, Dome A benefits from exceptional atmospheric conditions, such as good seeing, coldness and very low precipitable water vapour, which potentially makes it one of the best sites for an astronomical ground-based observatory. A Chinese team first landed on Dome A, in January 2005, and returned again to Dome A in 2008 (Yang *et al.* 2009). The so-called "PLATO" observatory has been installed in order to carry out a site testing campaign, which comprises a set of site testing instruments to characterize the Dome A site (Lawrence *et al.* 2008).

Two instruments have been set up to quantify the optical turbulence at Dome A. One is a Surface layer NOn-Doppler Acoustic Radar (SNODAR) to measure the turbulence within the near-surface layer (Bonner *et al.* 2010), the other is a Differential Image Motion Monitor (DIMM) to evaluate the seeing quality (Pei *et al.* 2010). We propose to develop a Single Star Scidar (SSS) in order to achieve a more detailed characterization of the optical turbulence within the boundary layer as well as the free atmosphere at Dome A. The SSS is a portable monitor, which can retrieve profiles of optical turbulence from the ground to the top of atmosphere. It has been running for more than one year at Dome C, as reported by Vernin *et al.* 2009 and Giordano *et al.* 2012.

2. Plan to install a SSS

The National Astronomical Observatories (NAOC) and the University of Nice Sophia Antipolis (UNSA) have been working together to develop a new SSS instrument for Dome A since 2009. We carried out a series of SSS experiments under real-sky conditions. We

Figure 1. First SSS experiment at NAOC during summer 2010.

performed the first experiment with the new SSS at the headquarters of NAOC (Fig.1), and turbulence measurements with the SSS have been successfully carried out at Xinglong Observatory in China. The same SSS has been moved to Ali observatory, Tibet, at an altitude of 5,100 m, and for the first time, we retrieved the turbulence and wind profiles above the Tibetan plateau in November, 2011 (Liu *et al.* 2012). We plan to install the SSS to Dome A after "antarcticization" of this instrument.

3. Improvement for Dome A

Due to the harsh environmental conditions in Antarctica, the SSS instrument robotization is important in order to guarantee continuous measurements during the Antarctic winter. An automatic guider has been developed to track the target star for a long time, as detailed in Liu *et al.* 2012. We are also committed to the development of remote control technology, which can monitor and diagnostics the SSS instrument via the Internet, and automatic download of the measurements. Cold temperatures, as low as -80°C, might be fatal for the instrument, and an antarcticization procedure is foreseen to ensure that the telescope mount and CCD camera will work well even at low temperatures.

4. Conclusions

We have developed a new SSS in order to characterize the optical turbulence at Dome A. The instrument was successfully tested under real-sky conditions in various observatories. We are committed to developing a fully robotic version of the SSS, which will be suitable for turbulence measurements even under extreme environment that prevail at Dome A.

Acknowledgements: This work is supported by the National Natural Science Foundation of China (NSFC, Grant Nos. 10903014 and 11073031).

References

Yang, H., Allen, G., Ashley, *et al.*, 2009, *PASP*, 121, 174-184.
J. S. Lawrence, G. R. Allen, M. C. B., Ashley, *et al.*, 2008, *Proc. SPIE*, 7012, 701227-701227-12.
C. S. Bonner, M. C. B., Ashley, X. Cui, *et al.*, 2010, *PASP*, 122, 1122-1131.
C. Pei, H. Chen, X. Yuan, *et al.*, 2010, *Proc. SPIE*, 7733, pp. 77334W-77334W-8.
Vernin, J., Chadid, M., Aristidi, E., *et al.*, 2009, *A&A*, 1271, 1276
Giordano, C., Vernin, J., Chadid, M., *et al.*, 2012, *PASP*, 494, 506
L. Liu, Y. Yao, J. Vernin, *et al.*, 2012, *Proc. SPIE*, 8444, pp. 844464 (1-7)

Astrophysics from Antarctica
Proceedings IAU Symposium No. 288, 2012
M. G. Burton, X. Cui & N. F. H. Tothill, eds.

© International Astronomical Union 2013
doi:10.1017/S1743921312017115

SONG China project
– participating in the global network

Licai Deng[1,*], Yu Xin[1], Xiaobin Zhang[1], Yan Li[2], Xiaojun Jiang[1], Guomin Wang[3], Kun Wang[1,4], Jilin Zhou[5], Zhengzhou Yan[4] and Zhiquan Luo[4]

[1]National Astronomical Observatories, CAS, 20A Datun Road, 100012 Beijing, China
*email: licai@bao.ac.cn

[2]Yunnan Astronomical Observatory, CAS, Kunming, China

[3]Nanjing Institute of Astronomical Optics and Technolog, CAS, China

[4]School of Physics and Electronic Information, China-West Normal University, China

[5]School of Astronomy and Space Science, Nanjing University, China

Abstract. SONG (Stellar Observations Network Goup) is a low-cost ground based international collaboration aimed at two cutting edge problems in contemporary astrophysics in the time-domain: 1) Direct diagnostics of the internal structure of stars and 2) looking for and studying extra solar planets, possibly in the habitable zone. The general plan is to set up a network of 1 m telescopes uniformly distributed in geographic latitude (in both hemispheres). China joined the collaboration (initiated by Danish astronomers) at the very beginning. In addition to SONG's original plan (http://song.phys.au.dk), the Chinese team proposed a parallel photometry subnet work in the northern hemisphere, namely 50BiN (50 cm Binocular Network, previously known as mini-SONG), to enable a large field photometric capability for the network, therefore maximising the potential of the network platform. The network will be able to produce nearly continuous time series observations of a number of selected objects with high resolution spectroscopy (SONG) and accurate photometry (50BiN), and to produce ultra-high accuracy photometry in dense field to look for micro-lensing events caused by planetary systems. This project has great synergy with Chinese Astronomical activities in Antarctica (Dome A), and other similar networks (e.g. LCOGT). The plan and current status of the project are overviewed in this poster.

1. The Chinese SONG node

SONG, as an international program with well defined science goals, is designed to operate as a single instrument on its mission (for details on science and instrumentations of SONG see song.phys.au.dk). Unlike previous global stellar observation networks, the telescopes and instruments of each SONG node follow the same design, even if not manufactured by the same company. The Chinese SONG node takes the same parameters as the prototype node in Denmark (now being commissioned at Tenerife, Spain); some of the original drawings were also accepted. The 1 m telescope and the spectrograph have been redesigned applying the same standard as the prototype, and are fabricated at NIAOT.

The site for the Chinese node is a key to realize the expected performance of the network. The other proposed sites in the northern hemisphere subnet all have very good quality. None of the existing optical sites in China (mostly near the eastern coast line) meet the requirements of SONG, therefore a search for a reasonably good site for the SONG node in west China is crucial. Suggested by the site survey team, we selected a number of radio sites with existing infrastructure that can host the node 3 years ago when we started the project. The radio observatory in Delingha has been selected for our project. The site quality in terms of supporting infrastructure, general weather pattern, optical observational condition (seeing, sky background, extinction) are optimal among all the candidate sites in west China.

The main components of the Chinese SONG node and corresponding current status are listed bellow:

(*a*) Site: Delingha with geographic coordinate of N37.378027, E97.732326. Qualification started 2009, foundation and observational buildings construction started Aug 2012;

(*b*) 1 m SONG telescope: Factory assembled August 2012, site installation will be in the end of April 2013;

(*c*) High resolution Spectrograph: in fabrication;

(*d*) Lucky-image camera system: in fabrication;

(*e*) Software systems (data pipelines, networking and operations): being developed.

2. 50BiN subnet of SONG

50BiN is a Chinese initiative that originated from common research interests of the whole stellar physics community in the country. China will fund the system as an add-on project of Chinese participation of SONG. We are obliged to consider any suggestions and contributions to 50BiN in terms of scientific provisions. Any type of participation in the program will be highly appreciated. Given the condition that 50BiN is going to share the infrastructure and network of SONG, a number of science goals matching all the goals of the SONG can be pre-defined.

The plan is to have a small photometry telescope installed at every SONG node site in the north, so that photometry for a selection of objects can be followed in the same long time based line and duty cycle as SONG.

Just like SONG, a long time-baseline high duty cycle photometry of large FOV will be offered by 50BiN, which facilitates researche in the following areas:

Primary science goals:

(*a*) Photometry of a sample of open clusters, determining basic physical parameters when measured in a uniform way;

(*b*) Time domain study of variable objects in selected clusters: long time base-line, high cadence and high duty cycle, high precision photometric observations.

• A complete survey of stellar variability along the main sequence (e.g. Cep, Scu & Dor) in selected open clusters;

• A complete survey of small and large amplitude red variables along the RGBs of selected open clusters;

• A complete survey of eclipsing binaries in selected clusters, including looking for the transit of exoplanet systems;

• Looking for flare type variabilities in selected clusters; Stellar rotation and spot activities.

(*c*) Time domain study of selected field areas.

The first node, the prototype of 50BiN, is contributed by the CWNU. It is a binocular equatorial mount system. Two parallel camera systems will be used so that simultaneous photometry in two channels will be realized. The FOV is 20 arcmin and is suited for most galactic open clusters. The prototype will be in commissioned by the end of 2012. We wish to find more partners for future, both domestic or international. We have started negotiations with host institutes of the SONG network, hopefully to complete the network within the next 2–3 years.

For further details on the project, please see http://song.bao.ac.cn.

Astrophysics from Antarctica
Proceedings IAU Symposium No. 288, 2012
M. G. Burton, X. Cui & N. F. H. Tothill, eds.

© International Astronomical Union 2013
doi:10.1017/S1743921312017127

Photometry of Variables from Dome A, Antarctica

Lingzhi Wang[1,2,3], L. M. Macri[3], L. Wang[3,4], M. C. B. Ashley[5], X. Cui[6], L. L. Feng[4], X. Gong[6], J. S. Lawrence[5,7], Q .Liu[1], D. Luong-Van[5], C. R. Pennypacker[8], Z. Shang[9], J. W. V. Storey[5], H. Yang[10], J. Yang[4], X. Yuan[6,11], D. G. York[12], X. Zhou[1,11], Z. Zhu[4] and Z. Zhu[2]

[1] National Astronomical Observatory of China, Chinese Academy of Sciences, Beijing 100012, China
email: wanglingzhi@bao.ac.cn

[2] Department of Astronomy, Beijing Normal University, Beijing, 100875, China
[3] Mitchell Institute for Fundamental Physics & Astronomy, Department of Physics & Astronomy, Texas A&M University, College Station, TX 77843, USA
[4] Purple Mountain Observatory, Chinese Academy of Sciences, Nanjing 210008, China
[5] School of Physics, University of New South Wales, NSW 2052, Australia
[6] Nanjing Institute of Astronomical Optics and Technology, Nanjing 210042, China
[7] Australian Astronomical Observatory, NSW 1710, Australia
[8] Center for Astrophysics, Lawrence Berkeley National Laboratory, Berkeley, CA, USA
[9] Tianjin Normal University, Tianjin 300074, China
[10] Polar Research Institute of China, Pudong, Shanghai 200136, China
[11] Chinese Center for Antarctic Astronomy, Nanjing 210008, China
[12] Department of Astronomy and Astrophysics and Enrico Fermi Institute, University of Chicago, Chicago, IL 60637, USA

Abstract. Dome A on the Antarctic plateau is likely one of the best observing sites on Earth (Saunders *et al.* 2009). We used the CSTAR telescope (Yuan *et al.* 2008) to obtain time-series photometry of 10^4 stars with $i < 14.5$ mag during 128 days of the 2008 Antarctic winter season (Wang *et al.* 2011). During the 2010 season we observed 2×10^4 stars with $i < 15$ mag for 183 days (Wang *et al.* 2012). We detected a total of 262 variables, a 6× increase relative to previous surveys of the same area and depth carried out from temperate sites (Pojmanski 2004). Our observations show that high-precision, long-term photometry is possible from Antarctica and that astronomically useful data can be obtained during 80% of the winter season.

1. Photometry and site statistics

We used CSTAR#1 (equipped with a SDSS i filter) to obtain over 6.3×10^5 images with exposure times ranging from 20 to 40s during 128 and 183 days of the 2008 & 2010 Antarctic winter seasons, respectively. The raw images were de-biased, flat-fielded and fringe-corrected. We carried out aperture photometry using DAOPHOT (Stetson 1987) and transformed the measurements to the standard system.

Our master frame contained $\sim 1.5 \times 10^5$ stars with $i < 21$ mag. We selected the brightest 10^4 objects in 2008 (2×10^4 in 2010) to search for variability and calculate site statistics. We determined that favourable conditions (defined as extinction due to clouds being below 0.4 mag) occur for 80% of the dark time(left panel of Fig. 1).

Representative phased light curves are presented in the right panel of Fig. 1.

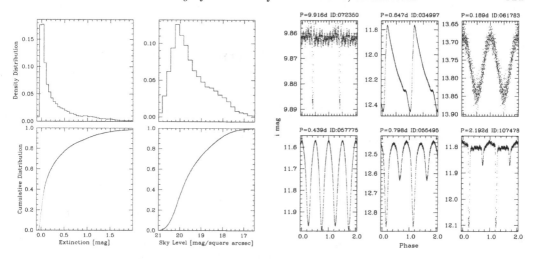

Figure 1: Left panel: Combined site statistics for the 2008 and 2010 Antarctic winter seasons based on CSTAR photometry (the total number of frames is 547,482). Note that the sky brightness histogram includes all Moon phases. Right panel: Phased light curves of six representative variables from Wang *et al.* (2011). Top row (left to right): exoplanet candidate; RR Lyrae; δ Scuti. Bottom row (left to right): contact; semi-detached and detached binaries.

References

Pojmański, G. 2004, *Astronomische Nachrichten*, 325, 553

Saunders, W., Lawrence, J. S., Storey, J. W. V., Ashley, M. C. B., Kato, S., Minnis, P., Winker, D. M., Liu, G., & Kulesa, C. 2009, *PASP*, 121, 976

Stetson, P. B. 1987 *PASP*, 99, 191

Wang, L., Macri, L. M., Krisciunas, K., Wang, L., Ashley, M. C. B., Cui, X., Feng, L.-L., Gong, X., Lawrence, J. S., Liu, Q., Luong-Van, D., Pennypacker, C. R., Shang, Z., Storey, J. W. V., Yang, H., Yang, J., Yuan, X., York, D. G., Zhou, X., Zhu, Z., & Zhu, Z. 2011, *AJ*, 142, 155

Wang, L., Macri, L. M., Wang, L., Ashley, M. C. B., Cui, X., Feng, L.-L., Gong, X., Lawrence, J. S., Liu, Q., Luong-Van, D., Pennypacker, C. R., Shang, Z., Storey, J. W. V., Yang, H., Yang, J., Yuan, X., York, D. G., Zhou, X., Zhu, Z., & Zhu, Z. 2012, *in preparation*

Yuan, X., Cui, X., Liu, G., Zhai, F., Gong, X., Zhang, R., Xia, L., Hu, J., Lawrence, J. S., Yan, J., Storey, J. W. V., Wang, L., Feng, L., Ashley, M. C. B., Zhou, X., Jiang, Z., & Zhu, Z. 2008, *SPIE*, 7012, 152

Astrophysics from Antarctica
Proceedings IAU Symposium No. 288, 2012
M. G. Burton, X. Cui & N. F. H. Tothill, eds.

Secular variation and fluctuation of GPS Total Electron Content over Antarctica

Rui Jin[1,2] and Shuanggen Jin[1]

[1]Shanghai Astronomical Observatory, Chinese Academy of Sciences, Shanghai 200030, China
[2]University of Chinese Academy of Sciences, Beijing 100049, China

Abstract. The total electron content (TEC) is an important parameters in the Earth's iono-sphere, related to various space weather and solar activities. However, understanding of the complex ionospheric environments is still a challenge due to the lack of direct observations, particularly in the polar areas, e.g., Antarctica. Now the Global Positioning System (GPS) can be used to retrieve total electron content (TEC) from dual-frequency observations. The con-tinuous GPS observations in Antarctica provide a good opportunity to investigate ionospheric climatology. In this paper, the long-term variations and fluctuations of TEC over Antarctica are investigated from CODE global ionospheric maps (GIM) with a resolution of $2.5° \times 5°$ every two hours since 1998. The analysis shows significant seasonal and secular variations in the GPS TEC. Furthermore, the effects of TEC fluctuations are discussed.

Keywords. GNSS, TEC, Ionosphere, Antarctica.

1. Introduction

The Earth's ionosphere is one of main parts in the Earth's upper atmosphere and plays a significant role in the space environment. Although the ionosphere represents less than 0.1% of the total mass of the Earth's atmosphere, it has a great effect on the global elec-tric circuit, the Earth's magnetic field and on electromagnetic wave propagation through the Earth's ionosphere due to its partially ionized gas. However, understanding of the complex ionospheric environments is still a challenge due to the lack of direct observa-tions, particularly in the polar areas, e.g., Antarctica. Antarctica is the southernmost and coldest continent on the Earth. It is well-known that Antarctica has special daily and seasonal variations due to the Earth's rotation and revolution, which also affect the Earth's ionospheric variation. Since the Earth's ionosphere is a dispersive medium, when electromagnetic wave signal propagates through the ionosphere, the signal will be de-layed. The ionospheric TEC can be calculated from dual-frequency GPS measurements. As more and more continuous operational GPS stations are set up, GPS has become a powerful tool to monitor global and region TEC. The international GNSS services (IGS) has routinely produced global vertical TEC maps every two hours for more than 10 years. Secular variations of vertical TEC over Antarctica can be investigated using these data. Here TEC maps provided by CODE are used to extract vertical TEC time series over the Antarctica. In section 2, the basic method on VTEC computation and data pro-cessing from CODE is described. Secular variations and fluctuations of GPS TEC over Antarctica are discussed and analyzed in section 3. Finally, the conclusions are given in Section 4.

2. Method and Data

As is well known, slant TEC can be derived from GPS geometry-free combination measurements. Usually, the high-order effects of the ionospheric refractivity are ignored

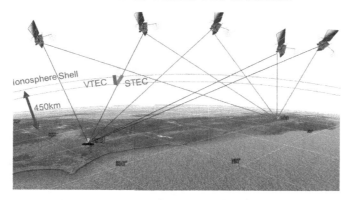

Figure 1. Geometry of GPS-TEC estimation.

as their effects are very small. Here we assumed all electrons are concentrated in an infinite thin shell in order to simplify expression of the slant TEC from GPS dual-frequency data. As shown in Figure 1 (here we just use two GPS stations (usud and mizu) to show the Geometry of GPS TEC estimate), the altitude of the ionosphere shell is adopted as 450 km, because the electron density peak of the Earths ionosphere is about 300–500 km. The vertical TEC is transformed from slant TEC using a cosine function of elevation.

$$\frac{vTEC}{cos(z)} = -\frac{f_1^2 f_2^2}{40.3(f_1^2 - f_2^2)}(P_1 - P_2 + B) = \frac{f_1^2 f_2^2}{40.3(f_1^2 - f_2^2)}(L_1 - L_2 - (N_1\lambda_1 - N_2\lambda_2) + b)$$

(2.1)

where, f_1 and f_2 are the GPS carrier wave frequencies, P_1 and P_2 are the GPS pseudorange measurements and carrier phase measurements, respectively, N_1 and N_2 are the ambiguity, λ_1 and λ_2 and are the wavelength of GPS signals. B and b are the differential codes biases and inter-frequencies differences, respectively.

In this paper, the GPS TEC time series over Antarctica are obtained from CODE. In order to avoid an error caused by the TEC computation algorithm, the recent data are used after March 16, 2002. These are results for the middle day of a 72-hour combination from the next day rather than 24-hour analysis. In this way, discontinuities at day boundaries can be minimized. Furthermore, a time-invariant quality level is achieved.

3. Result and Discussion

Although the GPS satellites orbit inclination angle is 55° and GPS satellites cannot fly over the South Pole, many GPS TEC measurements are provided by ground GPS stations located in Antarctica, such as vesl, syog, maw1, dav1, cas1. CODE provides global vertical TEC maps every two hours with a spatial resolution of 2.5°×5° in longitude and latitude. Vertical TEC values over Antarctica are extracted from ionox files released by CODE from day 076, 2002 to day 252, 2012. In order to check the characteristics of secular GPS TEC variations over the Antarctica, mean values of vertical TEC along the parallel of latitude are computed. As shown in Figure 2, obviously annual variation appears in long GPS TEC time series over Antarctica at latitudinal circles from S87.5° to S72.5°, and the mean values of the vertical TEC over Antarctica have a slight decreasing trend. The 2880 MHz solar flux variations during 2002 to 2012 are also shown in Figure 2. The vertical TEC variation shows agreement with the solar activities. Higher amplitudes

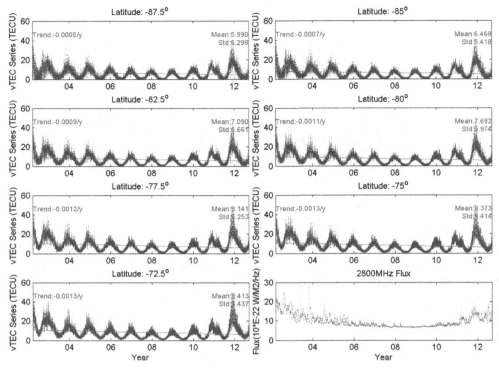

Figure 2. Long term variation and fluctuation of TEC from 2002 to 2012 over Antarctica.

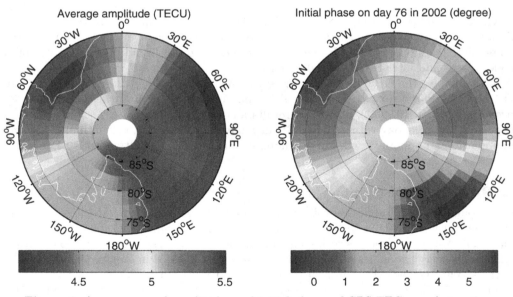

Figure 3. Average annual amplitudes and initial phases of GPS TEC over Antarctica.

of vertical TEC time series are found in around 2003 and 2011, which are higher solar activities years. Although, for the most time, the Sun is not visible in areas near the South Pole in Antarctica, the variation of amplitudes is obvious. In addition, the mean values and the standard deviations of about 10 years GPS TEC series over Antarctica increase as the latitude decreases. Probably, this phenomenon can be explained due to differences in solar radiation with latitude. Solar fluxes become smaller and smaller as the location gets closer to the South Pole. The Earth's ionosphere receives less energy for ionization near the South Pole. Lower vertical TEC and fewer fluctuations are shown in long-term vertical TEC in higher latitude area over Antarctica, which agrees with our inference. Higher amplitudes of vertical TEC time series are found around 2002 and 2012. Earth's ionosphere is affected by solar activity. Although, for the most time, the Sun is not visible in areas near the South Pole in Antarctica, the variation of amplitude is obvious. Figure 3 shows the distribution of the annual amplitudes and phases of GPS TEC times series from day 076, 2002 to day 252, 2012. The amplitudes and phases are both related to the longitude and latitude. The average amplitudes range from 4 to 6 TECU, and increase as the decrease of latitudes. And for W0° to W120°, the average amplitudes are relatively large. Initial phases also changes with the latitude variations exclude from E120° to E170°.

4. Conclusion

In this paper, the long-term variations and fluctuations of TEC over Antarctica are investigated from CODE global ionospheric maps (GIM) with a resolution of 2.5°*5° every two hours since 1998. The analysis results show significant seasonal and secular variations in GPS TEC. The long-term vertical TEC variations have good agreement with the solar activities. The annual amplitudes and phases are both related to the longitude and latitude. The average amplitudes range from 4 to 6 TECU over most Antarctic and phases are almost closer. The different patterns in West and East Antarctic are existed, which needs to be further investigated in the future.

References

A. Leick 1990 *GPS Satellite Surveying* (John Wiley and Sons).

Bassiri, S. and G. A. Hajj 1992 *Modeling the Global Positioning system signal Propagation Through the ionosphere*, TDA Progress Report 42-110.

Datta-Barua, S., T. Walter, *et al.* 2008 *Bounding higher-order ionosphere errors for the dual-frequency GPS user*, Radio Science 43, 5.

Hernandez-Pajares, M., J. M. Juan, *et al.* 2009 *The IGS VTEC maps: a reliable source of ionospheric information since 1998*, Journal of Geodesy 83(3-4): 263-275.

Liu Jiyu 2008 *The theory and algorithm of GPS satellite navigation positioning*, Science Press.

Michael C. Kelly 2009 *The earth's Ionosphere: Plasma Physics and Electrodynamics* (2nd Edition. Elsevier, Academic Press).

Mannucci, A. J., B. D. Wilson, *et al.* 1998 *A global mapping technique for GPS-derived ionospheric total electron content measurements*, Radio Science, 33(3): 565-582.

Norbert Jakowski, Christoph Mayer, Volker Wilken and Mohammed M. Hoque 2008 *Ionospheric impact on GNSS signals*, Fsica de la Tierra, vol 20.

Schaer, S. 1999 *Mapping and predicting the earth's ionosphere using the Global Positioning System*, Zurich, Institut fur Geodasie und Photogrammetrie, Eidg. Technische Hochschule Zurich.

Smith, D. A., E. A. Araujo-Pradere, *et al.* *A comprehensive evaluation of the errors inherent in the use of a two-dimensional shell for modeling the ionosphere* Radio Science, 43(6): RS6008.

Astrophysics from Antarctica
Proceedings IAU Symposium No. 288, 2012
M. G. Burton, X. Cui & N. F. H. Tothill, eds.

© International Astronomical Union 2013
doi:10.1017/S1743921312017140

Electron-antineutrino disappearance seen by Daya Bay reactor neutrino experiment

Ruiguang Wang, on behalf of the Daya Bay Collaboration

Institute of High Energy Physics, Beijing, 100049, China
email: wangrg@mail.ihep.ac.cn.

Abstract. The Daya Bay Reactor Neutrino Experiment has measured a non-zero value for the neutrino mixing angle θ_{13} with a significance of 7.7 standard deviations. Antineutrinos from six 2.9 GW_{th} reactors were detected in six antineutrino detectors deployed in two near and one far underground experimental halls. With a 116.8 kton-GW_{th}-day live-time exposure in 139 days, 28,909 (205,308) electron-antineutrino candidates were detected at the far hall (near hall). The ratio of the observed to expected number of antineutrinos at the far hall is R = 0.944 ± 0.007 ± 0.003 (syst). A rate-only analysis finds $\sin^2 2\theta_{13}$ = 0.089 ± 0.010 (stat) ± 0.005 (syst) in a three-neutrino framework.

Keywords. Neutrino oscillation, reactor neutrino, mixing angle θ_{13}, Daya Bay

1. Introduction

It is well known that the flavor of neutrinos oscillate as they propagate. Neutrino oscillations can be described by the three mixing angles ($\theta_{12}, \theta_{23}, \theta_{13}$) and a phase of the PMNS matrix, and two mass-squared differences (Δm_{32} and Δm_{21}). Of these mixing angles, θ_{13} is the least known. Some experiments, such as CHOOZ, T2K, MINOS and Double CHOOZ, have indicated that θ_{13} could be nonzero. In March 2012, the Daya Bay collaboration first announced the discovery of a non-zero θ_{13} with more than 5 σ (An *et al.* 2012a). Several months later this value was improved to 7.7 σ (An *et al.* 2013).

In reactor based experiments, the precise θ_{13} can be determined via the survival probability of the electron-antineutrino $\bar{\nu}_e$ at short distances from the reactors,

$$P_{\mathrm{sur}} \approx 1 - \sin^2 2\theta_{13} \sin^2(1.267 \Delta m_{31}^2 L/E), \qquad (1.1)$$

where $\Delta m_{31}^2 = \Delta m_{32}^2 \pm \Delta m_{21}^2$, E is the $\bar{\nu}_e$ energy in MeV and L is the distance in meters between the $\bar{\nu}_e$ source and the detector (baseline). The near-far arrangement of Daya Bay identical antineutrino detectors (ADs), as illustrated in Figure 1, allows for a relative measurement by comparing the observed $\bar{\nu}_e$ rates at various baselines with minimal uncertainties.

2. The experiment

The Daya Bay experiment sets up at the Daya Bay nuclear power complex in Shenzhen, south China. Six identical reactor cores spaced in pairs provide a total 17.4 GW_{th} thermal power. Three underground experimental halls (EHs) are connected with about 3,000 m horizontal tunnels. EHs' overburdens are 250, 265 and 860 (m.w.e). The baselines from six ADs to six cores are precisely measured by GPS. The baseline uncertainties have no effect on the oscillation results.

As shown in figure 2, each AD has three coaxial cylindrical vessels, 3 m inner (IAV) and 4 m outer (OAV) acrylic vessels and 5 m stainless steel vessel (SSV), filled with 20 t

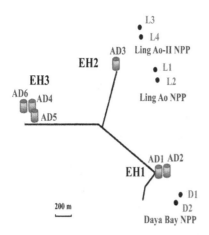

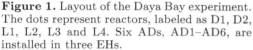

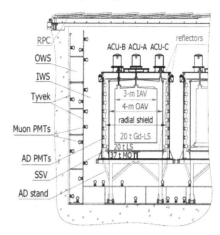

Figure 1. Layout of the Daya Bay experiment. The dots represent reactors, labeled as D1, D2, L1, L2, L3 and L4. Six ADs, AD1–AD6, are installed in three EHs.

Figure 2. The Daya Bay detectors, consisting of ADs (IAV, OAV,SSV), calibration units (ACU-A, ACU-B, and ACU-C), muon system (RPC, OWS, IWS) and PMTs.

of 0.1% by weight gadolinium-doped liquid scintillator (Gd-LS), 21 t of un-doped liquid scintillator (LS) and 37 t of mineral oil (MO), respectively. Three automated calibration units (ACU–A, ACU–B, and ACU–C) are mounted at the top of each SSV. Each ACU is equipped with a LED, a ^{68}Ge source, and a combined source of ^{241}Am–^{13}C and ^{60}Co. The muon detection system consists of a resistive plate chamber (RPC) tracker and a high-purity active inner (IWS) and outer (OWS) water shields. There are 192 8-inch Hamamatsu PMTs installed in each AD and 288 (384) PMTs installed in each near (far) hall water pool. A detailed description of the Daya Bay experiment can be found in (An *et al.* 2012b).

3. Data analysis and results

This analysis is based on data from December 24, 2011 to May 11, 2012. The $\bar{\nu}_e$s are detected via the inverse β-decay (IBD) reaction, $\bar{\nu}_e + p \rightarrow e^+ + n$, in Gd-LS. The coincidence of the prompt scintillation from the e^+ and the delayed neutron capture on Gd provides a distinctive $\bar{\nu}_e$ signature. The IBD selection includes: 1) prompt energy 0.7~ 12 MeV and delayed energy 6~12 MeV, 2) prompt-delayed signals time duration <200μs and 3) no triggers 200 μs before prompt or after delayed event. The contribution of IBD-like events is estimated less than 0.3%. The detector efficiencies as well as correlated and uncorrelated systematic uncertainties are also estimated as 78.8%, 1.9% and 0.2%, respectively.

The $\bar{\nu}_e$ rate in the far hall was predicted with a weighted combination of the two near hall measurements assuming no oscillation. The deficit observed at the far hall is $R = 0.944 \pm 0.007\,(\text{stat})\pm 0.003\,(\text{syst})$. The value of $\sin^2 2\theta_{13}$ was determined with a χ^2 constructed with pull terms accounting for the correlation of the systematic errors, which is $\sin^2 2\theta_{13} = 0.089 \pm 0.010\,(\text{stat.})\pm 0.005\,(\text{syst.})$ with a χ^2/NDF of 3.4/4. All best estimates of pull parameters are within its one standard deviation based on the corresponding systematic uncertainties. The no-oscillation hypothesis is excluded at 7.7 standard deviations (in figure 3). The observed $\bar{\nu}_e$ spectrum in the far hall was compared to a prediction based on the near hall measurements is shown in Figure 4. The distortion

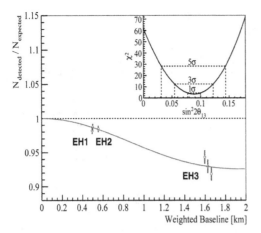

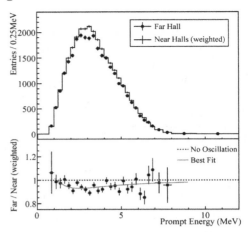

Figure 3. Ratio of measured versus expected signal in each detector, assuming no oscillation. The error bar is the uncorrelated uncertainty of each AD, including statistical, detector-related, and background-related uncertainties. The expected signal was corrected with the best-fit normalization parameter. Reactor and survey data were used to compute the flux-weighted average baselines. The oscillation survival probability at the best-fit value is given by the smooth curve. The χ^2 versus $\sin^2 2\theta_{13}$ is shown in the inset.

Figure 4. Top: Measured prompt energy spectrum of the far hall (sum of three ADs) compared with the no-oscillation prediction from the measurements of the two near halls. Bottom: The ratio of measured and predicted no-oscillation spectra. The red curve is the best-fit solution with $\sin^2 2\theta_{13} = 0.089$ obtained from the rate-only analysis. The dashed line is the no-oscillation prediction.

of the spectra is consistent with the expected one calculated with the best-fit θ_{13} obtained from the rate-only analysis, providing further evidence of neutrino oscillation.

Up to now Daya Bay experiment has obtained the most precise value of $\sin^2 2\theta_{13}$ in the world. After installation of final two ADs in the summer of 2012, running for several years with eight ADs will yield even better results.

References

An, F. P. *et al.* (Daya Bay Collaboration) 2012a, *Phys. Rev. Lett.*, 108, 171803
An, F. P. *et al.* (Daya Bay Collaboration) 2012b, *Nucl. Instr. and Meth.*, A685, 78
An, F. P. *et al.* (Daya Bay Collaboration) 2013, *Chinese Phys. C*, 3, 7, 011001

Astrophysics from Antarctica
Proceedings IAU Symposium No. 288, 2012   © International Astronomical Union 2013
M. G. Burton, X. Cui & N. F. H. Tothill, eds.   doi:10.1017/S1743921312017152

Test and Commissioning of the AST3-1 Control System

Xiaoyan Li[1,2,3] and Daxing Wang[1,3]

[1]National Astronomical Observatories/Nanjing Institute of Astronomical
Optics & Technology, Chinese Academy of Sciences,
Nanjing 210042, China
[2]Email: xyli@niaot.ac.cn

[3]Key Laboratory of Astronomical Optics & Technology, Nanjing Institute of Astronomical
Optics & Technology, Chinese Academy of Sciences,
Nanjing 210042, China

Abstract. The first of three Antarctic Survey Telescopes (AST3-1), a 50/68 cm Schmidt-like equatorial-mount telescope, is the first trackable Chinese telescope operating on the Antarctic plateau. It was installed at Dome A (80°22′S, 77°21′E, 4,093 m), the highest place on the Antarctic plateau, in 2012. The telescope is unmanned during night-time operations through the Austral winter. The telescope optics and mechanics, as well as the motors and position sensors, are exposed to a harsh environment. The mechanics is enclosed with a foldable tent-like dome to prevent snow, diamond dust and ice. While the control cabinet containing drive boxes, circuit board boxes, power converters and the Telescope Control Computer (TCC) is located inside the warm instrumental module. In about 15 weeks remote testing and commissioning, from January 24 when the expedition team left there to May 8, when the communication failed, we obtained images with the best FWHM of less than 2″. We also recorded the telescope movement performance and fine-tuned the dynamic properties of the telescope control system. Some experiences and lessons will be disscussed in this paper.

Keywords. AST3-1, Dome A, control system, movement performance

1. Introduction

The telescope control system of AST3-1, including the telescope movement system, focusing system and lens-heating system, consists of Telescope Control Software(TCS), Telescope Control Computer (TCC), driveboxes, motors, position sensors, current sensors and voltage sensors, etc. It has such features as:

Drive system: Gear drive with dual-motors per axis and drive boxes equipped with an anti-backlash control unit.

Motors: Customized AC servo PMSM (Permanent Magnet Synchronous Motor) with a two-pole resolver as the feedback sensor. Four such motors were mounted on the two axes of the telescope. Besides, a customized stepper motor was used in the focusing system. All these motors were not heated and they all have worked through lower than $-70°$C.

Resolvers: Multi-pole resolvers instead of commonly used photoelectric encoders were

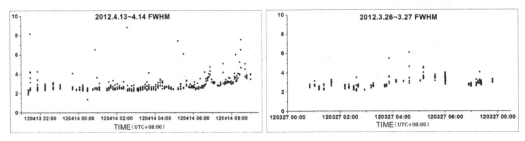

Figure 1. Average star FWHM from AST3-1 (one point stands for one image).

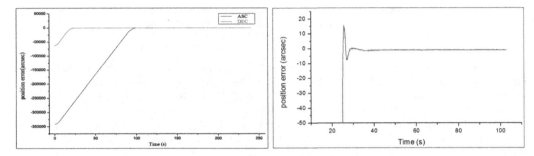

Figure 2. Asc/Dec position error.

used to obtain the position of the two telescope axes. Resolvers were not heated and they all worked well through the testing and commissioning stage.

Commands: About 50 commands provide comprehensive control and feedback of the telescope control system. The TCS, running on the TCC, receives and interprets these commands and manages the control system.

Protection: Various auxilliary and safety sensors help protect the telescope to keep it running safely.

2. Test and Commissioning Result

The control system of AST3-1 has worked through ambient temperatures lower than $-70°C$. The telescope continued running until the communication dropped on May 8. We have obtained many images in about 8-weeks of observation. Figure 1 shows the average star FWHM (Full Width at Half Maximum) of all images during one night of observing. Most of the FWHMs are between $2''$ and $3''$ demonstrate good seeing and tracking performance.

TCS recorded a quantitative measurement of the telescope movement. Accordingly, we can change the parameters of the hardware servo and software servo to fine tune the dynamic properties of the telescope. The left image of Figure 2 shows an example record of the Asc/Dec position error. When the position error get close to zero it means the telescope switches from pointing to tracking. Zoom into the intersection of pointing and tracking to obtain the right image in Figure 2. This shows a small overshoot of less than 1% and a fast transition of only one adjustment. The tracking following error of the Asc axis is less than $0.4''$ and for the Dec axis less than $0.6''$.

We also have obtained some bad images during observing on several nights. Some of these demonstrate that the weather conditions are poor at times. When the wind speed is high, the tent-like dome and the tube stiffness will affect the image quality.

Astrophysics from Antarctica
Proceedings IAU Symposium No. 288, 2012
M. G. Burton, X. Cui & N. F. H. Tothill, eds.

© International Astronomical Union 2013
doi:10.1017/S1743921312017164

Atmospheric calibration for submillimeter and terahertz observations

Xin Guan[1], Jürgen Stutzki[1] and Yoko Okada[1]

[1] I. Physikalisches Institut der Universität zu Köln,
Zülpicher Straße 77, 50937 Köln, Germany
email: **guan@ph1.uni-koeln.de**

Abstract. Submillimeter observations through the atmosphere can be affected by the complex spectroscopic features of the air. Calibration of astronomical observations in these frequencies requires proper modelling of the atmosphere. We analyzed sky observations from altitudes around 500 and 200 hPa respectively and found deficiencies in atmospheric models. Further research to improve the models are expected to help in future submillimeter observations.

Keywords. Atmospheric effects — Submillimeter: general

The earth's atmosphere causes significant attenuation to submillimeter radio signals. For this reason, submillimeter astronomical observations must be carried out either from space or from high altitudes where the preciptable water vapour is low.

Observing from spacecraft is an expensive task. Particularly, when the apparatus needs frequent maintenance or is simply too large to be launched. Then we must make use of ground based, airborne or balloon-borne observatories. For the first two options, atmospheric calibration can be essential to getting accurate scientific data.

When observing with a double-side-banded (DSB) receiver such as the GREAT (Heyminck *et al.* (2012)) and the SMART (Graf *et al.* (2003)) receivers, an atmospheric model (atm-model hereafter) is necessary to separate the contributions of the atmosphere from each sideband. Utilization of an atm-model can also improve signal-to-noise (s/n) ratio for spectroscopy because the s/n ratio directly obtained from limited observing time can be very low in the narrow frequency channels.

In order to model the atmosphere, firstly we modeled the vertical water vapour distribution (Guan *et al.* (2012)) according to meteorological measurements. Effective observation in submillimeter wavelengths requires very low precipitable water vapor (pwv). So the water mixing ratio is negligible. Therefore we can keep other parameters constant while varying the pwv, and prove that in our model the atmospheric opacity can be decomposited into wet and dry components (see Guan *et al.* (2012)):

$$\tau = b \times pwv + c.$$

Using Kirchhoff's law we can derive the atmosphere's opacity from its measured brightness temperature and then obtain pwv, the model parameter.

We made observations with the GREAT receiver on the SOFIA telescope from altitudes from 10 to 14 km, and with the SMART receiver on the NANTEN2 telescope at an altitude of 4.8 km. GREAT receives in several bands from 1.2 to 2.8 THz, while SMART is an array receiver of two bands (460/490 GHZ and 810 GHz) with 8 pixels each. SMART's unique property is that it is an array receiver and the results from its eight pairs of pixels can be cross-checked. The same cannot be easily done with GREAT/SOFIA.

Both of our GREAT and SMART observations showed inconsistencies with the atm-model. The fit model parameter "pwv" is often inconsistent between different bands.

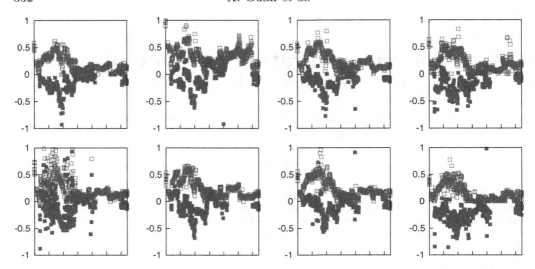

Figure 1. Fit pwv and dry-offset of SMART observations across 630 successive scans, in which pwv varied from a little below 1 mm to somewhere between 0.1 and 0.2 mm. "Common-fit" was used to fit these spectra so there is only a single pwv and dry-offset for each 460/810 GHz pairs. Red open squares are pwv's (millimeter) and blue filled squares are dry-offsets (neper). The horizontal axis is scan number. The eight subfigures show results from the eight pairs of 460/810 GHz pixels at corresponding positions in the focal plane.

Trying to find out the problem, we introduced the "dry-offset" parameter to the model:

$$\tau_\nu = b_\nu \, pwv + c_\nu - \begin{cases} o_{L1}, & \nu \text{ in L1} \\ o_{L2}, & \nu \text{ in L2} \end{cases}$$

in the hope of correcting possible errors of the model in predicting the quasi-continuum component of the dry atmospheric opacity from collision-induced-absorption. We tried fitting the dry-offsets of the 460 and 810 GHz data from all eight pairs of SMART pixels. The value of dry-offset is clearly correlated with the pwv. All eight pairs of 460/810 pixels show similar results in Fig. 1. In this example we fit a common set of pwv and dry-offset parameters for the 460/810 GHz data from each pairs of pixels. The absolute values of dry-offset are much larger when pwv is high, and nearly zero when pwv is low. Because of the large difference in the "b" coefficient between the 460 and 810 GHz bands, both of the pwv and dry-offset parameters should be determined and the dry-offset would have been zero if the atm-model were perfect.

The systematic deviation of the model from the observations indicates imperfection in the atm-model. The high frequency, high angular resolution data from astronomical observations has expanded the regime of meteorological studies, and will help to better understanding of the Earth's atmosphere, which will in turn allow astronomers to obtain more accurate submillimeter scientific data. We have already obtained more observing time on this issue and are expecting further developments in atmospheric calibration.

References

Graf, U. U., Heyminck, S., & Michael, E. A. 2003, *SPIE* 4855,322
Guan, X., Stutzki, J., Graf, U. U., *et al.* 2012, *A&A* 542L, 4
Heyminck, S., Graf, U. U., Güsten R. *et al.* 2012, *A&A* 542L, 1

Astrophysics from Antarctica
Proceedings IAU Symposium No. 288, 2012
M. G. Burton, X. Cui & N. F. H. Tothill, eds.

© International Astronomical Union 2013
doi:10.1017/S1743921312017176

Classification of Quasars and Stars by Supervised and Unsupervised Methods

Yanxia Zhang[1], Yongheng Zhao[1], Hongwen Zheng[2] and Xue-bing Wu[3]

[1] Key Laboratory of Optical Astronomy, National Astronomical Observatories, Chinese Academy of Sciences, Beijing 100012, China
email: zyx@bao.ac.cn, yzhao@bao.ac.cn

[2] Mathematics and Physics Department, North China Electronic Power University, Beijing 102206, China

[3] Department of Astronomy, Peking University, Beijing 100871, China

Abstract. Targeting quasar candidates is always an important task for large spectroscopic sky survey projects. Astronomers never give up thinking out effective approaches to separate quasars from stars. The previous methods on this issue almost belong to supervised methods or color-color cut. In this work, we compare the performance of a supervised method – Support Vector Machine (SVM)– with that of an unsupervised method one-class SVM. The performance of SVM is better than that of one-class SVM. But one-class SVM is an unsupervised algorithm which is helpful to recognize rare or mysterious objects. Combining supervised methods with unsupervised methods is effective to improve the performance of a single classifier.

Keywords. Classification, Astronomical databases: miscellaneous, Catalogs, Methods: data analysis, Methods: statistical

---

1. Introduction

Large samples of quasars are important tools in astrophysics and cosmology for several reasons. With them, we may not only study the quasar phenomenon itself, but also the numerous astrophysical applications they offer. The quasar phenomenon itself is related to the galaxy history and evolution, to star formation and possibly to the interaction with other galaxies. Quasars can be used to study the intervening intergalactic medium. Indeed, the mechanism which feeds the central black hole and triggers the nucleus active during a given period remains unclear. So far there have been many automated methods focusing on selecting quasar candidates.

2. Sample and results

The samples applied here were cross-identified from different survey catalogs: SDSS DR7 and UKIDSS DR7 catalogs. The star sample was adopted from the cross-identified pointed sample without brightness variation in Stripe 82. Finally we obtain 21,241 quasars and 154,739 stars.

Support Vector Machine (SVM) is a two-class algorithm (i.e. one needs negative as well as positive examples). One-class SVM focuses on positive data, and identifies outliers amongst the positive examples and uses them as negative examples. SVM and one-class SVM are compared to separate quasars from stars. The input pattern for SVM and one-class SVM is $i, u - g, g - r, r - i, i - z, z - Y, Y - J, J - H, H - K$. The classification results are shown in Table 1. Obviously the performance of SVM is superior to that of one-class SVM. The accuracy of quasars and stars with SVM is more than 99.9%, respectively. The accuracy of quasars with one-class SVM is more than 99.8% while the

333

Y. Zhang, Y. Zhao, H. Zheng & X. Wu

accuracy of stars is rather poor. Comparing the misclassified sources by the two methods, the misclassified quasars have no overlap. However misclassified stars by one-class SVM include misclassified stars by SVM. As a result, we propose a new classification scheme as indicated in Figure 1.

Table 1. The classification results with SVM and one-class SVM

method	SVM		one-class SVM	
classified↓known→	quasars	stars	quasars	stars
quasars	21,226	32	21,204	66,792
stars	15	154,707	37	87,948
accuracy	99.93%	99.98%	99.83%	56.80%

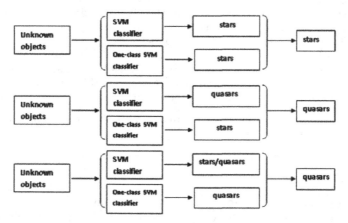

Figure 1. Classification scheme of combined SVM and one-class SVM.

Table 2. The classification result with combined SVM and one-class SVM

classified↓known→	quasars	stars
quasars	21,241	32
stars	0	154,707
accuracy	100.00%	99.98%

When combining the two approaches, the new objects are recognized as stars when SVM classifier and one-class SVM classifier both consider them as stars; the new objects are marked as quasars when SVM or one-class SVM classifies them as quasars. The new classification accuracy based on this scheme improves as shown in Table 2. In other words, ensemble different methods are helpful to enhance the performance of a single classifier.

Acknowledgements

This paper is funded by the National Natural Science Foundation of China under grant No.10778724, 11178021 and No. 11033001, the Natural Science Foundation of Education Department of Hebei Province under grant No. ZD2010127. We acknowledge use of the SDSS and UKIDSS databases.

Author Index

Printed in the United States
by Baker & Taylor Publisher Services